Homotopy Theory of Function Spaces and Related Topics

AF698611

CONTEMPORARY MATHEMATICS

519

Homotopy Theory of Function Spaces and Related Topics

Oberwolfach Workshop
April 5–11, 2009
Mathematisches Forschungsinstitut
Oberwolfach, Germany

Yves Félix
Gregory Lupton
Samuel B. Smith
Editors

American Mathematical Society
Providence, Rhode Island

Editorial Board

Dennis DeTurck, managing editor

George Andrews   Abel Klein   Martin J. Strauss

2000 *Mathematics Subject Classification.* Primary 55P15, 55P35, 55P45, 55P48, 55P50, 55P60, 55P62, 55Q52, 55R35, 46M20.

Library of Congress Cataloging-in-Publication Data

Oberwolfach Workshop on Homotopy Theory of Function Spaces and Related Topics (2009 : Mathematisches Forschungsinstitut Oberwolfach)

Homotopy theory of function spaces and related topics : Oberwolfach Workshop on Homotopy Theory of Function Spaces and Related Topics, April 5–11, 2009, Mathematisches Forschungsinstitut Oberwolfach, Germany / Yves Félix, Gregory Lupton, Samuel B. Smith, editors.

p. cm.

Includes bibliographical references.

ISBN 978-0-8218-4929-3 (alk. paper)

1. Homotopy Theory—Congresses. 2. Function spaces—Congresses. I. Félix, Y. (Yves) II. Lupton, Gregory, 1960– III. Smith, Samuel B., 1966– IV. Title.

QA612.7.024 2009
514′.24—dc22

2010009614

Copying and reprinting. Material in this book may be reproduced by any means for educational and scientific purposes without fee or permission with the exception of reproduction by services that collect fees for delivery of documents and provided that the customary acknowledgment of the source is given. This consent does not extend to other kinds of copying for general distribution, for advertising or promotional purposes, or for resale. Requests for permission for commercial use of material should be addressed to the Acquisitions Department, American Mathematical Society, 201 Charles Street, Providence, Rhode Island 02904-2294, USA. Requests can also be made by e-mail to `reprint-permission@ams.org`.

Excluded from these provisions is material in articles for which the author holds copyright. In such cases, requests for permission to use or reprint should be addressed directly to the author(s). (Copyright ownership is indicated in the notice in the lower right-hand corner of the first page of each article.)

© 2010 by the American Mathematical Society. All rights reserved.
The American Mathematical Society retains all rights
except those granted to the United States Government.
Copyright of individual articles may revert to the public domain 28 years
after publication. Contact the AMS for copyright status of individual articles.
Printed in the United States of America.

∞ The paper used in this book is acid-free and falls within the guidelines
established to ensure permanence and durability.
Visit the AMS home page at `http://www.ams.org/`

10 9 8 7 6 5 4 3 2 1   15 14 13 12 11 10

Contents

Problem List

Preface

This collection of articles is the proceedings volume for a conference entitled *Homotopy Theory of Function Spaces and Related Topics*, which was held 5th–11th April 2009 at the Mathematisches Forschungsinstitut Oberwolfach (MFO), Germany. The conference attracted an international group of 23 participants that included leading practitioners in the field. We would like to thank the MFO for including our conference in its workshop program for the 2009 year and for providing a congenial and productive atmosphere for the conference. As organizers, we were especially appreciative of the excellent level of facilities—technical, professional, and domestic—offered by the MFO.

Function spaces have been objects of central interest to homotopy theorists, and their study has seen steady activity for over sixty years. Current research in this area is remarkably diverse with connections to other areas ranging from geometry to analysis to robotics. The conference brought together researchers with expertise in a wide breadth of such topics. The lectures at the conference provided a snapshot of the current state-of-the-art of the subject, whilst the problem sessions suggested many promising directions for future work. We are grateful to all those who participated in the conference for a very stimulating week.

This volume contains 14 original research articles on function spaces and related topics. Each of the research articles was carefully refereed. We would like to thank the referees for their gracious acceptance and timely execution of this task. The volume also includes two general interest articles: a survey by SMITH and a problem list, curated by FÉLIX, which is an expanded and edited version of problems discussed in sessions held at the conference.

Several main themes of research in the area of homotopy theory of function spaces are represented by the selection of articles here. Also represented are a number of important connections to other areas. The survey article of SMITH gives a fairly complete picture of this landscape; we restrict ourselves to a brief summary here, and refer to that article for details and more extensive discussion.

Of course, the basic problem is to understand the homotopy type of a function space $\mathrm{map}(X, Y)$. Generally, this is a disconnected space and so one focuses on a path component, denoted here by $\mathrm{map}(X, Y; f)$, for some choice of map $f\colon X \to Y$. Different components generally display different homotopy types. In order to progress, hypotheses on the spaces X and Y are necessary: a popular choice is to restrict to the case in which X is a finite CW complex and Y is a nilpotent CW complex (e.g., a simply connected CW complex).

Classification of the homotopy types of components is clearly a deep and difficult problem. The subject has progressed through results that either focus on some particular aspect of the homotopy type of a component, or apply to particular choices of spaces X and Y. Again, the survey article contains a wealth of information about all this. Here, we briefly highlight some of those themes and directions represented by articles in this proceedings.

Rational homotopy theory has been used intensively in recent years to study function spaces. The articles of BUIJS and GATSINZI-KWASHIRA focus on the rational homotopy theory of a general component of a function space. Those of KURIBAYASHI and FÉLIX-TANRÉ use rational homotopy theory to study topics that have many points of contact with function spaces, namely classifying spaces of fibrations and configuration spaces—actually an extension of such, respectively.

As described in the survey article, many results about $\mathrm{map}(X, Y)$ have been proved in the cases in which either X or Y is a classifying space. Here, the article of KONO-TSUKUDA studies the case of $\mathrm{map}(X, BG)$. A connection to geometry and physics arises here due to a theorem of Gottlieb, which establishes an equivalence between the gauge group of a principal G-bundle over X and the loop space $\Omega\mathrm{map}(X, BG; h)$, where $h\colon X \to BG$ is the classifying map of the bundle. The article of SMITH-SCHOCHET is also on this subject; they focus on extending localization properties of the spaces concerned.

The free loop space $\mathrm{map}(S^1, Y)$ is an object of perennial interest from numerous points of view. The surge of activity around string topology has made it even more ubiquitous. The articles of CHATAUR, CRABB, GATSINZI, GIANSIRACUSA-SALVATORE, GOLASIŃSKI-GONÇALVES-WONG and WOCKEL all bear in some way on this very active area.

Although spaces of equivalences were not a direct focus of the conference, they are related to both the classifying space of a fibration and the Gottlieb groups of a space. Several articles have already been noted that bear on the former topic. The articles of GATSINZI and STROM-OPREA are concerned with the latter topic.

One further direction of application for the study of $\mathrm{map}(X, Y)$ is represented by the article of KOZLOWSKI-YAMAGUCHI. A number of seminal results establish equivalences of one sort or another—weak homotopy equivalences, equivalences after stabilization in a certain sense—between the space of maps that preserve some pertinent structure and the corresponding space of continuous maps. An emblematic result of this type is one of Segal that relates the space of based holomorphic maps $\mathrm{Hol}_*(S^2, \mathbb{C}P^n)$ with the based function space $\mathrm{map}_*(S^2, \mathbb{C}P^n)$ in this way. Such results then allow knowledge of the ordinary function space to be applied to yield information about the space of more structured maps, an object which, *a priori*, one might expect to be more difficult to analyze or requiring techniques different from those of the Homotopy Theory of Function Spaces.

We refer the reader to the survey article and to the problem list for further information about themes of research in the area and directions of application to, or connections with, other areas.

The AMS publications department has been very encouraging throughout the preparation of this volume. We would like especially to thank Christine Thivierge for her guidance at each stage of the process.

Y. Félix, G. Lupton, S. B. Smith

Conference Participants

Martin Arkowitz
Dartmouth College, U.S.A.

Méadhbh Boyle,
University of Aberdeen, Scotland

Urtzi Buijs
Universidad de Málaga, Spain

David Chataur
Université de Lille, France

Yves Félix
Université Catholique de Louvain, Belgium

Martin Fuchssteiner
TU Darmstadt, Germany

Jean-Baptiste Gatsinzi
University of Botswana, Botswana

Marek Golaziński
Nicolaus Copernicus University, Poland

Daniel Gottlieb
UCLA, U.S.A.

Katsuhiko Kuribayashi
Shinshu University, Japan

Andrey Lazarev
University of Leicester, England

Gregory Lupton
Cleveland State University, U.S.A.

John Oprea
Cleveland State University, U.S.A.

Paul-Eugène Parent
University of Ottawa, Canada

Paolo Salvatore
Universitá di Roma "Tor Vergata," Italy

Jonathan Scott
Cleveland State University, U.S.A.

Samuel Smith
St. Joseph's University, U.S.A.

Jeffrey Strom
University of Western Michigan, U.S.A.

Daniel Tanré
Université de Lille, France

Svjetlana Terzić
University of Montenegro, Montenegro

Shuichi Tsukuda
University of the Ryukus, Japan

Antonio Viruel
Universidad de Málaga, Spain

Christoph Wockel
Georg-August-Universität, Germany

Conference Presentations

In keeping with standard practice at MFO, the number of talks was kept relatively low, so as to allow generous amounts of time for discussion amongst participants.

The following four talks were expository (solicited as such by the organizers):

- David Chataur: *Division Functors and Mapping Spaces*
- Katsuhiko Kuribayashi: *Models for Function Spaces and Applications*
- John Oprea: *Gottlieb Groups, Evaluation Maps and Geometry*
- Shuichi Tsukuda: *Survey on Gauge Groups*

The remaining 13 were research talks on the presenters' work.

- Méadhbh Boyle: *An Algebraic Model for the Homology of Pointed Mapping Spaces out of a Closed Surface*
- Urtzi Buijs: *The Homotopy Lie Algebra of Function Spaces and Spaces of Sections (joint with A. Murillo)*
- Jean-Baptiste Gatsinzi: *Rational Homotopy Groups of Function Spaces*
- Marek Golaziński: *Fox and Gottlieb Groups and Whitehead Products (joint with D. Gonçalves, J. Mukai and P. Wong)*
- Daniel Gottlieb: *Self Coincidence Numbers and the Fundamental Group*
- Andrey Lazarev: *Characteristic Classes of Operadic Algebras*
- Paolo Salvatore: *Cyclic Formality of the Operad of Framed Little Discs, with Implications for Spaces of Knots*
- Jonathan Scott: *On the Geodesic Conjecture (joint with K. Hess)*
- Jeffrey Strom: *Miller Spaces*
- Svjetlana Terzić: *The Integral Pontrjagin Homology of the Based Loop Space on a Flag Manifold*
- Shuichi Tsukuda: *On the Configuration Space of a Certain n-arms Machine in Euclidean Space*
- Antonio Viruel: *Equivalences of a Product and Mal'cev Quasi-rings*
- Christoph Wockel: *Non-Integral Central Extension of Loop Groups via Gerbes*

Two problem sessions were also held, not included on this list.

Survey Article

Contemporary Mathematics
Volume **519**, 2010

The Homotopy Theory of Function Spaces: A Survey

Samuel Bruce Smith

ABSTRACT. We survey research on the homotopy theory of the space $map(X, Y)$ consisting of all continuous functions between two topological spaces. We summarize progress on various classification problems for the homotopy types represented by the path-components of $map(X, Y)$. We also discuss work on the homotopy theory of the monoid of self-equivalences $aut(X)$ and of the free loop space LX. We consider these topics in both ordinary homotopy theory as well as after localization. In the latter case, we discuss algebraic models for the localization of function spaces and their applications.

1. Introduction.

In this paper, we survey research in homotopy theory on function spaces treated as topological spaces of interest in their own right. We begin, in this section, with some general remarks on the topology of function spaces. We then give a brief historical sketch of work on the homotopy theory of function spaces. This sketch serves to introduce the basic themes around which the body of the paper is organized.

By work of BROWN [**38**, 1964] and STEENROD [**258**, 1967], the homotopy theory of function spaces may be studied in the "convenient category" of compactly generated Hausdorff spaces. Retopologizing is required, however. Given spaces X and Y in this category, let Y^X denote the space of all continuous functions with the compact-open topology. Define

$$map(X, Y) = k(Y^X)$$

to be the associated compactly generated space. Then $map(X, Y)$ satisfies the desired exponential laws and is a homotopy invariant of X and Y. The space $map(X, Y)$ is generally disconnected with path-components corresponding to the set of free homotopy classes of maps. We write $map(X, Y; f)$ for the path-component containing a given map $f : X \to Y$. Important special cases include: $map(X, Y; 0)$, the space of null-homotopic maps; $map(X, X; 1)$, the identity component; $aut(X)$,

---

2010 *Mathematics Subject Classification.* 55P15, 55P35, 55P48, 55P50, 55P60, 55P62, 55Q52, 55R35, 46M20.

Key words and phrases. Function space, monoid of self-equivalence, free loop space, space of holomorphic maps, gauge group, string topology, configuration space, section space, classifying space, Gottlieb group, localization, rational homotopy theory.

©2010 American Mathematical Society

the space of all homotopy self-equivalences of X; and $LX = map(S^1, X)$ the free loop space.

Concrete results on the path-components of $map(X,Y)$ often require much more restrictive hypotheses on X and Y. By MILNOR [**212**, 1959], when X is a compact, metric space and Y is a CW complex, the components $map(X,Y;f)$ are of CW homotopy type. A natural case to consider then is when X is a finite CW complex and Y is any CW complex. By KAHN [**151**, 1984], $map(X,Y)$ is also of CW type when X is any CW complex and Y has finitely many homotopy groups.

The space $map(X,Y)$ has two close relatives. If X and Y are pointed spaces, we have $map_*(X,Y)$ the space of basepoint preserving functions, with components $map_*(X,Y;f)$ for f a based map. Given a fibration $p\colon E \to X$, we have $\Gamma(p)$ the space of sections with components $\Gamma(p;s)$ for s a fixed section. Of course, $map(X,Y) \simeq \Gamma(p)$ when p fibre-homotopy trivial with fibre Y. Many theorems about $map(X,Y)$ generalize to $\Gamma(p)$ and many have related versions for $map_*(X,Y)$. For the sake of brevity, when possible we state theorems for the free function space and omit extensions and restrictions. Theorems stated for the based function space are then results that do not apply to $map(X,Y)$.

1.1. A Brief History. Function spaces are at the foundations of homotopy theory and appear in the literature dating back, at least, to Hurewicz's definition of the homotopy groups in the 1930s. Work focusing explicitly on the homotopy theory of a function space first appears in the 1940s. WHITEHEAD [**284**, 1946] introduced the problem of classifying the homotopy types represented by the path-components of a function space, focusing on the case $map(S^2,S^2)$. HU [**147**, 1946] showed

$$\pi_1(map(S^2,S^2;\iota_m)) \cong \mathbb{Z}/2|m|,$$

where ι_m is the map of degree m thus distinguishing components of different absolute degree.

A decade later, papers of THOM [**268**, 1957] and FEDERER [**87**, 1956] appeared giving dual methods for computing homotopy groups of components of $map(X,Y)$. Thom used a Postnikov decomposition of Y to indicate a method of calculation. Federer constructed a spectral sequence converging to these homotopy groups using a cellular decomposition of X. Both authors obtained the following basic identity:

$$\pi_q(map(X, K(\pi,n);0)) \cong H^{n-q}(X;\pi)$$

for X a CW complex and π an abelian group.

In the 1960s, the monoid $aut(X)$ of all homotopy self-equivalences of X emerged as a central object for the theory of fibrations. STASHEFF [**257**, 1963] constructed a universal fibration for CW fibrations with fibre of the homotopy type of a fixed finite CW complex X, building on work of DOLD-LASHOF [**71**, 1957]. His result implied the universal X-fibration is obtained, up to homotopy, by applying the Dold-Lashof classifying space functor to the evaluation fibration $\omega\colon map(X,X;1) \to X$. In this same period, GOTTLIEB [**107**, 1965] introduced and studied the evaluation subgroups or *Gottlieb groups*:

$$G_n(X) = \mathrm{im}\{\omega_\sharp\colon \pi_n(map(X,X;1)) \to \pi_n(X)\} \subseteq \pi_n(X)$$

initiating a rich literature on the evaluation map. Among many other properties, he showed the Gottlieb groups correspond to the image of the linking homomorphism in the long exact sequence of homotopy groups of the universal X-fibration. Thus the vanishing of a Gottlieb group $G_n(X)$ is equivalent to the vanishing of the linking homomorphism in degree n for every CW fibration with fibre X.

In the 1970s, HANSEN [**128**, 1974] began a systematic study of the homotopy classification problem for the path components of $map(X, Y)$. He completed the classification for $map(S^n, S^n)$ building on the methods of Whitehead, mentioned above. He and other authors obtained complete results in many special cases involving spheres, suspensions, projective spaces and certain manifolds.

The space of holomorphic maps $\mathrm{Hol}(M, N)$ between two complex manifolds offers a deep refinement of the homotopy classification problem for continuous maps with important interdisciplinary connections. SEGAL [**242**, 1979] initiated the study of the space $\mathrm{Hol}(M, N)$ in homotopy theory proving the inclusion

$$\mathrm{Hol}_k^*(S^2, \mathbb{C}P^m) \hookrightarrow map_*(S^2, \mathbb{C}P^m; \iota_k)$$

induces a homology equivalence through a range of degrees. Here $\mathrm{Hol}_k^*(S^2, \mathbb{C}P^m)$ denotes the space of based holomorphic maps of degree k. In fundamental work in complex geometry, GROMOV [**115**, 1989] identified the class of elliptic manifolds and proved they satisfy the "Oka Principle". As a consequence, he identified a large class of manifolds for which the inclusion $\mathrm{Hol}(M, N) \hookrightarrow map(M, N)$ is a weak equivalence. COHEN-COHEN-MANN-MILGRAM [**55**, 1991] described the full stable homotopy type of $\mathrm{Hol}_k^*(S^2, \mathbb{C}P^m)$, their description given in terms of configuration spaces. A related problem of stabilization for moduli spaces of connections is the subject of the famous "Atiyah-Jones conjecture" in mathematical physics ATIYAH-JONES [**14**, 1978].

The gauge groups provide a connection between the homotopy theory of function spaces and the theory of principal bundles. Let $P\colon E \to X$ be a principal G-bundle for G a connected topological group classified by a map $h\colon X \to BG$. The *gauge group* $\mathcal{G}(P)$ of P is defined to be the group of G-equivariant homeomorphisms $f\colon E \to E$ over X. ATIYAH-BOTT [**13**, 1983] used the gauge group in their celebrated study of Yang-Mills equations and principal bundles over a Riemann surface. They made use of Thom's theory and a multiplicative equivalence originally due to GOTTLIEB [**111**, 1972]

$$\mathcal{G}(P) \simeq \Omega map(X, BG; h)$$

to study the homotopy theory of $B\mathcal{G}(P)$. Gottlieb's identity, in turn, links the classification of gauge groups up to H-homotopy type, for fixed G and X, to the homotopy classification problem for $map(X, BG)$. CRABB-SUTHERLAND [**66**, 2000] proved that the gauge groups $\mathcal{G}(P)$ represent only finitely many homotopy types for G a compact Lie group and X a finite complex,. In contrast, the path-components of $map(X, BG)$ may represent infinitely many distinct homotopy types in this case by MASBAUM [**197**, 1991].

The advent of localization techniques introduced new depth to the study of function spaces while opening up a wide range of fundamental problems and applications. In his seminal paper on rational homotopy theory, SULLIVAN [**261**, 1977] sketched a construction for an algebraic model for components of $map(X, Y)$ for X and Y simply connected CW complexes with X finite, as an extension of Thom's ideas. Sullivan's construction was completed by HAEFLIGER [**122**, 1982]. Sullivan also identified the rational Samelson Lie algebra of $aut(X)$ for X a finite, simply connected CW complex via an isomorphism:

$$\pi_*(aut(X)) \otimes \mathbb{Q}, [\ ,\ ] \cong H_*(\mathrm{Der}(\mathcal{M}_X)), [\ ,\ ]$$

Here the latter space is the homology of the Lie algebra of degree lowering derivations of the Sullivan minimal model of X with the commutator bracket.

One of the early applications of Sullivan's rational homotopy theory was the proof by VIGUÉ-POIRRIER-SULLIVAN [**280**, 1976] of the unboundedness of the Betti numbers of the free loop space $LX = map(S^1, X)$ for certain simply connected CW complexes X. Combined with a famous result of GROMOLL-MEYER [**114**, 1969] in geometry, this calculation solved the "closed geodesic problem" for many manifolds. The calculation was deduced from a Sullivan model constructed for LX.

The p-local homotopy theory of a function space featured in a landmark result in algebraic topology, the proof of the Sullivan conjecture. MILLER [**211**, 1984] proved

$$\pi_n(map_*(B\pi, X; 0)) = 0 \quad \text{for all } n \geq 0$$

where π is any finite group and X any finite CW complex. Among many applications, this result was used by MCGIBBON-NEISENDORFER [**207**, 1984] to affirm Serre's conjecture: $\pi_m(X)$ contains a subgroup of order p for infinitely many m.

LANNES [**178**, 1987] constructed the T-functor which is left adjoint to the tensor product in the category of unstable modules over the Steenrod algebras. His construction provided a model for the mod p cohomology of the space $map(BV, X)$ where V is a p-group. Lannes' construction was adapted to the rational homotopy setting by BOUSFIELD-PETERSON-SMITH [**29**, 1989] and, later, BROWN-SZCZARBA [**36**, 1997] to give another model for the rational homotopy type of $map(X, Y; f)$. FRESSE [**101**, preprint] recently constructed a version of Lannes' functor in a category of operadic algebras giving a model for the integral homotopy type of certain function spaces.

The free loop space recently re-emerged as a central object for study in homotopy theory with the appearance of work of CHAS-SULLIVAN [**49**, preprint]. They constructed a product on the regraded homology

$$\mathbb{H}_*(LM^m) = H_{*+m}(LM^m)$$

for a simply connected, closed, oriented m-manifold M^m using intersection theory. They also defined a bracket on the equivariant homology of LM^m and a degree $+1$ operator giving $\mathbb{H}_*(LM^m)$ the structure of Batalin-Vilkovisky algebra. These structures have incarnations in diverse other settings. Their study, known as *string topology*, is now an active subfield in the intersection of homotopy theory and geometry.

1.2. Organization. In Section 2, we discuss work on the ordinary and stable homotopy theory, as opposed to the local homotopy theory of function spaces. We focus on the areas introduced above, namely: (i) the general path component $map(X, Y; f)$; (ii) the monoid $aut(X)$; and (iii) the free loop space LX. We also discuss work on the stable homotopy theory of these spaces and on spectral sequence calculations of their invariants. In Section 3, we discuss the localization of function spaces. We describe the algebraic models of Sullivan, and of later authors, for the general component, the monoid of self-equivalences and the free loop space in rational homotopy theory, and survey their applications. We also discuss the p-local homotopy theory of function spaces including the work of Miller, Lannes and others on the space of maps out of a classifying space, and algebraic models for function spaces in tame homotopy theory. The paper includes a rather extensive bibliography gathering together both papers directly focused on function spaces and papers giving significant applications and extensions.

2. Ordinary and Stable Homotopy Theory of Function Spaces.

We divide our discussion in this section according to the cases (i), (ii) and (iii) above. We then discuss some general results in stable homotopy theory and spectral sequence constructions for function spaces.

2.1. General Components. As mentioned in the introduction, the following open problem lies at the historical roots of the study of function spaces as objects in their own right.

PROBLEM 2.1. *Given spaces X and Y classify the path-components $map(X,Y;f)$ up to homotopy type for homotopy classes $f\colon X \to Y$.*

We consider a variety of cases here beginning with the most classical, mentioning progress on Problem 2.1, when appropriate.

2.1.1. *Maps from Spheres and Suspensions.* The components of $map(S^p, Y)$ correspond to the homotopy classes in $\pi_p(Y)$. The coproduct on S^p gives rise to an equivalence $map_*(S^p,Y;\alpha) \simeq map_*(S^p,Y;0)$, for any class $\alpha\colon S^p \to Y$. By adjointness, $\pi_n(map_*(S^p,Y;0)) \cong \pi_{n+p}(Y)$. These observations were made by WHITEHEAD [**284**, 1946] who gave the first algebraic method for computation. Whitehead identified the linking homomorphism in the long exact homotopy sequence for the evaluation fibration $map_*(S^p,Y;\alpha) \to map(S^p,Y;\alpha) \to Y$ obtaining:

$$\begin{array}{ccccccc} \cdots \longrightarrow & \pi_{n+1}(Y) & \xrightarrow{\partial} & \pi_n(map_*(S^p,Y;\alpha)) & \longrightarrow & \pi_n(map(S^p,Y;\alpha)) & \longrightarrow \cdots \\ & & \searrow^{W(\alpha)} & \downarrow \cong & & & \\ & & & \pi_{n+p}(Y) & & & \end{array}$$

where $W(\alpha)(\beta) = -[\alpha,\beta]_w$ denotes the Whitehead product map. Using this sequence, he proved $map(S^2,S^2;\iota) \not\simeq map(S^2,S^2;0)$ by comparing homotopy groups. HU [**147**, 1946] and KOH [**158**, 1960] computed $\pi_{2m-1}(map(S^{2m},S^{2m};\alpha))$ for small values of m. In these cases, the order of $\pi_{2m-1}(map(S^{2m},S^{2m};\alpha))$ depends on the absolute value of the degree of α and so distinguishes components with different absolute order. Since clearly $map(S^{2m},S^{2m};\alpha) \simeq map(S^{2m},S^{2m};-\alpha)$ the classification in these cases was complete with these calculations.

HANSEN [**128**, 1974] obtained the complete classification for self-maps of S^n. For even spheres, he proved

$$map(S^{2m},S^{2m};\alpha) \simeq map(S^{2m},S^{2m};\beta) \iff [\alpha,\iota]_w = \pm[\beta,\iota]_w.$$

Here $\iota \in \pi_{2m}(S^{2m})$ is the fundamental class. For odd spheres, the components of $map(S^{2m-1},S^{2m-1})$ are all homotopy equivalent for $m = 1,2,4$ due to the existence of a multiplication on S^{2m-1} in these cases. For $m \neq 1,2,4$, Hansen showed $map(S^{2m-1},S^{2m-1};\iota) \not\simeq map(S^{2m-1},S^{2m-1};0)$ and

$$map(S^{2m-1},S^{2m-1};\alpha) \simeq \begin{cases} map(S^{2m-1},S^{2m-1};\iota) & \text{if } \deg(\alpha) = \text{odd} \\ map(S^{2m-1},S^{2m-1};0) & \text{if } \deg(\alpha) = \text{even}. \end{cases}$$

Problem 2.1 remains open for $map(S^m,S^n)$ for $m > n$. YOON [**303**, 1995] observed a connection between the Gottlieb group $G_m(Y)$ and the homotopy classification problem for $map(S^m,Y)$ showing $map(S^m,Y;\alpha) \simeq map(S^m,Y;0)$ if and only if $\alpha \in G_m(Y)$. LUPTON-SMITH [**191**, 2008] extended this to a surjection of sets

$$\pi_m(Y)/G_m(Y) \twoheadrightarrow \frac{\{\text{components } map(S^m, Y; f)\}}{\text{homotopy equivalence}}.$$

Thus the complexity of the classification problem for $map(S^m, S^n)$ is roughly that of computing Gottlieb groups $G_m(S^n)$. Extensive, low-dimensional calculations of this group were recently made by GOLASIŃSKI-MUKAI [**105**, 2009]. LEE-MIMURA-WOO [**182**, 2004] calculated the Gottlieb groups for certain homogeneous spaces.

When $X = \Sigma A$ is a suspension, the fibres $map_*(\Sigma A, Y; f)$ of the various evaluation fibrations $\omega_f \colon map(\Sigma A, Y; f) \to Y$ are all homotopy equivalent to the space $map_*(\Sigma A, Y; 0)$ with homotopy groups

$$\pi_q(map_*(\Sigma A, Y; 0)) = [\Sigma^{q+1} A, Y].$$

LANG [**176**, 1973] extended Whitehead's exact sequence to this case replacing the Whitehead product in $\pi_*(Y)$ by the generalized Whitehead product in $[\Sigma^* A, Y]$. It is natural to consider, as Whitehead did, a stronger version of Problem 2.1, namely, the classification of the evaluation fibrations $\omega_f \colon map(X, Y; f) \to Y$ up to fibre homotopy type for homotopy classes $f \colon X \to Y$. HANSEN [**127**, 1974] defined $\omega_f \colon map(\Sigma A, \Sigma B; f) \to Y$ to be *strongly fibre homotopy equivalent* to $\omega_g \colon map(\Sigma A, \Sigma B; g) \to \Sigma B$ if the fibre homotopy equivalence is homotopic to the identity after (fixed) identification of the fibres with $map_*(\Sigma A, \Sigma B; 0)$. He proved:

$$\omega_f \text{ is strongly fibre homotopic to } \omega_g \iff [f, 1_{\Sigma B}] = [g, 1_{\Sigma B}]$$

where $[\ ,\ ]$ here denotes the generalized Whitehead product in $[\Sigma A, \Sigma B]$. MCCLENDON [**204**, 1981] showed that the evaluation fibrations $\omega_f \colon map(\Sigma A, Y; f) \to Y$ behave as principal fibrations and, in particular, are classified by maps $s \colon Y \to map(A, Y)$ determined by generalized Whitehead products.

2.1.2. *Maps into Eilenberg-Mac Lane Spaces.* The weak homotopy type of the space $map(X, K(\pi, n); f)$ may be described for any $f \colon X \to K(\pi, n)$ for π abelian. The ideas are due to THOM [**268**, 1957] with a refinement by HAEFLIGER [**122**, 1982]. First, observe that these components are all homotopy equivalent since $K(\pi, n)$ has the homotopy type of a topological group. A homotopy class $\alpha \in \pi_p(map(X, K(\pi, n; 0)))$ corresponds, by adjointness, to a map $A \colon S^p \times X \to K(\pi, n)$. On cohomology,

$$A^*(x_n) = 1 \otimes a_n + u_p \otimes a_{n-p}$$

where $a_n, a_{n-p} \in H^*(X; \pi)$ with subscripts indicating degree while $u_p \in H^p(S^p; \pi)$ and $x_n \in H^n(K(\pi, n); \pi)$ are the fundamental classes. Since A restricts to the constant map on $S^p \times *$ we see $a_n = 0$. The assignment $\alpha \mapsto a_{n-p}$ gives the identification

$$\pi_p(map(X, K(\pi, n); f)) \cong H^{n-p}(X; \pi),$$

mentioned in the introduction and leads to directly to a weak equivalence

$$map(X, K(\pi, n); f)) \simeq_w \prod_{p \geq 1} K(H^{n-p}(X; \pi), p).$$

Thom also indicated how the homotopy groups $\pi_p(map(X, Y; f))$ for Y a finite Postnikov piece are determined, up to extensions, by the k-invariants of Y and the groups $H^{n-p}(X, \pi_n(Y))$. This approach was encoded in Haefliger's construction of a Sullivan model for $map(X, Y; f)$, as discussed below.

GOTTLIEB [**109**, 1969] extended Thom's result to the case $n = 1$ and π any group. Here

$$map(X, K(\pi, 1); f) \simeq_w K(C(f_\sharp), 1)$$

where $C(f_\sharp)$ denotes the centralizer of the image of $f_\sharp\colon \pi_1(X) \to \pi$. MÖLLER [**216**, 1987] showed that when Y is a twisted Eilenberg-Mac Lane space, then $map(X, Y; f)$ is one also with homotopy groups determined by the cohomology groups of X with twisted coefficients in the homotopy groups of Y. Note that the weak equivalences above are homotopy equivalences by Whitehead's Theorem, when $map(X, K(\pi, n))$ is of CW type, e.g., when X is compact or a CW complex. In general, the study of the homotopy type of $map(X, Y; f)$ when Y has at least two nonvanishing homotopy groups is a difficult, open problem.

2.1.3. *Maps between Manifolds.* The homotopy theory of $map(M^m, N^n)$ for M^m and N^n closed manifolds is a topic of wide-ranging interest. In this case, important variations have been considered. Below we consider one such variation with direct ties to Problem 2.1, namely spaces of holomorphic maps. We begin with the space $map(M^m, N^n)$.

If T_g is an orientable surface, then $T_g \simeq K(\pi_1(T_g), 1)$ and the classification problem for $map(X, T_g)$ reduces to the computation of centralizers of homomorphisms into $\pi_1(T_g)$. For $g \geq 2$ this group is highly nonabelian and the only possible nontrivial centralizers are isomorphic to $\mathbb{Z}$ by HANSEN [**134**, 1983]. HANSEN [**130**, 1974] earlier considered the space $map(T_g, S^2)$. As a generalization of Whitehead's exact sequence, he showed an exact sequence

$$0 \to \mathbb{Z}/2|m| \to \pi_1(map(T_g, S^2; \iota_m)) \to \mathbb{Z}^{2g} \to 0$$

which gives the classification, in terms of degree, in this case. The fundamental group $\pi_1(map(T_g, S^2; \iota_m))$ was later completely determined by LARMORE-THOMAS [**180**, 1980].

HANSEN [**132**, 1981] extended his classification result for the space of self-maps of spheres to the case $map(M^m, S^m)$ where M^m is closed, oriented, connected manifold with vanishing first Betti number. Note that, by Hopf's Theorem, $[M^m, S^m] = \mathbb{Z}$ with maps classified by degree. When m is even and ≥ 4, Hansen showed the homotopy types of $map(M^m, S^m; \alpha)$ are classified by the absolute values of the degrees of the α. When m is odd and $m \neq 1, 4, 7$ there are two homotopy types corresponding to the distinct types $map(M^m, S^m; 0)$ and $map(M^m, S^m; \iota)$ where ι is of degree 1. Again in this case, components are classified by the parity of degree of the class α.

SUTHERLAND [**262**, 1983] extended Hansen's work eliminating the restriction on the first Betti number and dealing with the case M^m nonorientable. Note that in the latter case there are only two distinct classes $\alpha\colon M^m \to S^m$ and so the problem reduces to distinguishing these components for $m \neq 1, 4, 7$. Sutherland observed that the components of $map(M^m, S^m)$ all have the same homotopy type if there is a map $I\colon M^m \to map(S^m, S^m; \iota)$ making the diagram

$$\begin{array}{ccc} & & map(S^m, S^m; 1) \\ & \overset{I}{\nearrow} & \downarrow \omega \\ M^m & \xrightarrow{\quad \iota \quad} & S^m \end{array}$$

commute, where ι is of degree 1. Taking $M^m = \mathbb{R}P^m$, we have a lift I based on the lift $I' \colon \mathbb{R}P^m \to SO(n+1) \subseteq map(S^m, S^m; 1)$ of ι. Sutherland showed the components of $map(M^m, S^m)$ are all of the same homotopy type if there exists a map $f \colon M^m \to \mathbb{R}P^m$ of odd degree giving examples with $m \neq 1, 4, 7$ for which all the components are homotopy equivalent.

SASAO [**233**, 1974] studied the homotopy type of components of $map(\mathbb{C}P^m, \mathbb{C}P^n; i)$ for $m \leq n$ and $i \colon \mathbb{C}P^m \to \mathbb{C}P^n$ the inclusion. He constructed a map

$$\alpha_{m,n} \colon U(n+1)/\Delta(m+1) \times U(n-m) \to map(\mathbb{C}P^m, \mathbb{C}P^n; i)$$

where $\Delta(m+1) \subset U(m+1)$ denotes scalar multiplies of the identity. He proved $\alpha_{m,n}$ induces an isomorphism on rational homotopy groups and on ordinary homotopy groups through degree $4n - 4m + 1$. YAMAGUCHI [**287**, 1983] extended Sasao's analysis to the case of quaternionic projective spaces. MÖLLER [**213**, 1984] gave the complete classification for the components $map(\mathbb{C}P^m, \mathbb{C}P^n)$ showing the homotopy types are classified by the absolute value of the degree of a representative class. The result is a direct consequence of his calculation

$$H_{2n-2m+1}(map(\mathbb{C}P^m, \mathbb{C}P^n; \iota_k)) \cong \mathbb{Z}/d \text{ where } d = \binom{n+1}{m} |k|^m.$$

YAMAGUCHI [**290**, 2006] studied maps between real projective spaces. He defined the analogue of Sasao's map, here of the form

$$\beta_{m,n} \colon O(n+1)/\Delta(m+1) \times O(n-m) \to map(\mathbb{R}P^m, \mathbb{R}P^n; i)$$

and proved $\beta_{m,n}$ is an equivalence on ordinary and rational homotopy groups through certain ranges of degrees.

When G is a topological group (or group-like space) the path-components of $map(X, G)$ are all of the same homotopy type. Problem 2.1 thus reduces, in this case, to the study of the homotopy theory of the null-component $map(X, G; 0)$. Given Lie groups G and H, the calculation of homotopy invariants of $map(G, H)$ is a difficult open problem. Recently, MARUYAMA-ŌSHIMA [**196**, 2008] computed the homotopy groups of $map_*(G, G)$ for $G = SU(3), Sp(2)$ in degrees ≤ 8.

2.1.4. *Spaces of Holomorphic Maps.* SEGAL [**242**, 1979] proved a basic result on the homotopy theory of the space $\mathrm{Hol}(M, N)$. His work launched a vital subfield of research on the "stability" of the inclusion $\mathrm{Hol}(M, N) \hookrightarrow map(M, N)$. Segal proved

$$\mathrm{Hol}_k^*(T_g, \mathbb{C}P^n) \hookrightarrow map_*(T_g, ; \iota_k)$$

induces a homology isomorphism through dimension $(k - 2g)(2n - 1)$ where T_g is a Riemann surface of genus g. Specializing to the case of the sphere, he proved

$$\mathrm{Hol}_k^*(S^2, \mathbb{C}P^n) \hookrightarrow map_*(S^2, \mathbb{C}P^n; \iota_k)$$

induces a homotopy equivalence up to degree $2n - 1$.

Segal's work was extended by many authors. GUEST [**118**, 1984] proved the corresponding stability result on homology for $\mathrm{Hol}_k^*(S^2, F) \hookrightarrow map(S^2, F; \iota_k)$ for certain complex flag manifolds F. His proof involved developing the analogue of a Morse-theoretic result for the case of the energy functional on the space $C^\infty(S^2, F)$ of smooth maps. KIRWAN [**156**, 1986] extended Segal's result to the case the target is the complex Grassmannian manifold $G(n, n+m)$ of n-planes in $n+m$-space proving $\mathrm{Hol}_k^*(S^2, G(n, n+m)) \hookrightarrow map(S^2, G(n, n+m); \iota_k)$ induces a homology isomorphism in degrees depending on k, n and m. MANN-MILGRAM [**194**, 1991] considered this case as well, constructing a spectral sequence to analyze the homology of $\mathrm{Hol}_k^*(S^2, G(n, n+m))$. GRAVESON [**112**, 1989] studied holomorphic maps into space ΩG for G a complex, compact Lie group.

COHEN-COHEN-MANN-MILGRAM [**55**, 1991] and COHEN-SHIMAMOTA [**63**, 1991] described the stable homotopy type of $\mathrm{Hol}_k^*(S^2, \mathbb{C}P^n)$. They proved

$$\mathrm{Hol}_k^*(S^2, \mathbb{C}P^n) \simeq C_k(\mathbb{R}^2, S^{2n-1})$$

where $C_k(\mathbb{R}^2, S^{2n-1})$ is the configuration space of distinct points in $\mathbb{R}^2$ with labels in S^{2n-1} of length at most k. Cohen-Cohen-Mann-Milgram also computed the homology of $\mathrm{Hol}_k^*(S^2, \mathbb{C}P^n)$ with $\mathbb{Z}_p$-coefficients in terms of Dyer-Lashof operations. MANN-MILGRAM [**195**] used the stable homotopy decomposition above to prove the homology stability of the inclusion $\mathrm{Hol}_k^*(S^2, F) \hookrightarrow map_*(S^2, F)$ for F an $SL(n, \mathbb{C})$-flag-manifold. BOYER-MANN-HURTUBISE-MILGRAM [**32**, 1994] and HURTUBUISE [**148**, 1996] proved homology stabilization theorems for the space $\mathrm{Hol}_k^*(S^2, G/P)$ for certain complex homogeneous spaces G/P. BOYER-HURTUBISE-MILGRAM gave a configuration space description of $\mathrm{Hol}_k(T_g, M)$ for certain complex manifolds admitting nice Lie group actions extending the approach of Graveson.

Segal's stabilization problem has deep interdisciplinary connections. GROMOV [**115**, 1989] obtained general stability results as a consequence of his work on the *Oka Principle* in complex geometry. A complex manifold M satisfies the Oka principle if every continuous map $f\colon S \to M$ is homotopic to a holomorphic map where S is a Stein manifold. Gromov identified the class of "elliptic" manifolds and proved elliptic manifolds satisfied the Oka principle. Consequently, he obtained the inclusion

$$\mathrm{Hol}(S, M) \hookrightarrow map(S, M)$$

is a weak homotopy equivalence for S Stein and M elliptic. The class of elliptic manifolds includes complex Lie groups and their homogeneous spaces.

The problem of stabilization also has a famous incarnation in Yang-Mills theory and mathematical physics. ATIYAH-JONES [**14**, 1978] constructed a map

$$\theta_k\colon \mathcal{M}_k \to map_*(S^3, SU(2); \iota_k)$$

where $\mathcal{M}_k$ is a moduli space of connections on a principal $SU(2)$-bundle P_k over S^4 corresponding to a map $S^4 \to BSU(2)$ of degree k. They proved θ_k induces a homology surjection through a range of degrees and conjectured θ_k induces an equivalence in both homotopy and homology through a range depending on k. Work on the Atiyah-Jones conjecture includes TAUBES [**266**, 1989], GRAVESON [**112**, 1989] and BOYER-HURTUBISE–MANN-MILGRAM [**30**, 1993].

Many authors have studied related spaces of maps. VASSILIEV [**275**, 1992] proved a stable equivalence

$$\mathrm{Hol}_k^*(S^2, \mathbb{C}P^n) \simeq \mathrm{SP}_{n-1}^k(\mathbb{C})$$

where the latter is the space of monic complex polynomials of degree k with all roots of multiplicity $< n$. GUEST-KOZLOWSKI-YAMAGUCHI [**121**, 1994] extended Segal's result in a different direction, proving the inclusion

$$\mathrm{Hol}_k^*(S^2, X_n) \hookrightarrow map_*(S^2, X_n)$$

is a homotopy equivalence up to degree k where $X_n \subseteq \mathbb{C}P^{n-1}$ is the subspace of points with at most one coordinate zero. The cohomology of the space of basepoint-free holomorphic maps $\mathrm{Hol}_1(S^2, S^2)$ was studied by HAVLICEK [**139**, 1995] while the homotopy groups of $\mathrm{Hol}_k(S^2, S^2)$ were studied by GUEST-KOZLOWSKI-MURAYAMA-YAMAGUCHI [**120**, 1995]. KALLEL-MILGRAM [**152**, 1997] gave a complete calculation of the homology of $\mathrm{Hol}_k^*(T_g, \mathbb{C}P^m)$ for T_g an elliptic Riemann surface. The space of real rational functions was recently studied by KAMIYAMA [**155**, 2007].

KALLEL-SALVATORE [**153**, 2006] applied techniques from string topology to the study of spaces of maps between manifolds. Set

$$\mathbb{H}_*(map(M^m, N^n)) = H_{*+n}(map(M^m, N^n))$$

and similarly for $\mathrm{Hol}(S^2, N^n)$. When M^m, N^n are closed, compact and orientable, they proved $\mathbb{H}_*(map(S^m, N))$ has a ring structure corresponding to an intersection product and $\mathbb{H}_*(map(M^m, N^n))$ is a module over this ring. They used this structure to compute $\mathbb{H}_*(map(S^2, \mathbb{C}P^n; \iota_k))$ and $\mathbb{H}_*(\mathrm{Hol}_k(S^2, \mathbb{C}P^n))$ with $\mathbb{Z}_p$-coefficients. They also studied $\mathbb{H}_*(map(T_g, \mathbb{C}P^n; \iota_k); \mathbb{Z}_p))$ for T_g a compact Riemann surface proving, among other results, that these groups are isomorphic for all k when p divides n.

2.1.5. *Maps into a Classifying Space and Gauge Groups.* Let X be a space and G a connected topological group. Suppose $P\colon E \to X$ is a principal G-bundle. The gauge group $\mathcal{G}(P)$ is the topological group of bundle automorphisms of P. The gauge group featured in important work of ATIYAH-BOTT [**13**, 1983] in mathematical physics. They considered the action of $\mathcal{G}(P)$ on the moduli space $\mathcal{A}$ of Yang Mills connections on a principal $U(n)$-bundle $P\colon E \to M$ for M a Riemann manifold. Among other results, they proved $H^*(B\mathcal{G}(P))$ is torsion free and computed its Poincaré series. Their calculation depends on the identity:

$$\mathcal{G}(P) \simeq_H \Omega map(X, BG; h),$$

where $h\colon X \to BG$ is the classifying map of P, a result originally due to GOTTLIEB [**111**, 1972]. Here X is a finite CW complex. Thus $B\mathcal{G}(P) \simeq map(X, BG; h)$. By Bott periodicity, the loops and double loops on $BU(n)$ are torsion free. Atiyah-Bott used this fact and Thom's theory to make their calculations.

The classification of gauge groups for fixed X and G up to H-equivalence or, alternately, up to ordinary equivalence is the subject of active research. Gottlieb's identity implies the homotopy classification problem for $map(X, BG)$ refines the gauge group classification problem. MASBAUM [**197**, 1991] studied the homology of the components of the space $map(X, BSU(2))$ for X a 4-dimensional CW complex obtained by attaching a single 4-cell to a bouquet of 2-spheres. This case includes oriented, simply connected 4-dimensional manifolds. Using a cofibre sequence for X, he obtained, in particular, that the components of $map(S^4, BSU(2))$ represent infinitely many homotopy types. Using a related analysis, SUTHERLAND [**262**, 1992] considered the classification of components of $map(T_g, BU(n))$ for T_g an orientable surface of genus g. He obtained the calculatiuon

$$\pi_{2n-1}(map(T_g, BU(n); \iota_k)) \cong \mathbb{Z}^g \oplus \mathbb{Z}/d \quad \text{where } d = (n-1)!(k, n)$$

thus distinguishing the components corresponding to maps k and l with $(k, n) \neq (l, n)$. TSUKUDA [**274**, 2001] classified the homotopy types represented by the components of $map(S^4, BSU(2))$ showing $map(S^4, BSU(2); \iota_k) \simeq map(S^4, BSU(2); \iota_l)$ if and only if $k = \pm l$. KONO-TSUKUDA [**161**, 2000] generalized this result from $X = S^4$ to X a simply connected 4-dimensional manifold.

As regards the homotopy type of the gauge group, KONO [**159**, 1991] proved

$$\mathcal{G}(P_k) \simeq \mathcal{G}(P_l) \iff (12, k) = (12, l)$$

for $k, l \in \pi_4(BSU(2)) \cong \mathbb{Z}$. Thus the infinitely many distinct homotopy types represented by the path-components of $map(S^4, BSU(2))$ loop to only 6 distinct homotopy types. Kono's proof included the calculation

$$\pi_2(\mathcal{G}(P_k)) = \mathbb{Z}/(12, k)$$

using, essentially, Whitehead's exact sequence mentioned above. KONO-TSUKUDA [**161**, 1996] extended this result from $X = S^4$ to X a closed, simply connected manifold using a cofibre sequence for X to make the corresponding calculation. HAMANAKA-KONO [**126**, 2006] obtained a corresponding classification for $SU(3)$-bundles over S^4. They proved

$$\mathcal{G}(P_k) \simeq \mathcal{G}(P_l) \iff (120, k) = (120, l)$$

where P_k and P_l are principal $SU(3)$-bundles over S^6 with third Chern class equal to $2k$ and $2l$, respectively.

CRABB-SUTHERLAND [**66**, 2000] obtained a global result on the classification of gauge groups. They proved that, for any fixed finite CW complex X and compact Lie group G, there are only finitely many H-homotopy types represented by the gauge groups $\mathcal{G}(P)$ for P a principal G-bundle over X. Their proof is based on an alternate description of the gauge group:

$$\mathcal{G}(P) \cong \Gamma(\mathrm{Ad}(P))$$

where $\mathrm{Ad}(P) = E \times_G G^{\mathrm{ad}} \to X$ is the adjoint bundle associated to $P\colon E \to X$ and the space of sections has multiplication induced by G. A key step in their finiteness result is the proof that the fibrewise rationalization of $\mathrm{Ad}(P)$ is equivariantly trivial. They also classified the H-homotopy types of gauge groups of $SU(2)$-principal bundles over S^4 complementing Masbaum and Kono's work. Here

$$\mathcal{G}(P_k) \simeq_H \mathcal{G}(P_l) \iff (180, k) = (180, l).$$

Summarizing, the infinitely many distinct homotopy types represented by the components of $map(S^4, BSU(2))$ loop to 6 distinct homotopy types and 18 distinct H-homotopy types.

2.2. Spaces of Self-Equivalences. The space of equivalences $aut(X)$ of a space X with some additional structure admits many important refinements. When M is a Riemannian manifold, we have the chain of subspaces

$$\mathrm{Isom}(M) \hookrightarrow \mathrm{Diff}(M) \hookrightarrow \mathrm{Homeo}(M) \hookrightarrow aut(M)$$

given by spaces of isometries, diffeomorphisms and homeomorphisms, respectively. Each of these spaces is the subject of active research in homotopy theory and geometric topology. SMALE [**245**, 1959] proved the inclusion

$$\mathrm{Isom}(S^2) \hookrightarrow \mathrm{Diff}(S^2)$$

is a homotopy equivalence. Since $\mathrm{Isom}(S^2) \simeq O(3)$ this determines the homotopy type of $\mathrm{Diff}(S^2)$. The (Generalized) Smale Conjecture asserts that $\mathrm{Isom}(M) \simeq \mathrm{Diff}(M)$ for M a 3-manifold of constant, positive curvature. The Smale Conjecture was affirmed for $M = S^3$ by HATCHER [**136**, 1983]. GABAI [**103**, 2003] proved the corresponding result for M a closed, hyperbolic 3-dimensional manifold.

The first result on the H-homotopy type of $aut(X)$ is due, essentially, to THOM [**268**, 1957]. By his results mentioned above, we have

$$aut(K(\pi, n)) \simeq_H aut(\pi) \rtimes K(\pi, n)$$

for π abelian. When $n = 1$, Gottlieb's extension of Thom's results leads to an identification

$$aut(K(\pi, 1)) \simeq_H \mathrm{Out}(\pi) \rtimes K(\pi, 1)$$

where $\mathrm{Out}(\pi)$ denotes the group of outer automorphisms. Note that these results include a description of $\pi_0(aut(X))$, the group of free homotopy self-equivalences of X. This group is, in general, quite complicated even for simple X. See ARKOWITZ [**11**, 1990] and RUTTER [**231**, 1997] for surveys of the extensive literature on this group.

Since the path-components of $aut(X)$ are all of the same homotopy type, we focus on the component of the identity which we denote $aut(X)_\circ$. Thus

$$aut(X)_\circ = map(X, X; 1)$$

is the identity component in the space of self-maps.

HANSEN [**135**, 1990] identified the homotopy type of $aut(S^2)_\circ$ by comparing the evaluation fibration for this space with the fibre sequence $SO(2) \to SO(3) \to S^2$. He proved

$$aut(S^2)_\circ \simeq_H SO(3) \times \widetilde{aut_*(S^2)_\circ}$$

where $\widetilde{Z}$ denotes the universal cover. Combined with Smale's result, this shows $\mathrm{Diff}(S^2) \hookrightarrow aut(S^2)$ is not a homotopy equivalence. YAMANOSHITA [**302**, 1993] obtained a related result proving

$$aut(\mathbb{R}P^2)_\circ \simeq_H SO(3) \times \left(aut_*(\mathbb{R}P^2)_\circ)/O(2)\right).$$

This result implies $\mathrm{Diff}(\mathbb{R}P^2) \simeq O(3)$ is not homotopy equivalent to $aut(\mathbb{R}P^2)$. YAMANOSHITA [**299**, 1985] also obtained a general result

$$aut(X \times Y) \simeq aut(X) \times aut(Y) \times map_*(Y, aut(X)) \times map_*(X, aut(Y))$$

provided the dimension of X is less than the connectivity of Y. In particular:

$$aut(S^1 \times Y) \simeq O(2) \times aut(Y) \times \Omega aut_\circ(Y)$$

for simply connected Y.

MCCULLOUGH [**205**, 1981] computed $\pi_q(aut(M)_\circ)$ for $1 \leq q \leq n-3$ for M a connected sum of closed, aspherical manifolds of dimension ≤ 3 proving the groups $\pi_{n-2}(aut(M)_\circ)$ are not finitely generated. He used this result to give examples of closed 3-manifolds M such that the fundamental group of $\mathrm{Homeo}(M)$ is not finitely generated. DIDIERJEAN [**69**, 1990] and [**70**, 1992] used a Postnikov decomposition of X to construct a spectral sequence converging to the homotopy groups of $aut(X)$. She determined low degree homotopy groups of $aut(X)$ for $X = SU(3)$ and $X = Sp(2)$ up to extensions.

Given a fibration $p\colon E \to B$, we may consider the monoid $aut(p)$ of fibre-homotopy equivalences $f\colon E \to E$ covering the identity of B. BOOTH-HEATH-MORGAN-PICCININI [**25**, 1984] extended Gottlieb's result for the gauge group to prove an H-equivalence

$$aut(p) \simeq_H \Omega map(B, Baut(F); h))$$

where F is the fibre of p and $h\colon B \to Baut(F)$ is the classifying map. A simplicial version of this result was earlier obtained by DROR-DWYER-KAN [**72**, 1980]. DIDIERJEAN [**68**, 1987] extended Thom's result to the fibrewise setting, proving

$$\pi_q(aut(p)) \cong H^{n-q}(B; \pi)$$

for $p\colon E \to B$ a principal fibration with fibre $K(\pi, n)$ and made calculations of the homotopy groups of $aut(p)$ when F has two nonzero homotopy groups. She constructed a spectral sequence converging to the homotopy groups of $aut(p)$, expanding on work of LEGRAND [**183**, 1983].

2.3. The Free Loop Space. The space of maps

$$LX = map(S^1, X)$$

is the subject of intensive research in diverse branches of mathematics. Given a compact Lie group G, the space of smooth loops on G is an infinite-dimensional Lie group, called the "loop group" of G. The representation theory of loop groups has deep connections to mathematical physics (cf. PRESSLEY-SEGAL [**228**, 1986]). GROMOLL-MEYER [**114**, 1969] linked the closed geodesic problem for a compact Riemannian manifold M to the homotopy theory of the free-loop space LM. They proved M admits infinitely many closed, prime geodesics if the Betti numbers of LM grow without bound. VIGUÉ-POIRRIER-SULLIVAN [**280**, 1976] proved that the Betti numbers of LM are unbounded when the rational cohomology of M requires at least two generators. Their calculation was facilitated by a Sullivan model for LM, described in Section 3, below.

More recently, CHAS-SULLIVAN [**49**, preprint] and [**50**, 2004] unearthed a wealth of structure on the (regraded) homology of LM of a closed, oriented, smooth m-manifold M. Setting $\mathbb{H}_*(LM) = H_{*+m}(LM)$ they defined a graded-commutative and associative product $\bullet$ on $\mathbb{H}_*(LM)$ and a related Lie bracket on the equivariant homology. The pair give $\mathbb{H}_*(LM)$ the structure of Gerstenhaber algebra. A degree $+1$ operator Δ give $\mathbb{H}_*(LM)$ the structure of Batalin-Vilkovisky (BV) algebra. These structures were obtained by geometric methods using intersection theory and transversality arguments.

A homotopy theoretic construction of the string topology structures was given by COHEN-JONES [**60**, 2002] using Thom spectra. They also proved, for M simply connected, an isomorphism of graded algebras

$$\mathbb{H}_*(LM; \mathbb{F}) \cong HH^*(S^*(M), S^*(M); \mathbb{F})$$

where the latter is the Hochschild cohomology of the algebra of singular cochains of M. Here $\mathbb{F}$ is a field. The Cohen-Jones construction was extended to more general ring spectra by GRUHER-SALVATORE [**117**, 2008]. COHEN-KLEIN-SULLIVAN [**62**, 2008], CRABB [**64**, 2008], and Gruher-Salvatore independently proved the homotopy invariance of the loop product and bracket, a significant advance since the original constructions depended on the smooth structure of M. CHATAUR [**51**, 2005], HU [**146**, 2006] and KALLEL-SALVATORE [**153**, 2006] considered generalizations of string operations from $LM = map(S^1, M)$ to $map(S^n, M)$. The work of these various authors include constructions of the string topology operations in the frameworks of ring spectra (Cohen-Jones and Gruher-Salvatore]), bordism theory (Chataur) and fibrewise homotopy theory (Crabb).

As regards the ordinary homotopy theory of the free loop space, HANSEN [**129**, 1974] gave an example of an aspherical space X with $\pi_1(LX)$ not finitely generated. Note that when X is an H-space $LX \simeq X \times \Omega X$. AGUADÉ [**2**, 1987] made a general study of spaces X, called T-*spaces*, for which the evaluation fibration

$$\Omega X \to LX \to X$$

is fibre-homotopically trivial. He obtained a refinement of the notion of H-space via a sequence of classes $T = T_1 \subset T_2 \subset \ldots \subset T_\infty = H$-spaces with separating examples. WOO-YOON [**286**, 1995] proved that when X is a T-space the components of $map(\Sigma A, X)$ are all homotopy equivalent. FADELL-HUSSEINI [**86**, 1989] proved LX has infinite L.S. category for X a simply connected CW complex with finitely generated, nontrivial rational cohomology.

SMITH [**246**, 1981] and [**247**, 1984] constructed an Eilenberg-Moore spectral sequence for the cohomology of a free loop space. Starting with the pull-back square

$$\begin{array}{ccc} LX & \xrightarrow{\omega} & X \\ {\scriptstyle\omega}\downarrow & & \downarrow{\scriptstyle\Delta} \\ X & \xrightarrow{\Delta} & X \times X \end{array}$$

he obtained a spectral sequence

$$E_2^{*,*} = \mathrm{Tor}_{H^*(X \times X;\mathbb{F})}(H^*(X;\mathbb{F}), H^*(X;\mathbb{F})) \implies H^*(LX;\mathbb{F}).$$

He proved collapsing results for this spectral sequence and obtained calculations of $H^*(LM;\mathbb{F})$ for $\mathbb{F}$ of characteristic 2 and 0. KURIBAYASHI [**167**, 1991] used the Eilenberg-Moore spectral sequence to prove the fibre is totally noncohomologous to zero in the fibration $\Omega M \to LM \to M$ for M a Grassmann or Stiefel manifold and mod p cohomology for certain primes p.

MCCLEARY-ZILLER [**203**, 1987] proved the Betti numbers of LM are unbounded for M a compact, simply connected homogeneous space not equivalent to a rank one symmetric space using spectral sequence methods and extending earlier work of ZILLER [**305**, 1977] who used Morse theory. ROOS [**230**, 1988] studied the Poincaré-Betti series for LX for X a wedge of spheres using local algebra. He proved the series for $X = S^2 \vee S^2$ is not rational. HALPERIN-VIGUÉ-POIRRIER [**125**, 1991] proved the $\mathbb{F}$-Betti numbers are unbounded for a field $\mathbb{F}$ of positive characteristic k provided $H^*(X;\mathbb{F})$ requires at least 2 generators and under certain restrictions on k and X. MCCLEARY-MCLAUGHLIN [**202**, 1992] studied the free loop space in the context of Morava K-theory while OTTOSEN [**224**, 2003] considered the Borel cohomology of the free loop space. LAMBRECHTS [**175**, 2001] proved the Betti numbers of the free loop space are unbounded for certain connected sums.

BURGHELEA-FIEDOROWICZ [**43**, 1984] and GOODWILLIE [**106**, 1985] proved an isomorphism of graded spaces

$$H^*(LX) \cong HH^*(S_*(\Omega X), S_*(\Omega X))$$

where the latter space is the Hochschild cohomology. MENICHI [**209**, 2001], DUPONT-HESS [**76**, 2002] and NDOMBOL-THOMAS [**221**, 2002] independently proved this is an isomorphism of algebras. Menichi also made calculations of the graded algebra $H^*(LX;\mathbb{Z}_p)$ for X a suspension and $X = \mathbb{C}P^n$ while Ndombol-Thomas used Hochschild cohomology to make calculations of $H^*(LX;\mathbb{Z}_p)$ for $X = S^m, \mathbb{C}P^m$ and $\Sigma\mathbb{C}P^m$. KURIBAYASHI-YAMAGUCHI [**171**, 1997] made complete calculations of $H^*(LX;\mathbb{Z}_p)$ when X is simply connected with mod p cohomology an exterior algebra on few generators using Hochschild cohomology to obtain information on the E_2-term in the Eilenberg-Moore spectral sequence. Recently, SEELIGER [**241**, 2008] used the Serre spectral sequence applied to the evaluation fibration to make calculations of $H^*(L\mathbb{C}P^m)$.

Since the appearance of the paper of Chas-Sullivan, the structure of $\mathbb{H}_*(LM)$ has seen an explosion of research with many partial descriptions of the loop product, the Gerstenhaber algebra and the BV algebra structure in special cases. We mention a sampling of these results here. COHEN-JONES-YAN [**61**, 2004] constructed a spectral sequence of algebras

$$E^2_{*,*} = H^*(M; H_*(\Omega M)) \Longrightarrow \mathbb{H}_*(LM)$$

and used this to calculate the loop product for $M = S^2, \mathbb{C}P^n$. TAMANOI [**264**, 2006] computed the BV algebra structure of $\mathbb{H}_*(LM)$ for $M = SU(n)$ and M a complex Stiefel manifold. GRUHER-SALVATORE [**117**, 2008] extended the string operations to the case $M = BG$ for G a compact Lie group. MENICHI [**210**, 2009] proved the Cohen-Jones isomorphism

$$\mathbb{H}_*(LS^2; \mathbb{F}_2) \cong HH^*(S^*(S^2; \mathbb{F}_2), S^*(S^2; \mathbb{F}_2))$$

mentioned above is an isomorphism of Gerstenhaber algebras but, surprisingly, not an isomorphism of BV algebras.

2.4. Spectral Sequences and Stable Decompositions. We here discuss some general results on homotopy invariants and the stable homotopy type of function spaces not covered by the preceding discussion.

FEDERER [**87**, 1956] constructed a spectral sequence converging to the homotopy groups of $map(X, Y; f)$ for X any finite CW complex and Y a simple CW complex. He defined an exact couple from the long exact homotopy sequences of the restriction maps $\rho_n\colon map(X_n, Y; f_n) \to map(X_{n-1}, Y; f_{n-1})$. He identified the homotopy groups of the fibre of ρ_n with cellular cochain groups of X with coefficients in $\pi_*(Y)$ and obtained a spectral sequence

$$E_2^{p,q} = H^q(X; \pi_{p+q}(Y)) \implies \pi_p(map(X, Y; f)).$$

DYER [**85**, 1966] applied the Federer spectral sequence to calculate low degree homotopy and homology group of components of $map(X, Y)$ when $\dim X$ is less than the connectivity of Y. SCHULTZ [**240**, 1973] and MÖLLER [**217**, 1990] constructed equivariant versions of this spectral sequence.

BORSUK [**26**, 1952] proved that if X is a finite CW complex of dimension k with nonzero kth Betti number then $map_*(X, S^n)$ has nonzero $(n-k)$th Betti number. MOORE [**220**, 1956] extended this analysis and asked for a spectral sequence relating the cohomology of X to the homology of $map_*(X, S^n)$. SPANIER [**256**, 1959] showed the functor

$$F_n(X) = \varinjlim_k map(X, \Omega^k \mathrm{SP}^{n+k} S^{n+k}),$$

converts cohomology groups to homotopy groups. Here SP^{n+k} is the symmetric product functor. ANDERSON [**6**, 1972] constructed an Eilenberg-Moore spectral sequence for the cohomology of $map_*(X, Y)$ using the cobar construction in the category of cosimplicial spaces. LEGRAND [**184**, 1986] constructed a spectral sequence in the spirit of Moore's problem for $map_*(X, Y)$ using a Postnikov decomposition of Y. PATRAS-THOMAS [**227**, 2003] proved the Anderson spectral sequence converges when the dimension of X is no bigger than the connectivity of Y. CHATAUR-THOMAS [**53**, 2004] gave a related E_∞-model for function spaces. Applied to the free loop space, their model gives an operadic version of Hochschild cohomology.

The stable homotopy type of the based function space $map_*(X, Y)$ has been described in many cases. When $X = S^n$, this is just the loop space $\Omega^n Y$. Stable decompositions of this space are central structural results in homotopy theory. SNAITH [**255**, 1974] proved a stable decomposition for $\Omega^n Y$ in terms of configuration spaces of j-tuples of distinct points of $\mathbb{R}^n$. His proof was obtained by analyzing the stable homotopy type of approximations due to MAY [**198**, 1972] for iterated loop spaces. BÖDIGHEIMER [**22**, 1987] proved a generalization of Snaith's splitting result. He showed

$$\Omega^\infty \Sigma^\infty map_*(M, \Sigma^n Y) \simeq \Omega^\infty \Sigma^\infty \left(C(M, \partial M; n) \wedge_{\Sigma_n} Y^n\right)$$

where M is a compact manifold with boundary and $C(M, \partial M; n)$ is the configuration space. ARONE [**12**, 1999] described the Goodwillie tower of $\Omega^\infty \Sigma^\infty map_*(X, \Sigma^n Y)$ for more general X recovering the previous splittings. He obtained an Eilenberg-Moore spectral sequence from this description. AHEARN-KUHN [**5**, 2002] and KUHN [**165**, 2006] studied this spectral sequence proving it a spectral sequence of graded algebras and studying functorial properties.

CAMPBELL-COHEN-PETERSON-SELICK [**44**, 1987] studied $map_*(P_m(2^r), S^n)$ where $P_m(2^r) = S^{m-1} \cup e^m$ is a Moore space with attaching map of degree 2^r. They gave a partial description of the mod 2 Steenrod operations and proved that $map_*(P_3(2), S^n)$ is not decomposable as a product except, perhaps, for finitely many n. WESTERLAND [**282**, 2006] obtained a stable splitting for components of the space $map_*(T_g, S^2)$ where T_g is a surface of genus $g > 0$. COHEN [**59**, 1987] and BÖDIGHEIMER [**22**, 1987] independently obtained a stable splitting of the free loop space $L\Sigma X$ in terms of configuration spaces. Other results on the stable homotopy of the free loop space include splitting results for $L\mathbb{R}P^n$ by BAUER-CRABB-SPREAFICO [**15**, 2001] and YAMAGUCHI [**292**, 2005].

3. Localization of Function Spaces.

In this section, we survey work on function spaces after localization. Recall a *nilpotent space* X is a connected CW complex such that $\pi_1(X)$ is a nilpotent group and the standard action of $\pi_1(X)$ on the higher homotopy groups of X is a nilpotent action. By SULLIVAN [**260**, 1971] and HILTON-MISLIN-ROITBERG [**141**, 1975], a nilpotent space X admits a *P-localization* $\ell_X \colon X \to X_P$ which is a map inducing P-localization on homotopy groups. When $P = \{p\}$ we write X_p for the p-localization and when P is empty we write $X_{\mathbb{Q}}$ for the rationalization of X.

Under reasonable hypotheses on X and Y, function spaces behave well with respect to localization. Hilton-Mislin-Roitberg proved that if X is a finite CW complex and Y is a nilpotent space then the path-components of $map(X, Y)$ are nilpotent spaces. The components of $map(X, Y)$ are of CW type in this case by MILNOR [**212**, 1959]. Hilton-Mislin-Roitberg also proved that composition by ℓ_Y gives a P-localization map

$$(\ell_Y)_* \colon map(X, Y; f) \to map(X, Y_P; \ell_Y \circ f).$$

Below we write $f_P = \ell_Y \circ f$. HILTON-MISLIN-ROITBERG-STEINER [**142**, 1978] obtained the same results if, alternately, X is a finite type CW complex and Y is a nilpotent Postnikov piece. Here the CW type result is due to KAHN [**151**, 1984].

MÖLLER [**215**, 1987] extended these results to the function space of relative liftings

$$\begin{array}{ccc} A & \xrightarrow{u} & E \\ {\scriptstyle i}\downarrow & \nearrow & \downarrow{\scriptstyle p} \\ B & \xrightarrow{f} & X \end{array}$$

where here i is closed cofibration and p is a fibration. In this case, P-localization is obtained by passing from $p\colon E \to X$ to its fibrewise P-localization $p_{(P)}\colon E_{(P)} \to X$ as constructed by MAY [**199**, 1980]. In particular, fibrewise localization induces P-localization $\Gamma(p; s) \to \Gamma(p_{(P)}; s')$ for X finite CW and $F = p^{-1}(*)$ a nilpotent space. MÖLLER [**217**, 1990] proved the corresponding nilpotence and localization results for certain equivariant function spaces. KLEIN-SCHOCHET-SMITH [**157**, 2009] extended the nilpotence and localization results from the case when X is finite CW to the case X is compact metric provided the corresponding function or section space is known *a priori* to be nilpotent.

BOUSFIELD-KAN [**28**, 1972] introduced a more general localization theory for subrings $R \subset \mathbb{Q}$ including homotopy completions $Y \to R_\infty Y$. They proved that R-completion (respectively, R-localization) induces R-completion (respectively, R-localization) on the based function spaces $map_*(X, Y; 0)$ when X is a finite CW complex and Y is nilpotent. Further significant results on the behavior of function spaces under Bousfield-Kan localization and completion are discussed below.

3.1. Rational Homotopy Theory of Function Spaces. QUILLEN [**229**, 1969] constructed an equivalence between the homotopy category of simply connected rational CW complexes and a homotopy category of connected, differential graded Lie algebras (DGLAs) over $\mathbb{Q}$ initiating rational homotopy theory. The Quillen minimal model of a simply connected space X is a *minimal* DGLA $(\mathcal{L}(X), d_X)$ which means $\mathcal{L}(X) = \mathbb{L}(V)$ is a free GLA and d_X satisfies $d_X(V) \subseteq [\mathbb{L}(V), \mathbb{L}(V)]$. The rational homology and homotopy Lie algebra of X are recovered via isomorphisms

$$V \cong s^{-1}\widetilde{H}(X;\mathbb{Q}) \text{ and } \pi_*(\Omega X)\otimes\mathbb{Q}, [\,,\,] \cong H_*(\mathcal{L}(X)), [\,,\,].$$

SULLIVAN [**261**, 1977] constructed another categorical equivalence, here between the homotopy theory of simply connected CW complexes and a homotopy category of connected differential graded algebras (DGAs) over $\mathbb{Q}$. The Sullivan minimal model of a space X is a minimal DGA $(\mathcal{M}(X), d_X)$ where $\mathcal{M}(X) = \Lambda V$ is a free DGA with $d_X(V) \subseteq \Lambda^+ V \cdot \Lambda^+ V$ with

$$V \cong \mathrm{Hom}(\pi_*(X), \mathbb{Q}) \text{ and } H_*(\mathcal{M}(X)) \cong H^*(X;\mathbb{Q}).$$

More generally, a Sullivan model $(\mathcal{A}(X), d)$ for X is a DGA admitting a chain equivalence to the de Rahm complex of rational PL forms on X. In particular, $H_*(\mathcal{A}(X)) \cong H^*(X;\mathbb{Q})$ and $(\mathcal{A}(X), d) \simeq (\mathcal{M}(X), d_X)$ are homotopy equivalent in Sullivan's DGA category. Comprehensive treatments of the subject were given by TANRÉ [**265**, 1983] and FÉLIX-HALPERIN-THOMAS [**93**, 2001].

Sullivan described separate models for the general path-component of a function space, $map(X, Y; f)$, the space of self-equivalences $aut(X)_\circ$ and the free loop space LX each within his framework of DGAs. We discuss these models and their extensions and applications now.

3.1.1. *General Components.* Following the sketch by Sullivan, HAEFLIGER [**122**, 1982] constructed a (non-minimal) Sullivan model for the rational homotopy type of $map(X,Y;f)$ where $f\colon X\to Y$ is a map of nilpotent spaces with X finite. The construction builds on the ideas of Thom, described above. Let $p_r\colon Y_r\to Y_{r-1}$ with fibre $K(G_r,n_r)$ be a term in the principal refinement of the Postnikov tower of $Y_{\mathbb{Q}}$ with k invariant $k_{r-1}\colon Y_{r-1}\to K(G_r,n_r+1)$. We then obtain a pullback diagram

$$\begin{array}{ccc} map(X,Y_r;(f_{\mathbb{Q}})_r) & \longrightarrow & Pmap(X,K(G_r,n_r+1);0) \\ {\scriptstyle (p_r)_*}\downarrow & & \downarrow \\ map(X,Y_{r-1};(f_{\mathbb{Q}})_{r-1}) & \xrightarrow[(k_{r-1}\circ(f_{\mathbb{Q}})_{r-1})_*]{} & map(X,K(G_r,n_r+1);0) \end{array}$$

where the right fibration is the path/loop fibration. Let $V=\bigoplus_r \mathrm{Hom}(G_r,\mathbb{Q})$. Since X is finite, X admits a finite model (A,d). Write $\underline{A}=\mathrm{Hom}(A,\mathbb{Q})$ for the dual to A and grade $\underline{A}$ in negative degrees. Thom's calculation of $\pi_q(map(X,K(G,n);0))$ is then reflected in the grading on the ordinary tensor product $\underline{A}\otimes V$. Let I denote the ideal of $\Lambda(\underline{A}\otimes V)$ generated by elements of degree ≤ 0. Haefliger described a differential d on $\Lambda(\underline{A}\otimes V)/I$ in terms of the "k-invariants" $(k_{r-1}\circ(f_{\mathbb{Q}})_{r-1})_*$ above and proved the result is a Sullivan model for $map(X,Y;f)$.

BOUSFIELD-PETERSON-SMITH [**29**, 1989] gave an alternate construction, motivated by seminal work of LANNES [**178**, 1987] in p-local homotopy theory, discussed below. The construction makes use of the fact that $map(X,\_\_)$ defines a functor on topological spaces that is right adjoint to the product functor $\_\_\times X$. In the category of DGAs, this corresponds to the fact that $\mathrm{Hom}(A,\_\_)$ is right adjoint to the tensor product functor $\_\_\otimes A$. In this setting, $\_\_\otimes A$ has an left adjoint, as well, provided A is finite. The construction is a version of Lannes' T-functor. It is conveniently written here as $(\_\_:A)$. Given a map $\psi\colon B\to A$ define $(B:A)_\psi$ to be the connected DGA determined by ψ. Assume X and Y are nilpotent spaces with X finite. Let A be a finite Sullivan model for X. Bousfield-Peterson-Smith proved $(\mathcal{M}(Y):A)_\psi$ is a Sullivan model for $map(X,Y;f)$ where $\psi\colon\mathcal{M}(Y)\to A$ is a Sullivan model for $f\colon X\to Y$.

BROWN-SZCZARBA [**36**, 1998] and [**37**, 1998] expanded on the work of Haefliger and Bousfield-Peterson-Smith. They constructed a model $(\Lambda W,d)$ for $map(X,Y;f)$ where

$$W^q=\left(\sum_n \pi_n(Y)\otimes H_{n-q}(X;\mathbb{Q})\right)/K^q$$

for certain subspaces K^q. The differential d was described explicitly in terms of the coproduct on $H_*(X;\mathbb{Q})$. Here X is a finite CW complex and Y is nilpotent. They deduced descriptions of the rational homotopy groups of $map(X,Y;f)$. When f is trivial they proved $K_q=0$ thus obtaining an isomorphism of graded spaces

$$\pi_*(map(X,Y;0))\cong(H_*(X;\mathbb{Q}))\otimes(\pi_*(Y)\otimes\mathbb{Q}).$$

Again, the space $H_*(X;\mathbb{Q})$ is assumed to be negatively graded. This last result was earlier proved by SMITH [**248**, 1994].

Applications of the Haefliger model include the following results: VIGUÉ-POIRRIER [**278**, 1986] identified the rational homotopy Lie algebra of $map(X,Y;0)$ for X nilpotent of dimension strictly less than the degree of the first nontrivial homotopy group of Y via an isomorphism

$$\pi_*(\Omega map(X,Y;0)) \otimes \mathbb{Q}, [\,,\,] \cong (H^*(X;\mathbb{Q})) \otimes (\pi_*(\Omega Y) \otimes \mathbb{Q})), [\,,\,].$$

Here $H^*(X;\mathbb{Q})$ is negatively graded and the tensor product has the GLA structure induced by the product and bracket on the terms. MÖLLER-RAUSSEN [**219**, 1986] studied the rational homotopy classification problem for components of $map(X,Y)$ with $Y = S^n, \mathbb{C}P^n$ for X nilpotent and suitably rationally co-connected. They obtained complete classifications in these cases including descriptions of the rational homotopy types. FÉLIX [**88**, 1990] proved the rational L.S. category of $map(X,Y;0)$ is often infinite. SMITH [**251**, 1997] studied the rational homotopy classification problem for $map(G_1/T_1, G_2/T_2)$, where G_1, G_2 are classical compact Lie groups and T_1, T_2 maximal tori, identifying the rational type of certain components as generalized flag manifolds. SMITH [**252**, 1999] gave an explicit description of the Haefliger model for X and Y *elliptic* spaces (simply connected spaces having finite-dimensional rational homotopy and homology) with evenly graded rational cohomology obtaining examples of components of $map(X,Y)$ of finite L.S. category.

KOTANI [**163**, 2004] used the Brown-Szczarba model to give necessary and sufficient conditions for the space $map_*(X,Y)$ to be a rational H-space for X a formal, nilpotent CW complex of dimension $\leq$ the connectivity of Y. FÉLIX-TANRÉ [**97**, 2005] generalized this result replacing the formality of X by a condition involving L.S. category. BUIJS-MURILLO [**42**, 2006] constructed the Brown-Szczarba model within the simplicial category framework for rational homotopy theory due to BOUSFIELD-GUGENHEIM [**27**, 1976]. They obtained a functorial version of the Brown-Szczarba model in this setting and used this model, in BUIJS-MURILLO [**42**, 2008], to identify the rational homotopy Lie algebra of components $map(X,Y;f)$ with X, Y restricted, as usual, to nilpotent spaces with X finite. KURIBAYASHI-YAMAGUCHI [**172**, 2006] combined the Haefliger and Brown-Sczarba approaches to obtain a rational splitting of $map_*(X \cup_\alpha e^{k+1}, Y;0)$ where α is an attaching map. Under certain restrictions on X, Y and α they proved

$$map_*(X \cup_\alpha e^{k+1}, Y;0) \simeq_{\mathbb{Q}} map_*(X,Y;0) \times \Omega^{k+1}Y.$$

HIRATO-KURIBAYASHI-ODA [**145**, 2008] applied the Brown-Szczarba model to the study of the rational evaluation subgroups, i.e., the image of the map induced on rational homotopy groups by the evaluation map $\omega_f \colon map(X,Y;f) \to Y$. BUIJS-FÉLIX-MURILLO [**40**, 2009] used the Brown-Szczarba model to study the rational homotopy type of the homotopy fixed point of a circle action.

The higher rational homotopy groups of $map(X,Y;f)$ for suitable X and Y can be described directly in terms the homology of certain DG space of derivations. In the DGA setting, given a map $\psi\colon (A,d) \to (B,d)$ of DGAs let $\mathrm{Der}_n(A,B;\psi)$ denote the space of linear maps $\theta\colon A^* \to B^{*-n}$ satisfying, for $x,y \in A$, the identity $\theta(xy) = \theta(x)\psi(y) + (-1)^{n|x|}\psi(x)\theta(y)$. In the DGLA setting, let $\mathrm{Der}_*(L,K;\psi)$ denote the space of degree raising linear maps satisfying the corresponding identity. In both cases, a degree -1 differential is given by $D(\theta) = d \circ \theta - (-1)^n \theta \circ d$. When X and Y nilpotent spaces with X finite

$$\pi_n(map(X,Y;f)) \cong H_n(\mathrm{Der}(\mathcal{M}(Y), \mathcal{M}(X); \mathcal{M}(f))).$$

for $n \geq 2$. This result is due to Sullivan for the case $f = 1$ as discussed below. The general result was proved, independently, by BLOCK-LAZAREV [**21**,

2005], LUPTON-SMITH [**189**, 2007] and BUIJS-MURILLO [**42**, 2008]. The rationalization of the fundamental group $\pi_1(map(X,Y;f))$ is, in general, nonabelian. LUPTON-SMITH [**188**, 2007] proved the rank of $\pi_1(map(X,Y;f))_{\mathbb{Q}}$ is the dimension of $H_1(\mathrm{Der}(\mathcal{M}(Y),\mathcal{M}(X);\mathcal{M}(f)))$. BUIJS-MURILLO [**42**, 2008] extended this to an identification of the Malc'ev completion of $\pi_1(map(X,Y;f))_{\mathbb{Q}}$.

Within the framework of Quillen minimal models, we have an isomorphism

$$\pi_n(map(X,Y;f))\otimes\mathbb{Q}\cong H_n(\mathrm{Rel}(\mathrm{ad}_{\mathcal{L}(f)}))$$

for $n\geq 2$ for X and Y simply connected CW with X finite. Here $\mathrm{Rel}_*(\mathrm{ad}_{\mathcal{L}(f)})$ is the mapping cone of the chain map

$$\mathrm{ad}_{\mathcal{L}(f)}\colon \mathcal{L}(Y)\to \mathrm{Der}_*(\mathcal{L}(X),\mathcal{L}(Y);\mathcal{L}(f))$$

given by $\mathrm{ad}_{\mathcal{L}(f)}(y)(x)=[\mathcal{L}(f)(x),y]$ for $x\in\mathcal{L}(X), y\in\mathcal{L}(Y)$. This result was proved for the identity component by TANRÉ [**265**, 1983] and SCHLESSINGER-STASHEFF [**239**, preprint]. The result for the general component was proved by LUPTON-SMITH [**190**, 2007]. LUPTON-SMITH [**192**, 2010] identified rational Whitehead products in terms of this identification. BUIJS-FÉLIX-MURILLO [**39**, 2009] described a Quillen model for function spaces and obtained a result on the exponential growth of rational homotopy groups of function spaces.

The homotopy classification problem for gauge groups corresponding to principal G-bundles $P\colon E\to X$ is trivial after rationalization for X finite CW and G a compact Lie group. In this case, BG is a rational H-space. As mentioned above, CRABB-SUTHERLAND [**66**, 2000] used this fact to prove the fibrewise localization of the universal G-adjoint bundle $\mathrm{Ad}(P_G)\colon E_G\times_G G^{\mathrm{ad}}$ is equivariantly trivial. Their result implies a rational H-equivalence

$$(\mathcal{G}(P)_\circ)_{\mathbb{Q}}\simeq_H map(X,G_{\mathbb{Q}};0).$$

The rational homotopy groups of $\mathcal{G}(P)$ may thus be computed as

$$\pi_*(\mathcal{G}(P))\otimes\mathbb{Q}\cong H_*(X;\mathbb{Q})\otimes(\pi_*(G)\otimes\mathbb{Q})$$

since $G_{\mathbb{Q}}$ is a product of Eilenberg-Mac Lane spaces where here, as above, $H_*(X;\mathbb{Q})$ is negatively graded.

LUPTON-PHILLIPS-SCHOCHET-SMITH [**187**, 2009] proved a related result in the context of commutative Banach algebras. Let A be unital, commutative Banach algebra and $\mathrm{GL}_n(A)$ the group of $n\times n$ invertible matrices with coefficients in A. Then

$$\mathrm{GL}_n(A)_\circ\simeq_{\mathbb{Q}}\prod K(V_n,n)\ \text{ where }\ V=\check{H}_*(\mathrm{Max}(A);\mathbb{Q})\otimes\Lambda(s_1,\ldots,s_{2n-1}).$$

Here $\mathrm{Max}(A)$ is the maximal ideals space and s_{2i-1} is of degree $2i-1$. KLEIN-SCHOCHET-SMITH [**157**, 2009] extended this result to the group of unitaries UA_ζ where A_ζ is the C^*-algebra sections of a complex n-matrix bundle ζ over a compact metric space X. The latter result is based on an extension of the result $(\mathcal{G}(P)_\circ)_{\mathbb{Q}}\simeq_H map(X,G_{\mathbb{Q}};0)$ from the case X finite CW to the case X compact metric.

3.1.2. *Spaces of Self-Equivalences.* The rational homotopy type of a connected grouplike space G is completely determined by isomorphism type of the Samelson Lie algebra $\pi_*(G)\otimes\mathbb{Q},[\,,\,]$ (c.f. SCHEERER [**235**, 1985]). SULLIVAN [**261**, 1977] identified the rational Samelson Lie algebra of the space $aut(X)_\circ$ for X a simply connected finite CW complex via an isomorphism:

$$\pi_*(aut(X)_\circ)\otimes\mathbb{Q},[\,,\,]\cong H_*(\mathrm{Der}(\mathcal{M}(X))),[\,,\,].$$

A corresponding identity in Quillen's DGLA framework for rational homotopy theory was given by TANRÉ [**265**, 1983] and SCHLESSINGER-STASHEFF [**239**, preprint]:

$$\pi_*(aut(X)_\circ) \otimes \mathbb{Q}, [\, , \,] \cong H_*(\mathrm{Rel}(\mathrm{ad}_{\mathcal{L}(X)})), [\, , \,].$$

Here the bracket on the mapping cone of $\mathrm{ad}_{\mathcal{L}(X)}$ is induced from that on $\mathcal{L}(X)$.

Sullivan's identity connects the monoid $aut(X)$ to a fundamental open conjecture in rational homotopy theory. Let X be a simply connected elliptic CW complex with evenly graded rational cohomology. We refer to such spaces as *F_0-spaces.* The class includes (products of) spheres, complex projective spaces and homogeneous spaces G/H with G a compact Lie group and $H \subset G$ a closed subgroup of maximal rank. Motivated by this last case, HALPERIN [**123**, 1978] conjectured that the rational Serre spectral sequence collapses at the E_2 term for all orientable fibrations with fibre an F_0-space. THOMAS [**269**, 1981] and MEIER [**208**, 1981] independently proved that Halperin's conjecture is equivalent to the condition $H_{\mathrm{even}}(\mathrm{Der}(\mathcal{M}(X))) = 0$ for an F_0-space X. Thus, by Sullivan's identity, Halperin's conjecture holds for X if and only if $aut(X)_\circ$ is rationally equivalent to a product of odd spheres.

The Halperin conjecture has been affirmed in several special cases including for Kählerian manifolds by Meier, for homogeneous spaces of maximal rank pairs by SHIGA-TEZUKA [**244**, 1987] and for F_0-spaces with rational cohomology generated by ≤ 3 generators by LUPTON [**186**, 1990]. HAUSCHILD [**137**, 1993] [**138**, 2001] computed rational homotopy groups of $aut(X)$ for X an F_0-space and of $aut(p)$ for p a fibration with fibre X. FÉLIX-THOMAS [**90**, 1994] studied the rational homotopy of $aut(X)$ for various homogeneous spaces giving explicit calculations. GRIVEL [**113**, 1994] proved that for, X an F_0-space,

$$H_{\mathrm{even}}(\mathrm{Der}(\mathcal{M}(X))) \cong \mathrm{Der}_{\mathrm{even}}(H(\mathcal{M}(X)))$$

and used this result to give a formula for $\pi_{\mathrm{odd}}(aut(X)_\circ) \otimes \mathbb{Q}$.

SALVATORE [**232**, 1997] proved the nilpotency of the Lie algebra $H_*(\mathrm{Der}(\mathcal{M}(X)))$ coincides with the *rational homotopical nilpotency* of the monoid $aut(X)_\circ$ — the least integer n such that the n-fold commutator for $aut(X)_\circ$ is rationally trivial. He calculated the rational homotopical nilpotency of $aut(X)_\circ$ for X a rational two-stage Postnikov system and proved the monoid $aut(S^{2n-1} \vee S^{2n-1})_\circ$ is not rationally homotopy nilpotent. SMITH [**253**, 2001] computed the rational homotopy nilpotency of $aut(X)_\circ$ for certain spaces X admitting a two-stage Sullivan model. Recently, FÉLIX-LUPTON-SMITH [**95**, preprint] obtained a formula, in the spirit of Sullivan's above, for the rational Samelson Lie algebra of the monoid $aut(p)_\circ$ of fibre-homotopy self-equivalences of a fibration $p\colon E \to B$ of simply connected finite CW complexes:

$$\pi_*(aut(p)_\circ) \otimes \mathbb{Q}, [\, , \,] \cong H_*(\mathrm{Der}_{\Lambda V}(\Lambda W \otimes \Lambda V)), [\, , \,].$$

Here $(\Lambda W, d_B) \to (\Lambda W \otimes \Lambda V, D)$ is the Koszul-Sullivan model of the fibration and $\mathrm{Der}_{\Lambda V}(\Lambda W \otimes \Lambda V)$ denotes the DGLA of derivations vanishing on ΛV.

As for the rational homotopy of Gottlieb group $G_n(X)$ and the evaluation map $\omega\colon aut(X)_\circ \to X$, LANG [**177**, 1975] proved $G_*(X)_{\mathbb{Q}} \cong G_*(X_{\mathbb{Q}})$ for X a finite simply connected CW complex. FÉLIX-HALPERIN [**92**, 1982] identified the rationalized Gottlieb groups $G_n(X) \otimes \mathbb{Q}$ for these X in terms of the Sullivan identification $\pi_*(X) \otimes \mathbb{Q} \cong V$ where $\mathcal{M}(X) = \Lambda V$. Here an element $v \in V_n$ corresponds to a rational Gottlieb element if the dual map $v \mapsto 1$ extends to a derivation cycle in

$\mathrm{Der}_n(\mathcal{M}(X))$. They used this result to prove two global results on the rationalized Gottlieb groups of a simply connected CW complex X of finite rational L. S. category:

$$G_{\mathrm{even}}(X) \otimes \mathbb{Q} = 0 \quad \text{and} \quad \dim(G_{\mathrm{odd}}(X) \otimes \mathbb{Q}) \leq \mathrm{cat}_{\mathbb{Q}}(X).$$

A Quillen model description of the rationalized Gottlieb group was given by TANRÉ [**265**, 1983]:

$$G_n(X)_{\mathbb{Q}} \cong \ker\{H(\mathrm{ad}_{\mathcal{L}(X)})\colon \mathcal{L}(X) \to \mathrm{Der}(\mathcal{L}(X))\}.$$

Rationalized Gottlieb groups have been calculated by many authors using various means including SMITH [**249**, 1996], LUPTON-SMITH [**189**, 2007] and [**190**, 2007], HIRATO-KURIBAYASHI-ODA [**145**, 2008] and YAMAGUCHI [**296**, 2008]. FÉLIX-LUPTON [**94**, 2007] proved the evaluation map $\omega\colon \mathit{aut}(X)_\circ \to X$ is rationally homotopy trivial if and only if it is trivial on rational homotopy groups for X a finite, simply connected CW complex.

3.1.3. *The Free Loop Space.* VIGUÉ-POIRRIER-SULLIVAN [**280**, 1976] constructed a Sullivan model for LX when X is a simply connected CW complex. Let $(\Lambda V, d)$ be the minimal model for X. Their model for LX is given by

$$(\Lambda V \otimes \Lambda sV, \delta) \text{ with } \delta(v) = dv \text{ and } \delta(sv) = -sd(v) \text{ for } v \in V$$

where s is the degree -1 derivation of $\Lambda V \otimes \Lambda sV$ defined by setting $s(v) = sv$ and $s(sV) = 0$. HALPERIN [**124**, 1981] constructed a related model in the non-simply-connected case under restrictions on the component. As discussed above, Vigué-Poirrier-Sullivan used their model to prove the Betti numbers of LX are unbounded if $H^*(X;\mathbb{Q})$ requires at least two generators. In fact, they showed the Betti numbers of LX grow exponentially in this case. VIGUÉ-POIRRIER [**277**, 1984] proved that the same is true for wedges of spheres and manifolds of L.S. category less than 2. She conjectured the Betti numbers of LX grow exponentially for all finite, simply connected X with infinite dimensional rational homotopy. She gave examples of spaces X with finite-dimensional rational homotopy for which

$$\limsup \frac{\log \sum_{i=1}^n \beta_i}{\log n} = \dim(\pi_{\mathrm{odd}}(X) \otimes \mathbb{Q}).$$

LAMBRECHTS [**175**, 2001] affirmed Vigué-Poirrier's conjecture for the class of simply connected, coformal, finite complexes.

DUPONT-VIGUÉ-POIRRIER [**78**, 1998] proved a basic result concerning the question of formality for the free loop space. Given a simply connected CW complex X with Noetherian rational cohomology, the space LX is formal if and only if X is a rational H-space. YAMAGUCHI [**294**, 2000] generalized this to the function space $\mathit{map}(X, Y; 0)$ for X simply connected CW of dimension less than the connectivity of Y an elliptic space. He showed that, again, $\mathit{map}(X, Y; 0)$ is formal if and only if Y is a rational H-space. VIGUÉ-POIRRIER [**279**, 2007] proved a further result in this vein showing, with the same dimension/connectivity hypotheses, that $\mathit{map}(X, Y; 0)$ is formal if and only if Y is a rational H-space provided the odd rational Hurewicz homomorphism of X is nontrivial.

FÉLIX-THOMAS-VIGUÉ-POIRRIER [**100**, 2007] studied the string topology operations on $\mathbb{H}_*(LM;\mathbb{Q})$ within the framework of Sullivan minimal models. They gave a description of the loop-product and the string-bracket in this setting making explicit computations. They also proved the isomorphism of graded spaces due to

COHEN-JONES [**60**, 2002]

$$\mathbb{H}_*(LM;\mathbb{Q}) \cong HH^*(C^*(M), C^*(M);\mathbb{Q})$$

is an isomorphism of Gerstenhaber algebras. FÉLIX-THOMAS [**99**, 2008] extended this last result proving the above is an isomorphism of BV-algebras.

3.2. Function Spaces and p-Localization. Function spaces are central to the theory of homotopy localizations and completions with respect to subrings $R \subset \mathbb{Q}$. As mentioned above, BOUSFIELD-KAN [**28**, 1972] proved that their R-completion functor induces R-completion on the based function space $map_*(X, Y; 0)$ for X finite CW and Y nilpotent. This result was used in the proof of the fracture lemma for R-completions. A key step in Miller's proof of the Sullivan conjecture was the proof of a weak equivalence

$$map_*(X, Y; 0) \simeq_w map_*(X, R_\infty Y; 0)$$

where $R = \mathbb{F}_p$ and R_∞ is the R-completion functor. Here Y is a nilpotent space and X is a connected, $\mathbb{Z}[\frac{1}{p}]$-acyclic space. Function spaces also feature in the theory of homotopy localization and cellularization. Given a map $f\colon X \to Y$, a space Z is defined to be f-local if the induced map of function spaces $f^*\colon map_*(Y, Z) \to map_*(X, Z)$ is a weak homotopy equivalence. DROR-FARJOUN [**74**, 1996] constructed f-localizations and showed the known localization functors all occur as special cases, for suitable choices of the map f. As usual, we consider work here which focuses explicitly on the homotopy type of function spaces.

3.2.1. *Maps out of a Classifying Space.* The function spaces $map(B\pi, X)$ and $map_*(B\pi, X)$ for π a finite group appear in major developments in homotopy theory in the p-local category. In celebrated work, MILLER [**211**, 1984] affirmed the Sullivan conjecture, proving

$$\pi_n(map_*(B\pi, X)) = 0 \text{ for all } \ n \geq 0,$$

for π a locally finite group and X a connected, finite CW complex. Using the R-completion theorem for $map_*(X, Y)$ mentioned above, the problem reduces to proving the weak triviality of spaces $map_*(B\mathbb{Z}_p, R_\infty Y)$ where R_∞ is the Bousfield-Kan p-completion functor. Miller proved the latter fact by establishing the vanishing of certain Ext-sets in a category of unstable modules over the mod p Steenrod algebra.

Miller's Theorem had many important consequences for function spaces. ZABRODSKY [**304**, 1991] connected the result to the study of phantom maps. He also obtained the following extension. Let W be a connected CW complex with finitely many, locally finite homotopy groups. Then

$$\pi_n(map_*(W, X)) = 0 \text{ for all } \ n \geq 0,$$

for X any connected, finite CW complex. Zabrodsky also proved that, if $P\colon E \to B$ is a principal G-bundle with $map_*(G, Y)$ contractible, then $map_*(B, Y) \to map_*(E, Y)$ is a homotopy equivalence. Miller's result had equivariant generalizations in terms of fixed point and homotopy fixed point set. In particular, if X is a π-space for π a p-group then

$$R_\infty(X^\pi) = (R_\infty X)^{h\pi}$$

where $R = \mathbb{Z}/p$ and X^π is the fixed point set while $X^{h\pi}$ is the homotopy fixed point sets (c.f. CARLSSON [**46**, 1991]). DWYER-ZABRODSKY [**79**, 1986] applied this latter result to obtain a decomposition

$$map(B\pi, BG) \simeq \coprod_{\rho} BC(\rho).$$

Here π is a finite group, G is a compact Lie group and the disjoint union is over G-conjugacy classes of homomorphisms $\rho\colon \pi \to G$. As usual, $C(\rho)$ denotes the centralizer of the image of ρ in G. FRIEDLANDER-MISLIN [**102**, 1986] gave conditions on a Lie group G such that $map_*(BG, R_\infty X)$ is weakly trivial. Here X is nilpotent and R_∞ is p-completion. MCGIBBON [**206**, 1996] proved $map_*(W, R_\infty X)$ is weakly contractible for W a connected infinite loop space with torsion fundamental group. STROM [**259**, 2005] proved that if $map_*(X, S^n)$ is weakly contractible for all sufficiently large n then $map_*(X, Y)$ is actually weakly contractible for any nilpotent, finite CW complex Y. He thus obtained a method for recognizing spaces X satisfying the conclusion of Miller's Theorem.

LANNES [**178**, 1987] and [**179**, 1992] complemented and extended Miller's work. Let $\mathcal{U}$ and $\mathcal{K}$ denote, respectively, the category of unstable modules and algebras over $\mathcal{A}$, the mod p Steenrod algebra. Lannes constructed the T-functor which is left-adjoint to the tensor product functor $\_\_ \otimes_{\mathcal{U}} H^*(B\mathbb{Z}_p; \mathbb{Z}_p)$ on $\mathcal{U}$. He showed, among other properties, that T is exact, preserves tensor products and restricts to a functor on $\mathcal{K}$. These results are used to prove the natural map

$$\Theta_X\colon T(H^*(X; \mathbb{Z}_p)) \to H^*(map(B\mathbb{Z}_p, X); \mathbb{Z}_p)$$

is an isomorphism of unstable $\mathcal{A}$-algebras whenever $T(H^*(X; \mathbb{Z}_p))$ is trivial in degree 1 and of finite type.

AGUADÉ [**3**, 1989] computed $T(M)$ for certain $\mathcal{A}$-algebras M including subalgebras of $\mathbb{Z}_p[x_1, \ldots, x_n]$ invariant under the general linear action. Here p is an odd prime. DROR-SMITH [**73**, 1990] constructed an Eilenberg-Moore spectral sequence for computing Θ_X above and gave a geometric interpretation of the T-functor. DWYER-WILKERSON [**84**, 1990] proved if π is a locally finite group and X is a simply connected p-complete space with $H^*(X; \mathbb{Z}_p)$ finitely generated as an algebra, then $map_*(B\pi, X; 0)$ is weakly contractible. KUHN-WINSTEAD [**166**, 1996] proved

$$\widetilde{H}^*(X; \mathbb{Z}_p) = 0 \implies \widetilde{H}^*(map(B\mathbb{Z}_p, X); \mathbb{Z}_p) = 0.$$

More generally, they showed the image of Θ_X consists of $\mathbb{Z}_p^\wedge$-integral classes if $H^*(X; \mathbb{Z}_p)$ does where a $\mathbb{Z}_p^\wedge$-integral class in $H^*(\_\_; \mathbb{Z}_p)$ is a class in the reduction from $H^*(\_\_; \mathbb{Z}_p^\wedge)$. DEHON-LANNES [**67**, 2000] proved that if X is p-complete and $H^*(X; \mathbb{Z}_p)$ is Noetherian and generated in even degrees then $H^*(map(B\mathbb{Z}_p, X); \mathbb{Z}_p)$ is Noetherian and $map(B\mathbb{Z}_p, X)$ is p-complete. AGUADÉ-BROTO-SAUMELL [**4**, 2004] introduced the notion of T-representability for a space X and proved it a sufficient condition for Θ_X to be an isomorphism. They gave an example of a p-complete space X for which Θ_X is not an isomorphism.

We mention three more results falling under the current heading. DWYER-MISLIN [**83**, 1987] identified the homotopy type of the nontrivial components of $map_*(BS^3, BS^3)$. The null component is contractible by ZABRODSKY [**304**, 1991], as mentioned above. BLANC-NOTBOHM [**20**, 1993] proved

$$map(BG, BH; f)_p^\wedge \simeq map(BG, (BH)_p^\wedge; f_p^\wedge)$$

for G, H compact Lie groups. BROTO-LEVI [**34**, 2002] proved

$$aut((B\pi)_p^\wedge)_\circ \simeq K(Z(\pi/O_{p'}(\pi)), 1)$$

for π a finite group. Here $O_{p'}(\pi)$ denotes the maximal normal p'-subgroup of π. The proof uses the Bousfield-Kan spectral sequence to prove asphericality and the $\mathbb{Z}^*$-theorem from group theory to compute the fundamental group.

3.2.2. *Algebraic Models.* In this final section, we discuss results on modeling the p-local homotopy theory of function spaces.

DWYER [**80**, 1979] proved that a version of Quillen's rational homotopy theory extends to the homotopy theory of *tame spaces* X which are $(r-1)$-connected CW complexes, $r \geq 3$, with $\pi_{r+k}(X)$ uniquely p-divisible for all primes p with $2p-3 \leq k$. He proved the homotopy category of tame spaces endowed with an appropriate model structure is equivalent to a homotopy category of $(r-1)$-reduced integral DGLAs. Tame homotopy theory admits a version in the spirit of Sullivan's approach to rational homotopy theory by CENKL-PORTER [**48**, 1983]. SCHEERER-TANRÉ [**237**, 1988] identified homotopy invariants, e.g., homology and the homotopy Lie algebra, in Dwyer's framework. ANICK [**9**, 1989] and [**10**, 1990] gave an alternate approach using a classical construction of ADAMS-HILTON [**1**, 1955] on the Pontryagin algebra $H_*(\Omega X; R)$ for $R \subset \mathbb{Q}$. He constructed a DGLA model over R for the category CW_r^m consisting of $(r-1)$-connected complexes of dimension m when all primes p with $m > pr$ are invertible in R.

The description of function spaces in tame homotopy theory has been undertaken in several works. ANICK-DROR-FARJOUN [**8**, 1990] described a simplicial skeleton of the space $map_*(X,Y)$ for R-local spaces $X, Y \in \mathrm{CW}_r^m$. Given suitably reduced DGLAs L and K over R they constructed a function complex $\mathrm{Hom}(L,K)$ giving an explicit description through a range of dimensions. SCHEERER-TANRÉ [**238**, 1992] described R-local homotopy theory in a suitable category of DG coalgebras of R. They constructed an adjoint to the wedge functor in this context and used it to give a model for the space $map_*(X,Y)$ for R-local $X, Y \in \mathrm{CW}_r^m$. As an example, they described a model for the R-localization of $map_*(\mathbb{H}P^2, M_t)$, where M_t denotes a tamed Moore space and R is suitably chosen. FÉLIX-THOMAS [**89**, 1993] obtained a p-local decomposition

$$map_*(\Sigma X, Y) \simeq_p \prod_{i=2}^{n} (\Omega^{i+1} Y)^{\beta_i(X)}$$

for X simply connected with torsion-free homology of dimension $n < 2p$ and Y $(r-1)$-connected with $r > n+1$. SCHEERER [**236**, 1994] recovered this result as a special case of a corresponding decomposition for $map_*(C,Y)$ for C a co-H-space. DUPONT-HESS [**75**, 1999], [**76**, 2002] and [**77**, 2003] used Anick's framework to obtain a model for the mod p cohomology of the free loop space LX for a simply connected space $X \in \mathrm{CW}_r^m$ and prime p with $m \leq pr$. They constructed a DGA over $\mathbb{Z}_p$, built from Anick's model, and proved the homology of this algebra is isomorphic to $H^*(LX; \mathbb{Z}_p)$.

DWYER-KAN [**81**, 1980] identified function complexes in a simplicial homotopy theory category $L^H(\mathbf{M})$ of a general Quillen model category $\mathbf{M}$. Here L^H is their "hammock" localization functor. They showed $L^H(\mathbf{M}(X,Y))$ for $X, Y \in \mathbf{M}$ behaves properly with respect to simplicial and cosimplicial resolutions. This approach yielded, in particular, a good model for the monoid of self-equivalences for arbitrary model categories. In DWYER-KAN [**82**, 1983], they identified the homotopy type of the function complex as a homotopy inverse limit for X cofibrant and Y fibrant.

Recently, FRESSE [**101**, preprint] constructed an algebraic model for the homotopy type of $map(X,Y)$ for X a finite complex and $\pi_n(Y)$ a finite p-group for all n. Let $N^*(Y;\overline{\mathbb{Z}}_p)$ denote the normalized cochain complex where $\overline{\mathbb{Z}}_p$ is the algebraic closure of the field of p-elements. Then $N^*(Y;\overline{\mathbb{Z}}_p)$ is an E_∞-algebra and, more generally, an algebra over the Barrett-Eccles operad $\mathcal{E}$. MANDELL [**193**, 2001] proved this algebra structure determines the homotopy type of Y. Fresse constructed a Lannes T-functor left adjoint to the tensor product in the category of algebras over $\mathcal{E}$. As a consequence, he obtained a model $(N^*(Y;\overline{\mathbb{Z}}_p) : N^*(X;\overline{\mathbb{Z}}_p))$ for $map(X,Y)$. CHATAUR-KURIBAYASHI [**52**, 2007] extended this result to the case Y is connected and nilpotent of finite type and made calculations with the resulting spectral sequence.

References

1. J. F. Adams and P. J. Hilton, *On the chain algebra of a loop space*, Comment. Math. Helv. **30** (1956), 305–330. MR0077929 (17,1119b)
2. Jaume Aguadé, *Decomposable free loop spaces*, Canad. J. Math. **39** (1987), no. 4, 938–955. MR915024 (88m:55012)
3. ———, *Computing Lannes T functor*, Israel J. Math. **65** (1989), no. 3, 303–310. MR1005014 (90h:55026)
4. Jaume Aguadé, Carles Broto, and Laia Saumell, *The functor T and the cohomology of mapping spaces*, Categorical decomposition techniques in algebraic topology (Isle of Skye, 2001), Progr. Math., vol. 215, Birkhäuser, Basel, 2004, pp. 1–20. MR2039756 (2005e:55033)
5. Stephen T. Ahearn and Nicholas J. Kuhn, *Product and other fine structure in polynomial resolutions of mapping spaces*, Algebr. Geom. Topol. **2** (2002), 591–647 (electronic). MR1917068 (2003j:55009)
6. D. W. Anderson, *A generalization of the Eilenberg-Moore spectral sequence*, Bull. Amer. Math. Soc. **78** (1972), 784–786. MR0310889 (46 #9987)
7. M. H. Andrade Claudio and M. Spreafico, *Homotopy type of gauge groups of quaternionic line bundles over spheres*, Topology Appl. **156** (2009), no. 3, 643–651. MR2492312
8. D. J. Anick and E. Dror Farjoun, *On the space of maps between R-local CW complexes*, Astérisque (1990), no. 191, 5, 15–27, International Conference on Homotopy Theory (Marseille-Luminy, 1988). MR1098963 (92c:55008)
9. David J. Anick, *Hopf algebras up to homotopy*, J. Amer. Math. Soc. **2** (1989), no. 3, 417–453. MR991015 (90c:16007)
10. ———, *R-local homotopy theory*, Homotopy theory and related topics (Kinosaki, 1988), Lecture Notes in Math., vol. 1418, Springer, Berlin, 1990, pp. 78–85. MR1048177 (91c:55008)
11. Martin Arkowitz, *The group of self-homotopy equivalences—a survey*, Groups of self-equivalences and related topics (Montreal, PQ, 1988), Lecture Notes in Math., vol. 1425, Springer, Berlin, 1990, pp. 170–203. MR1070585 (91i:55001)
12. Greg Arone, *A generalization of Snaith-type filtration*, Trans. Amer. Math. Soc. **351** (1999), no. 3, 1123–1150. MR1638238 (99i:55011)
13. M. F. Atiyah and R. Bott, *The Yang-Mills equations over Riemann surfaces*, Philos. Trans. Roy. Soc. London Ser. A **308** (1983), no. 1505, 523–615. MR702806 (85k:14006)
14. M. F. Atiyah and J. D. S. Jones, *Topological aspects of Yang-Mills theory*, Comm. Math. Phys. **61** (1978), no. 2, 97–118. MR503187 (80j:58021)
15. Sven Bauer, Michael Crabb, and Mauro Spreafico, *The space of free loops on a real projective space*, Groups of homotopy self-equivalences and related topics (Gargnano, 1999), Contemp. Math., vol. 274, Amer. Math. Soc., Providence, RI, 2001, pp. 33–38. MR1817001 (2001m:55028)
16. H. J. Baues, *Algebraic homotopy*, Cambridge Studies in Advanced Mathematics, vol. 15, Cambridge University Press, Cambridge, 1989. MR985099 (90i:55016)
17. H. J. Baues and J.-M. Lemaire, *Minimal models in homotopy theory*, Math. Ann. **225** (1977), no. 3, 219–242. MR0431172 (55 #4174)
18. Martin Bendersky and Sam Gitler, *The cohomology of certain function spaces*, Trans. Amer. Math. Soc. **326** (1991), no. 1, 423–440. MR1010881 (92d:55005)

19. David Blanc, *Mapping spaces and M-CW complexes*, Forum Math. **9** (1997), no. 3, 367–382. MR1441926 (98i:55015)
20. David Blanc and Dietrich Notbohm, *Mapping spaces of compact Lie groups and p-adic completion*, Proc. Amer. Math. Soc. **117** (1993), no. 1, 251–258. MR1112487 (93c:55016)
21. J. Block and A. Lazarev, *André-Quillen cohomology and rational homotopy of function spaces*, Adv. Math. **193** (2005), no. 1, 18–39. MR2132759 (2006a:55014)
22. C.-F. Bödigheimer, *Stable splittings of mapping spaces*, Algebraic topology (Seattle, Wash., 1985), Lecture Notes in Math., vol. 1286, Springer, Berlin, 1987, pp. 174–187. MR922926 (89c:55011)
23. C.-F. Bödigheimer, F. R. Cohen, and M. D. Peim, *Mapping class groups and function spaces*, Homotopy methods in algebraic topology (Boulder, CO, 1999), Contemp. Math., vol. 271, Amer. Math. Soc., Providence, RI, 2001, pp. 17–39. MR1831345 (2002f:55036)
24. Marcel Bökstedt and Iver Ottosen, *A splitting result for the free loop space of spheres and projective spaces*, Q. J. Math. **56** (2005), no. 4, 443–472. MR2182460 (2006f:55008)
25. Peter Booth, Philip Heath, Chris Morgan, and Renzo Piccinini, *H-spaces of self-equivalences of fibrations and bundles*, Proc. London Math. Soc. (3) **49** (1984), no. 1, 111–127. MR743373 (85k:55013)
26. Karol Borsuk, *Concerning the homological structure of the functional space* S_m^X, Fund. Math. **29** (1952), 25–37 (1953). MR0056285 (15,51g)
27. A. K. Bousfield and V. K. A. M. Gugenheim, *On* PL *de Rham theory and rational homotopy type*, Mem. Amer. Math. Soc. **8** (1976), no. 179, ix+94. MR0425956 (54 #13906)
28. A. K. Bousfield and D. M. Kan, *Homotopy limits, completions and localizations*, Lecture Notes in Mathematics, Vol. 304, Springer-Verlag, Berlin, 1972. MR0365573 (51 #1825)
29. A. K. Bousfield, C. Peterson, and L. Smith, *The rational homology of function spaces*, Arch. Math. (Basel) **52** (1989), no. 3, 275–283. MR989883 (90d:55020)
30. C. P. Boyer, J. C. Hurtubise, B. M. Mann, and R. J. Milgram, *The topology of instanton moduli spaces. I. The Atiyah-Jones conjecture*, Ann. of Math. (2) **137** (1993), no. 3, 561–609. MR1217348 (94h:55010)
31. C. P. Boyer, J. C. Hurtubise, and R. J. Milgram, *Stability theorems for spaces of rational curves*, Internat. J. Math. **12** (2001), no. 2, 223–262. MR1823576 (2002g:55010)
32. C. P. Boyer, B. M. Mann, J. C. Hurtubise, and R. J. Milgram, *The topology of the space of rational maps into generalized flag manifolds*, Acta Math. **173** (1994), no. 1, 61–101. MR1294670 (95h:55007)
33. Charles P. Boyer, H. Blaine Lawson, Jr., Paulo Lima-Filho, Benjamin M. Mann, and Marie-Louise Michelsohn, *Algebraic cycles and infinite loop spaces*, Invent. Math. **113** (1993), no. 2, 373–388. MR1228130 (95a:55021)
34. Carles Broto and Ran Levi, *Loop structures on homotopy fibers of self maps of spheres*, Amer. J. Math. **122** (2000), no. 3, 547–580. MR1759888 (2001c:55005)
35. ———, *On spaces of self-homotopy equivalences of p-completed classifying spaces of finite groups and homotopy group extensions*, Topology **41** (2002), no. 2, 229–255. MR1876889 (2002j:55013)
36. Edgar H. Brown, Jr. and Robert H. Szczarba, *On the rational homotopy type of function spaces*, Trans. Amer. Math. Soc. **349** (1997), no. 12, 4931–4951. MR1407482 (98c:55015)
37. ———, *Some algebraic constructions in rational homotopy theory*, Topology Appl. **80** (1997), no. 3, 251–258. MR1473920 (98h:55017)
38. R. Brown, *Function spaces and product topologies*, Quart. J. Math. Oxford Ser. (2) **15** (1964), 238–250. MR0165497 (29 #2779)
39. Urtzi Buijs, Yves Félix, and Aniceto Murillo, *Lie models for the components of sections of a nilpotent fibration*, Trans. Amer. Math. Soc. **361** (2009), no. 10, 5601–5614. MR2515825
40. ———, *Rational homotopy of the (homotopy) fixed point sets of circle actions*, Adv. Math. **222** (2009), no. 1, 151–171. MR2531370
41. Urtzi Buijs and Aniceto Murillo, *Basic constructions in rational homotopy theory of function spaces*, Ann. Inst. Fourier (Grenoble) **56** (2006), no. 3, 815–838. MR2244231 (2007h:55009)
42. ———, *The rational homotopy Lie algebra of function spaces*, Comment. Math. Helv. **83** (2008), no. 4, 723–739. MR2442961 (2009f:55009)
43. D. Burghelea and Z. Fiedorowicz, *Hermitian algebraic K-theory of topological spaces*, Algebraic K-theory, number theory, geometry and analysis (Bielefeld, 1982), Lecture Notes in Math., vol. 1046, Springer, Berlin, 1984, pp. 32–46. MR750675 (86c:18005)

44. H. E. A. Campbell, F. R. Cohen, F. P. Peterson, and P. S. Selick, *The space of maps of Moore spaces into spheres*, Algebraic topology and algebraic K-theory (Princeton, N.J., 1983), Ann. of Math. Stud., vol. 113, Princeton Univ. Press, Princeton, NJ, 1987, pp. 72–100. MR921473 (89a:55012)
45. G. E. Carlsson, R. L. Cohen, T. Goodwillie, and W. C. Hsiang, *The free loop space and the algebraic K-theory of spaces*, K-Theory **1** (1987), no. 1, 53–82. MR899917 (88i:55002)
46. Gunnar Carlsson, *Equivariant stable homotopy and Sullivan's conjecture*, Invent. Math. **103** (1991), no. 3, 497–525. MR1091616 (92g:55007)
47. Carles Casacuberta and José L. Rodríguez, *On weak homotopy equivalences between mapping spaces*, Topology **37** (1998), no. 4, 709–717. MR1607716 (98k:55006)
48. Bohumil Cenkl and Richard Porter, *Differential forms and torsion in the fundamental group*, Adv. in Math. **48** (1983), no. 2, 189–204. MR700985 (85k:57003)
49. Moira Chas and Dennis Sullivan, *String topology*, preprint, arXiv:math.GT/9911159.
50. ______, *Closed string operators in topology leading to Lie bialgebras and higher string algebra*, The legacy of Niels Henrik Abel, Springer, Berlin, 2004, pp. 771–784. MR2077595 (2005f:55007)
51. David Chataur, *A bordism approach to string topology*, Int. Math. Res. Not. (2005), no. 46, 2829–2875. MR2180465 (2007b:55009)
52. David Chataur and Katsuhiko Kuribayashi, *An operadic model for a mapping space and its associated spectral sequence*, J. Pure Appl. Algebra **210** (2007), no. 2, 321–342. MR2320001 (2008h:55017)
53. David Chataur and Jean-Claude Thomas, *E_∞-model for a mapping space*, Topology Appl. **145** (2004), no. 1-3, 191–204. MR2100872 (2005k:55011)
54. Younggi Choi, *Homology of gauge groups associated with special unitary groups*, Topology Appl. **155** (2008), no. 12, 1340–1349. MR2423972 (2009f:55014)
55. F. R. Cohen, R. L. Cohen, B. M. Mann, and R. J. Milgram, *The topology of rational functions and divisors of surfaces*, Acta Math. **166** (1991), no. 3-4, 163–221. MR1097023 (92k:55011)
56. ______, *The homotopy type of rational functions*, Math. Z. **213** (1993), no. 1, 37–47. MR1217669 (94a:55006)
57. F. R. Cohen and L. R. Taylor, *The homology of function spaces*, Proceedings of the Northwestern Homotopy Theory Conference (Evanston, Ill., 1982) (Providence, RI), Contemp. Math., vol. 19, Amer. Math. Soc., 1983, pp. 39–50. MR711041 (85d:55016)
58. ______, *Homology of function spaces*, Math. Z. **198** (1988), no. 3, 299–316. MR946606 (89d:55017)
59. Ralph L. Cohen, *A model for the free loop space of a suspension*, Algebraic topology (Seattle, Wash., 1985), Lecture Notes in Math., vol. 1286, Springer, Berlin, 1987, pp. 193–207. MR922928 (89d:55018)
60. Ralph L. Cohen and John D. S. Jones, *A homotopy theoretic realization of string topology*, Math. Ann. **324** (2002), no. 4, 773–798. MR1942249 (2004c:55019)
61. Ralph L. Cohen, John D. S. Jones, and Jun Yan, *The loop homology algebra of spheres and projective spaces*, Categorical decomposition techniques in algebraic topology (Isle of Skye, 2001), Progr. Math., vol. 215, Birkhäuser, Basel, 2004, pp. 77–92. MR2039760 (2005c:55016)
62. Ralph L. Cohen, John R. Klein, and Dennis Sullivan, *The homotopy invariance of the string topology loop product and string bracket*, J. Topol. **1** (2008), no. 2, 391–408. MR2399136 (2009h:55004)
63. Ralph L. Cohen and Don H. Shimamoto, *Rational functions, labelled configurations, and Hilbert schemes*, J. London Math. Soc. (2) **43** (1991), no. 3, 509–528. MR1113390 (93c:55009)
64. M. C. Crabb, *Loop homology as fibrewise homology*, Proc. Edinb. Math. Soc. (2) **51** (2008), no. 1, 27–44. MR2391632 (2009c:55018)
65. M. C. Crabb and W. A. Sutherland, *Function spaces and Hurwitz-Radon numbers*, Math. Scand. **55** (1984), no. 1, 67–90. MR769026 (86d:58014)
66. ______, *Counting homotopy types of gauge groups*, Proc. London Math. Soc. (3) **81** (2000), no. 3, 747–768. MR1781154 (2001m:55024)
67. François-Xavier Dehon and Jean Lannes, *Sur les espaces fonctionnels dont la source est le classifiant d'un groupe de Lie compact commutatif*, Inst. Hautes Études Sci. Publ. Math. (1999), no. 89, 127–177 (2000). MR1793415 (2001m:55038)
68. Geneviève Didierjean, *Homotopie de l'espace des équivalences d'homotopie fibrées*, Ann. Inst. Fourier (Grenoble) **35** (1985), no. 3, 33–47. MR810666 (87e:55008)

69. ———, *Homotopie des espaces d'équivalences*, Groups of self-equivalences and related topics (Montreal, PQ, 1988), Lecture Notes in Math., vol. 1425, Springer, Berlin, 1990, pp. 32–39. MR1070573 (91j:55008)
70. ———, *Homotopie de l'espace des équivalences d'homotopie*, Trans. Amer. Math. Soc. **330** (1992), no. 1, 153–163. MR986023 (92f:55017)
71. Albrecht Dold and Richard Lashof, *Principal quasi-fibrations and fibre homotopy equivalence of bundles.*, Illinois J. Math. **3** (1959), 285–305. MR0101521 (21 #331)
72. E. Dror, W. G. Dwyer, and D. M. Kan, *Automorphisms of fibrations*, Proc. Amer. Math. Soc. **80** (1980), no. 3, 491–494. MR581012 (81h:55012)
73. E. Dror Farjoun and J. Smith, *A geometric interpretation of Lannes' functor* **T**, Astérisque (1990), no. 191, 6, 87–95, International Conference on Homotopy Theory (Marseille-Luminy, 1988). MR1098968 (92h:55013)
74. Emmanuel Dror Farjoun, *Cellular spaces, null spaces and homotopy localization*, Lecture Notes in Mathematics, vol. 1622, Springer-Verlag, Berlin, 1996. MR1392221 (98f:55010)
75. Nicolas Dupont and Kathryn Hess, *How to model the free loop space algebraically*, Math. Ann. **314** (1999), no. 3, 469–490. MR1704545 (2000e:18011)
76. ———, *An algebraic model for homotopy fibers*, Homology Homotopy Appl. **4** (2002), no. 2, part 1, 117–139 (electronic), The Roos Festschrift volume, 1. MR1918186 (2003e:55012)
77. ———, *Commutative free loop space models at large primes*, Math. Z. **244** (2003), no. 1, 1–34. MR1981874 (2004c:55016)
78. Nicolas Dupont and Micheline Vigué-Poirrier, *Formalité des espaces de lacets libres*, Bull. Soc. Math. France **126** (1998), no. 1, 141–148. MR1651384 (99i:55018)
79. W. Dwyer and A. Zabrodsky, *Maps between classifying spaces*, Algebraic topology, Barcelona, 1986, Lecture Notes in Math., vol. 1298, Springer, Berlin, 1987, pp. 106–119. MR928826 (89b:55018)
80. W. G. Dwyer, *Tame homotopy theory*, Topology **18** (1979), no. 4, 321–338. MR551014 (81a:55020)
81. W. G. Dwyer and D. M. Kan, *Function complexes in homotopical algebra*, Topology **19** (1980), no. 4, 427–440. MR584566 (81m:55018)
82. ———, *Function complexes for diagrams of simplicial sets*, Nederl. Akad. Wetensch. Indag. Math. **45** (1983), no. 2, 139–147. MR705421 (85e:55038)
83. W. G. Dwyer and G. Mislin, *On the homotopy type of the components of* $\mathrm{map}_*(BS^3, BS^3)$, Algebraic topology, Barcelona, 1986, Lecture Notes in Math., vol. 1298, Springer, Berlin, 1987, pp. 82–89. MR928824 (89e:55018)
84. William G. Dwyer and Clarence W. Wilkerson, *Spaces of null homotopic maps*, Astérisque (1990), no. 191, 6, 97–108, International Conference on Homotopy Theory (Marseille-Luminy, 1988). MR1098969 (92b:55004)
85. Micheal N. Dyer, *The influence of* $\pi_1(Y)$ *on the homology of* $M(X, Y)$, Illinois J. Math. **10** (1966), 648–651. MR0203728 (34 #3577)
86. E. Fadell and S. Husseini, *A note on the category of the free loop space*, Proc. Amer. Math. Soc. **107** (1989), no. 2, 527–536. MR984789 (90a:55008)
87. Herbert Federer, *A study of function spaces by spectral sequences*, Trans. Amer. Math. Soc. **82** (1956), 340–361. MR0079265 (18,59b)
88. Y. Félix, *Rational category of the space of sections of a nilpotent bundle*, Comment. Math. Helv. **65** (1990), no. 4, 615–622. MR1078101 (91j:55012)
89. Y. Félix and Jean-Claude Thomas, *Homology and homotopy groups of mapping spaces at large primes*, Topology **32** (1993), no. 1, 1–4. MR1204401 (93m:55018)
90. ———, *The monoid of self-homotopy equivalences of some homogeneous spaces*, Exposition. Math. **12** (1994), no. 4, 305–322. MR1297839 (95i:55013)
91. Y. Félix, Jean-Claude Thomas, and M. Vigué-Poirrier, *Free loop spaces of finite complexes have infinite category*, Proc. Amer. Math. Soc. **111** (1991), no. 3, 869–875. MR1025277 (91f:55003)
92. Yves Félix and Stephen Halperin, *Rational LS category and its applications*, Trans. Amer. Math. Soc. **273** (1982), no. 1, 1–38. MR664027 (84h:55011)
93. Yves Félix, Stephen Halperin, and Jean-Claude Thomas, *Rational homotopy theory*, Graduate Texts in Mathematics, vol. 205, Springer-Verlag, New York, 2001. MR1802847 (2002d:55014)

94. Yves Félix and Gregory Lupton, *Evaluation maps in rational homotopy*, Topology **46** (2007), no. 5, 493–506. MR2337558 (2008e:55014)
95. Yves Félix, Gregory Lupton, and Samuel B. Smith, *The rational homotopy type of the space of self-equivalences of a fibration*, arXiv:0903.1470v1.
96. Yves Félix and John Oprea, *Rational homotopy of gauge groups*, Proc. Amer. Math. Soc. **137** (2009), no. 4, 1519–1527. MR2465678 (2009h:55011)
97. Yves Félix and Daniel Tanré, *H-space structure on pointed mapping spaces*, Algebr. Geom. Topol. **5** (2005), 713–724 (electronic). MR2153111 (2006a:55020)
98. Yves Félix and Jean-Claude Thomas, *Monoid of self-equivalences and free loop spaces*, Proc. Amer. Math. Soc. **132** (2004), no. 1, 305–312 (electronic). MR2021275 (2004k:55007)
99. ———, *Rational BV-algebra in string topology*, Bull. Soc. Math. France **136** (2008), no. 2, 311–327. MR2415345 (2009c:55015)
100. Yves Félix, Jean-Claude Thomas, and Micheline Vigué-Poirrier, *Rational string topology*, J. Eur. Math. Soc. (JEMS) **9** (2007), no. 1, 123–156. MR2283106 (2007k:55009)
101. Benoit Fresse, *Derived division functors and mapping spaces*, preprint, arXiv:math/0208091.
102. Eric M. Friedlander and Guido Mislin, *Locally finite approximation of Lie groups. I*, Invent. Math. **83** (1986), no. 3, 425–436. MR827361 (87i:55038)
103. David Gabai, *The Smale conjecture for hyperbolic 3-manifolds:* $\mathrm{Isom}(M^3) \simeq \mathrm{Diff}(M^3)$, J. Differential Geom. **58** (2001), no. 1, 113–149. MR1895350 (2003c:57016)
104. Tudor Ganea, *Lusternik-Schnirelmann category and cocategory*, Proc. London Math. Soc. (3) **10** (1960), 623–639. MR0126278 (23 #A3574)
105. Marek Golasiński and Juno Mukai, *Gottlieb groups of spheres*, Topology **47** (2008), no. 6, 399–430. MR2427733 (2009j:55017)
106. Thomas G. Goodwillie, *Cyclic homology, derivations, and the free loopspace*, Topology **24** (1985), no. 2, 187–215. MR793184 (87c:18009)
107. Daniel Henry Gottlieb, *A certain subgroup of the fundamental group*, Amer. J. Math. **87** (1965), 840–856. MR0189027 (32 #6454)
108. ———, *On fibre spaces and the evaluation map*, Ann. of Math. (2) **87** (1968), 42–55. MR0221508 (36 #4560)
109. ———, *Covering transformations and universal fibrations*, Illinois J. Math. **13** (1969), 432–437. MR0239595 (39 #952)
110. ———, *Evaluation subgroups of homotopy groups*, Amer. J. Math. **91** (1969), 729–756. MR0275424 (43 #1181)
111. ———, *Applications of bundle map theory*, Trans. Amer. Math. Soc. **171** (1972), 23–50. MR0309111 (46 #8222)
112. Jens Gravesen, *On the topology of spaces of holomorphic maps*, Acta Math. **162** (1989), no. 3-4, 247–286. MR989398 (90g:32023)
113. Pierre-Paul Grivel, *Algèbres de Lie de dérivations de certaines algèbres pures*, J. Pure Appl. Algebra **91** (1994), no. 1-3, 121–135. MR1255925 (95a:55024)
114. Detlef Gromoll and Wolfgang Meyer, *Periodic geodesics on compact riemannian manifolds*, J. Differential Geometry **3** (1969), 493–510. MR0264551 (41 #9143)
115. M. Gromov, *Oka's principle for holomorphic sections of elliptic bundles*, J. Amer. Math. Soc. **2** (1989), no. 4, 851–897. MR1001851 (90g:32017)
116. Karsten Grove, Stephen Halperin, and Micheline Vigué-Poirrier, *The rational homotopy theory of certain path spaces with applications to geodesics*, Acta Math. **140** (1978), no. 3-4, 277–303. MR496895 (80g:58024)
117. Kate Gruher and Paolo Salvatore, *Generalized string topology operations*, Proc. Lond. Math. Soc. (3) **96** (2008), no. 1, 78–106. MR2392316 (2009e:55013)
118. M. A. Guest, *Topology of the space of absolute minima of the energy functional*, Amer. J. Math. **106** (1984), no. 1, 21–42. MR729753 (85h:58047)
119. ———, *The topology of the space of rational curves on a toric variety*, Acta Math. **174** (1995), no. 1, 119–145. MR1310847 (95k:58021)
120. M. A. Guest, A. Kozlowski, M. Murayama, and K. Yamaguchi, *The homotopy type of the space of rational functions*, J. Math. Kyoto Univ. **35** (1995), no. 4, 631–638. MR1365252 (96j:58020)
121. M. A. Guest, A. Kozlowski, and K. Yamaguchi, *The topology of spaces of coprime polynomials*, Math. Z. **217** (1994), no. 3, 435–446. MR1306670 (95i:55014)

122. André Haefliger, *Rational homotopy of the space of sections of a nilpotent bundle*, Trans. Amer. Math. Soc. **273** (1982), no. 2, 609–620. MR667163 (84a:55010)
123. Stephen Halperin, *Rational fibrations, minimal models, and fibrings of homogeneous spaces*, Trans. Amer. Math. Soc. **244** (1978), 199–224. MR0515558 (58 #24264)
124. ———, *Modèle de l'espace des lacets libres dans un espace à groupe fondamental non nul*, C. R. Acad. Sci. Paris Sér. I Math. **293** (1981), no. 1, 79–81. MR633568 (82i:55012)
125. Stephen Halperin and Micheline Vigué-Poirrier, *The homology of a free loop space*, Pacific J. Math. **147** (1991), no. 2, 311–324. MR1084712 (92e:55012)
126. Hiroaki Hamanaka and Akira Kono, *Homotopy type of gauge groups of* SU(3)*-bundles over* S^6, Topology Appl. **154** (2007), no. 7, 1377–1380. MR2310471 (2008c:57047)
127. Vagn Lundsgaard Hansen, *Equivalence of evaluation fibrations*, Invent. Math. **23** (1974), 163–171. MR0368000 (51 #4242)
128. ———, *The homotopy problem for the components in the space of maps on the n-sphere*, Quart. J. Math. Oxford Ser. (2) **25** (1974), 313–321. MR0362396 (50 #14838)
129. ———, *On the fundamental group of a mapping space. An example*, Compositio Math. **28** (1974), 33–36. MR0341477 (49 #6228)
130. ———, *On the space of maps of a closed surface into the 2-sphere*, Math. Scand. **35** (1974), 149–158. MR0385919 (52 #6778)
131. ———, *Decomposability of evaluation fibrations and the brace product operation of James*, Compositio Math. **35** (1977), no. 1, 83–89. MR0494104 (58 #13035)
132. ———, *On spaces of maps of n-manifolds into the n-sphere*, Trans. Amer. Math. Soc. **265** (1981), no. 1, 273–281. MR607120 (82c:55014)
133. ———, *Spaces of maps into Eilenberg-Mac Lane spaces*, Canad. J. Math. **33** (1981), no. 4, 782–785. MR634136 (83a:58022)
134. ———, *The homotopy groups of a space of maps between oriented closed surfaces*, Bull. London Math. Soc. **15** (1983), no. 4, 360–364. MR703761 (84f:55012)
135. ———, *The space of self-maps on the 2-sphere*, Groups of self-equivalences and related topics (Montreal, PQ, 1988), Lecture Notes in Math., vol. 1425, Springer, Berlin, 1990, pp. 40–47. MR1070574 (91j:55009)
136. Allen E. Hatcher, *A proof of the Smale conjecture,* Diff$(S^3) \simeq$ O(4), Ann. of Math. (2) **117** (1983), no. 3, 553–607. MR701256 (85c:57008)
137. Volker Hauschild, *Deformations and the rational homotopy of the monoid of fiber homotopy equivalences*, Illinois J. Math. **37** (1993), no. 4, 537–560. MR1226781 (94h:55017)
138. ———, *Fibrations, self homotopy equivalences and negative derivations*, Groups of homotopy self-equivalences and related topics (Gargnano, 1999), Contemp. Math., vol. 274, Amer. Math. Soc., Providence, RI, 2001, pp. 169–182. MR1817009 (2001m:55033)
139. John W. Havlicek, *The cohomology of holomorphic self-maps of the Riemann sphere*, Math. Z. **218** (1995), no. 2, 179–190. MR1318152 (96c:58029)
140. Kathryn Hess and Ran Levi, *An algebraic model for the loop space homology of a homotopy fiber*, Algebr. Geom. Topol. **7** (2007), 1699–1765. MR2366176 (2008j:57056)
141. Peter Hilton, Guido Mislin, and Joe Roitberg, *Localization of nilpotent groups and spaces*, North-Holland Publishing Co., Amsterdam, 1975, North-Holland Mathematics Studies, No. 15, Notas de Matemática, No. 55. [Notes on Mathematics, No. 55]. MR0478146 (57 #17635)
142. Peter Hilton, Guido Mislin, Joseph Roitberg, and Richard Steiner, *On free maps and free homotopies into nilpotent spaces*, Algebraic topology (Proc. Conf., Univ. British Columbia, Vancouver, B.C., 1977), Lecture Notes in Math., vol. 673, Springer, Berlin, 1978, pp. 202–218. MR517093 (80c:55007)
143. Yasumasa Hirashima and Nobuyuki Oda, *Pairings of function spaces*, Topology Appl. **154** (2007), no. 12, 2412–2424. MR2333796 (2008g:54021)
144. Koichi Hirata and Kohhei Yamaguchi, *Spaces of polynomials without 3-fold real roots*, J. Math. Kyoto Univ. **42** (2002), no. 3, 509–516. MR1967220 (2004c:55013)
145. Yoshihiro Hirato, Katsuhiko Kuribayashi, and Nobuyuki Oda, *A function space model approach to the rational evaluation subgroups*, Math. Z. **258** (2008), no. 3, 521–555. MR2369043 (2008j:55012)
146. Po Hu, *Higher string topology on general spaces*, Proc. London Math. Soc. (3) **93** (2006), no. 2, 515–544. MR2251161 (2007f:55007)

147. Sze-tsen Hu, *Concerning the homotopy groups of the components of the mapping space Y^{S^p}*, Nederl. Akad. Wetensch., Proc. **49** (1946), 1025–1031 = Indagationes Math. 8, 623–629 (1946). MR0019920 (8,481a)
148. J. C. Hurtubise, *Stability theorems for moduli spaces*, Canadian Mathematical Society. 1945–1995, Vol. 3, Canadian Math. Soc., Ottawa, ON, 1996, pp. 153–171. MR1661615 (2000c:14016)
149. Norio Iwase, *On the splitting of mapping spaces between classifying spaces. I*, Publ. Res. Inst. Math. Sci. **23** (1987), no. 3, 445–453. MR905020 (89i:55004)
150. John D. S. Jones, *Cyclic homology and equivariant homology*, Invent. Math. **87** (1987), no. 2, 403–423. MR870737 (88f:18016)
151. Peter J. Kahn, *Some function spaces of CW type*, Proc. Amer. Math. Soc. **90** (1984), no. 4, 599–607. MR733413 (85j:55027)
152. Sadok Kallel and R. James Milgram, *The geometry of the space of holomorphic maps from a Riemann surface to a complex projective space*, J. Differential Geom. **47** (1997), no. 2, 321–375. MR1601616 (98m:58014)
153. Sadok Kallel and Paolo Salvatore, *Rational maps and string topology*, Geom. Topol. **10** (2006), 1579–1606 (electronic). MR2284046 (2007k:58018)
154. Sadok Kallel and Denis Sjerve, *On the topology of fibrations with section and free loop spaces*, Proc. London Math. Soc. (3) **83** (2001), no. 2, 419–442. MR1839460 (2002c:55021)
155. Yasuhiko Kamiyama, *Remarks on spaces of real rational functions*, Rocky Mountain J. Math. **37** (2007), no. 1, 247–257. MR2316447 (2008c:55012)
156. Frances Kirwan, *On spaces of maps from Riemann surfaces to Grassmannians and applications to the cohomology of moduli of vector bundles*, Ark. Mat. **24** (1986), no. 2, 221–275. MR884188 (88h:14014)
157. John R. Klein, Claude L. Schochet, and Samuel B. Smith, *Continuous trace C^*-algebras, gauge groups and rationalization*, J. Topol. Anal. **1** (2009), no. 3, 261–288. MR2574026
158. S. S. Koh, *Note on the homotopy properties of the components of the mapping space X^{S^p}*, Proc. Amer. Math. Soc. **11** (1960), 896–904. MR0119201 (22 #9967)
159. Akira Kono, *A note on the homotopy type of certain gauge groups*, Proc. Roy. Soc. Edinburgh Sect. A **117** (1991), no. 3-4, 295–297. MR1103296 (92b:55005)
160. Akira Kono and Kazumoto Kozima, *The adjoint action of a Lie group on the space of loops*, J. Math. Soc. Japan **45** (1993), no. 3, 495–510. MR1219882 (94h:57053)
161. Akira Kono and Shuichi Tsukuda, *A remark on the homotopy type of certain gauge groups*, J. Math. Kyoto Univ. **36** (1996), no. 1, 115–121. MR1381542 (97k:57040)
162. ———, *4-manifolds X over $BSU(2)$ and the corresponding homotopy types* $\mathrm{Map}(X, BSU(2))$, J. Pure Appl. Algebra **151** (2000), no. 3, 227–237. MR1776430 (2001k:55020)
163. Yasusuke Kotani, *Note on the rational cohomology of the function space of based maps*, Homology Homotopy Appl. **6** (2004), no. 1, 341–350 (electronic). MR2084591 (2005e:55014)
164. Andrzej Kozlowski and Kohhei Yamaguchi, *Topology of complements of discriminants and resultants*, J. Math. Soc. Japan **52** (2000), no. 4, 949–959. MR1774637 (2001e:55024)
165. Nicholas J. Kuhn, *Mapping spaces and homology isomorphisms*, Proc. Amer. Math. Soc. **134** (2006), no. 4, 1237–1248 (electronic), With an appendix by Greg Arone and the author. MR2196061 (2006h:55010)
166. Nicholas J. Kuhn and Mark Winstead, *On the torsion in the cohomology of certain mapping spaces*, Topology **35** (1996), no. 4, 875–881. MR1404914 (97d:55013)
167. Katsuhiko Kuribayashi, *On the mod p cohomology of the spaces of free loops on the Grassmann and Stiefel manifolds*, J. Math. Soc. Japan **43** (1991), no. 2, 331–346. MR1096437 (92c:55010)
168. ———, *Module derivations and non triviality of an evaluation fibration*, Homology Homotopy Appl. **4** (2002), no. 1, 87–101 (electronic). MR1937960 (2003h:55025)
169. ———, *Eilenberg-Moore spectral sequence calculation of function space cohomology*, Manuscripta Math. **114** (2004), no. 3, 305–325. MR2075968 (2005d:55025)
170. ———, *A rational model for the evaluation map*, Georgian Math. J. **13** (2006), no. 1, 127–141. MR2242333 (2007b:18015)
171. Katsuhiko Kuribayashi and Toshihiro Yamaguchi, *The cohomology algebra of certain free loop spaces*, Fund. Math. **154** (1997), no. 1, 57–73. MR1472851 (98j:55007)

172. ———, *A rational splitting of a based mapping space*, Algebr. Geom. Topol. **6** (2006), 309–327 (electronic). MR2220679 (2007g:55011)
173. Katsuhiko Kuribayashi and Masaaki Yokotani, *Iterated cyclic homology*, Kodai Math. J. **30** (2007), no. 1, 19–40. MR2319074 (2008a:55009)
174. Pascal Lambrechts, *The Betti numbers of the free loop space of a connected sum*, J. London Math. Soc. (2) **64** (2001), no. 1, 205–228. MR1840780 (2002f:55022)
175. ———, *On the Betti numbers of the free loop space of a coformal space*, J. Pure Appl. Algebra **161** (2001), no. 1-2, 177–192. MR1834084 (2002d:55015)
176. George E. Lang, Jr., *The evaluation map and* ehp *sequences*, Pacific J. Math. **44** (1973), 201–210. MR0341484 (49 #6235)
177. ———, *Localizations and evaluation subgroups*, Proc. Amer. Math. Soc. **50** (1975), 489–494. MR0367986 (51 #4228)
178. Jean Lannes, *Sur la cohomologie modulo p des p-groupes abéliens élémentaires*, Homotopy theory (Durham, 1985), London Math. Soc. Lecture Note Ser., vol. 117, Cambridge Univ. Press, Cambridge, 1987, pp. 97–116. MR932261 (89e:55037)
179. ———, *Sur les espaces fonctionnels dont la source est le classifiant d'un p-groupe abélien élémentaire*, Inst. Hautes Études Sci. Publ. Math. (1992), no. 75, 135–244, With an appendix by Michel Zisman. MR1179079 (93j:55019)
180. L. L. Larmore and E. Thomas, *On the fundamental group of a space of sections*, Math. Scand. **47** (1980), no. 2, 232–246. MR612697 (82h:55014)
181. A. Lazarev, *The Stasheff model of a simply-connected manifold and the string bracket*, Proc. Amer. Math. Soc. **136** (2008), no. 2, 735–745 (electronic). MR2358516 (2008k:55025)
182. Kee-Young Lee, Mamoru Mimura, and Moo Ha Woo, *Gottlieb groups of homogeneous spaces*, Topology Appl. **145** (2004), no. 1-3, 147–155. MR2100869 (2005g:55024)
183. Claude Legrand, *Sur l'homologie des espaces fonctionnels*, C. R. Acad. Sci. Paris Sér. I Math. **297** (1983), no. 7, 429–432. MR732851 (86a:55022)
184. ———, *Homologie du spectre fonctionnel*, Bull. Soc. Math. Belg. Sér. B **39** (1987), no. 3, 331–346. MR925276 (88m:55010)
185. Jiayuan Lin, *Rational homotopy stability for the spaces of algebraic maps*, New Zealand J. Math. **38** (2008), 179–186. MR2515373
186. Gregory Lupton, *Note on a conjecture of Stephen Halperin's*, Topology and combinatorial group theory (Hanover, NH, 1986/1987; Enfield, NH, 1988), Lecture Notes in Math., vol. 1440, Springer, Berlin, 1990, pp. 148–163. MR1082989 (92a:55012)
187. Gregory Lupton, N. Christopher Phillips, Claude L. Schochet, and Samuel B. Smith, *Banach algebras and rational homotopy theory*, Trans. Amer. Math. Soc. **361** (2009), no. 1, 267–295. MR2439407 (2009k:46086)
188. Gregory Lupton and Samuel Bruce Smith, *Rank of the fundamental group of any component of a function space*, Proc. Amer. Math. Soc. **135** (2007), no. 8, 2649–2659 (electronic). MR2302588 (2008h:55027)
189. ———, *Rationalized evaluation subgroups of a map. I. Sullivan models, derivations and G-sequences*, J. Pure Appl. Algebra **209** (2007), no. 1, 159–171. MR2292124 (2008c:55017)
190. ———, *Rationalized evaluation subgroups of a map. II. Quillen models and adjoint maps*, J. Pure Appl. Algebra **209** (2007), no. 1, 173–188. MR2292125 (2008c:55018)
191. ———, *Criteria for components of a function space to be homotopy equivalent*, Math. Proc. Cambridge Philos. Soc. **145** (2008), no. 1, 95–106. MR2431641 (2009e:55009)
192. ———, *Whitehead products in function spaces: Quillen model formulae*, J. Japan. Math. Soc. **62** (2010), no. 1, 49–81.
193. Michael A. Mandell, *E_∞ algebras and p-adic homotopy theory*, Topology **40** (2001), no. 1, 43–94. MR1791268 (2001m:55025)
194. Benjamin M. Mann and R. James Milgram, *Some spaces of holomorphic maps to complex Grassmann manifolds*, J. Differential Geom. **33** (1991), no. 2, 301–324. MR1094457 (93e:55022)
195. ———, *On the moduli space of* SU(n) *monopoles and holomorphic maps to flag manifolds*, J. Differential Geom. **38** (1993), no. 1, 39–103. MR1231702 (95c:58031)
196. Ken-ichi Maruyama and Hideaki Ōshima, *Homotopy groups of the spaces of self-maps of Lie groups*, J. Math. Soc. Japan **60** (2008), no. 3, 767–792. MR2440413 (2009h:55014)
197. Gregor Masbaum, *On the cohomology of the classifying space of the gauge group over some 4-complexes*, Bull. Soc. Math. France **119** (1991), no. 1, 1–31. MR1101938 (92a:55009)

198. J. P. May, *The geometry of iterated loop spaces*, Springer-Verlag, Berlin, 1972, Lectures Notes in Mathematics, Vol. 271. MR0420610 (54 #8623b)
199. ———, *Fibrewise localization and completion*, Trans. Amer. Math. Soc. **258** (1980), no. 1, 127–146. MR554323 (81f:55005)
200. John McCleary, *On the mod p Betti numbers of loop spaces*, Invent. Math. **87** (1987), no. 3, 643–654. MR874040 (88i:57018)
201. ———, *Homotopy theory and closed geodesics*, Homotopy theory and related topics (Kinosaki, 1988), Lecture Notes in Math., vol. 1418, Springer, Berlin, 1990, pp. 86–94. MR1048178 (91e:57060)
202. John McCleary and Dennis A. McLaughlin, *Morava K-theory and the free loop space*, Proc. Amer. Math. Soc. **114** (1992), no. 1, 243–250. MR1079897 (92c:55011)
203. John McCleary and Wolfgang Ziller, *On the free loop space of homogeneous spaces*, Amer. J. Math. **109** (1987), no. 4, 765–781. MR900038 (88k:58023)
204. J. F. McClendon, *On evaluation fibrations*, Houston J. Math. **7** (1981), no. 3, 379–388. MR640979 (83g:55012)
205. Darryl McCullough, *Homotopy groups of the space of self-homotopy-equivalences*, Trans. Amer. Math. Soc. **264** (1981), no. 1, 151–163. MR597873 (83e:55004)
206. C. A. McGibbon, *A note on Miller's theorem about maps out of classifying spaces*, Proc. Amer. Math. Soc. **124** (1996), no. 10, 3241–3245. MR1328362 (96m:55033)
207. C. A. McGibbon and J. A. Neisendorfer, *On the homotopy groups of a finite-dimensional space*, Comment. Math. Helv. **59** (1984), no. 2, 253–257. MR749108 (86b:55015)
208. W. Meier, *Rational universal fibrations and flag manifolds*, Math. Ann. **258** (1981/82), no. 3, 329–340. MR649203 (83g:55009)
209. Luc Menichi, *The cohomology ring of free loop spaces*, Homology Homotopy Appl. **3** (2001), no. 1, 193–224 (electronic). MR1854644 (2002f:55023)
210. ———, *String topology for spheres*, Comment. Math. Helv. **84** (2009), no. 1, 135–157, With an appendix by Gerald Gaudens and Menichi. MR2466078 (2009k:55017)
211. Haynes Miller, *The Sullivan conjecture on maps from classifying spaces*, Ann. of Math. (2) **120** (1984), no. 1, 39–87. MR750716 (85i:55012)
212. John Milnor, *On spaces having the homotopy type of a CW-complex*, Trans. Amer. Math. Soc. **90** (1959), 272–280. MR0100267 (20 #6700)
213. Jesper Michael Møller, *On spaces of maps between complex projective spaces*, Proc. Amer. Math. Soc. **91** (1984), no. 3, 471–476. MR744651 (85f:55004)
214. ———, *On the homology of spaces of sections of complex projective bundles*, Pacific J. Math. **116** (1985), no. 1, 143–154. MR769828 (86k:55019)
215. ———, *Nilpotent spaces of sections*, Trans. Amer. Math. Soc. **303** (1987), no. 2, 733–741. MR902794 (88j:55007)
216. ———, *Spaces of sections of Eilenberg-Mac Lane fibrations*, Pacific J. Math. **130** (1987), no. 1, 171–186. MR910659 (89d:55048)
217. ———, *On equivariant function spaces*, Pacific J. Math. **142** (1990), no. 1, 103–119. MR1038731 (91a:55024)
218. ———, *Samelson products in spaces of self-homotopy equivalences*, Canad. J. Math. **42** (1990), no. 1, 95–108. MR1043513 (91a:55015)
219. Jesper Michael Møller and Martin Raussen, *Rational homotopy of spaces of maps into spheres and complex projective spaces*, Trans. Amer. Math. Soc. **292** (1985), no. 2, 721–732. MR808750 (86m:55019)
220. J. C. Moore, *On a theorem of Borsuk*, Fund. Math. **43** (1956), 195–201. MR0083123 (18,662d)
221. Bitjong Ndombol and Jean-Claude Thomas, *On the cohomology algebra of free loop spaces*, Topology **41** (2002), no. 1, 85–106. MR1871242 (2002h:57052)
222. Joseph Neisendorfer, *Lie algebras, coalgebras and rational homotopy theory for nilpotent spaces*, Pacific J. Math. **74** (1978), no. 2, 429–460. MR494641 (80b:55010)
223. Dietrich Notbohm and Larry Smith, *Rational homotopy of the space of homotopy equivalences of a flag manifold*, Algebraic topology (San Feliu de Guíxols, 1990), Lecture Notes in Math., vol. 1509, Springer, Berlin, 1992, pp. 301–312. MR1185980 (94c:55012)
224. Iver Ottosen, *On the Borel cohomology of free loop spaces*, Math. Scand. **93** (2003), no. 2, 185–220. MR2009581 (2004h:55004)

225. Iver Ottosen and Marcel Bökstedt, *String cohomology groups of complex projective spaces*, Algebr. Geom. Topol. **7** (2007), 2165–2238. MR2366191 (2008m:55015)
226. James Michael Parks, *A note on the monoid of self-equivalences*, Houston J. Math. **7** (1981), no. 3, 403–406. MR640982 (83c:55009)
227. Frédéric Patras and Jean-Claude Thomas, *Cochain algebras of mapping spaces and finite group actions*, Topology Appl. **128** (2003), no. 2-3, 189–207. MR1956614 (2004b:55028)
228. Andrew Pressley and Graeme Segal, *Loop groups*, Oxford Mathematical Monographs, The Clarendon Press Oxford University Press, New York, 1986, Oxford Science Publications. MR900587 (88i:22049)
229. Daniel Quillen, *Rational homotopy theory*, Ann. of Math. (2) **90** (1969), 205–295. MR0258031 (41 #2678)
230. Jan-Erik Roos, *Homology of free loop spaces, cyclic homology and nonrational Poincaré-Betti series in commutative algebra*, Algebra—some current trends (Varna, 1986), Lecture Notes in Math., vol. 1352, Springer, Berlin, 1988, pp. 173–189. MR981826 (90f:55020)
231. John W. Rutter, *Spaces of homotopy self-equivalences*, Lecture Notes in Mathematics, vol. 1662, Springer-Verlag, Berlin, 1997, A survey. MR1474967 (98f:55005)
232. Paolo Salvatore, *Rational homotopy nilpotency of self-equivalences*, Topology Appl. **77** (1997), no. 1, 37–50. MR1443426 (98d:55011)
233. Seiya Sasao, *The homotopy of* Map (CP^m, CP^n), J. London Math. Soc. (2) **8** (1974), 193–197. MR0346783 (49 #11507)
234. ———, *On the homotopy of certain mapping spaces*, Kodai Math. J. **11** (1988), no. 2, 306–315. MR949137 (89h:55025)
235. Hans Scheerer, *On rationalized H- and co-H-spaces. With an appendix on decomposable H- and co-H-spaces*, Manuscripta Math. **51** (1985), no. 1-3, 63–87. MR788673 (86m:55016)
236. ———, *An application of algebraic R-local homotopy theory*, J. Pure Appl. Algebra **91** (1994), no. 1-3, 329–332. MR1255936 (95b:55008)
237. Hans Scheerer and Daniel Tanré, *Exploration de l'homotopie modérée de W. G. Dwyer. I*, C. R. Acad. Sci. Paris Sér. I Math. **307** (1988), no. 14, 783–785. MR972081 (89k:55020a)
238. ———, *Homotopie modérée et tempérée avec les coalgèbres. Applications aux espaces fonctionnels*, Arch. Math. (Basel) **59** (1992), no. 2, 130–145. MR1170636 (93k:55016)
239. M. Schlessinger and J. Stasheff, *Deformation theory and rational homotopy type*, preprint.
240. Reinhard Schultz, *Homotopy decompositions of equivariant function spaces. I*, Math. Z. **131** (1973), 49–75. MR0407866 (53 #11636)
241. Nora Seeliger, *Cohomology of the free loop space of a complex projective space*, Topology Appl. **155** (2007), no. 3, 127–129. MR2370366 (2008m:55002)
242. Graeme Segal, *The topology of spaces of rational functions*, Acta Math. **143** (1979), no. 1-2, 39–72. MR533892 (81c:55013)
243. Katsuyuki Shibata, *On Haefliger's model for the Gel′fand-Fuks cohomology*, Japan. J. Math. (N.S.) **7** (1981), no. 2, 379–415. MR729444 (85k:57033)
244. H. Shiga and M. Tezuka, *Rational fibrations, homogeneous spaces with positive Euler characteristics and Jacobians*, Ann. Inst. Fourier (Grenoble) **37** (1987), no. 1, 81–106. MR894562 (89g:55019)
245. Stephen Smale, *Diffeomorphisms of the 2-sphere*, Proc. Amer. Math. Soc. **10** (1959), 621–626. MR0112149 (22 #3004)
246. Larry Smith, *On the characteristic zero cohomology of the free loop space*, Amer. J. Math. **103** (1981), no. 5, 887–910. MR630771 (83k:57035)
247. ———, *The Eilenberg-Moore spectral sequence and the mod 2 cohomology of certain free loop spaces*, Illinois J. Math. **28** (1984), no. 3, 516–522. MR748960 (86d:57026)
248. Samuel B. Smith, *Rational homotopy of the space of self-maps of complexes with finitely many homotopy groups*, Trans. Amer. Math. Soc. **342** (1994), no. 2, 895–915. MR1225575 (94j:55012)
249. ———, *Rational evaluation subgroups*, Math. Z. **221** (1996), no. 3, 387–400. MR1381587 (97a:55014)
250. ———, *A based Federer spectral sequence and the rational homotopy of function spaces*, Manuscripta Math. **93** (1997), no. 1, 59–66. MR1446191 (98f:55011)
251. ———, *Rational classification of simple function space components for flag manifolds*, Canad. J. Math. **49** (1997), no. 4, 855–864. MR1471062 (98e:55012)

252. ———, *Rational L.S. category of function space components for F_0-spaces*, Bull. Belg. Math. Soc. Simon Stevin **6** (1999), no. 2, 295–304. MR1705124 (2000g:55016)
253. ———, *The rational homotopy Lie algebra of classifying spaces for formal two-stage spaces*, J. Pure Appl. Algebra **160** (2001), no. 2-3, 333–343. MR1836007 (2002f:55031)
254. Jaka Smrekar, *Compact open topology and CW homotopy type*, Topology Appl. **130** (2003), no. 3, 291–304. MR1978893 (2004c:55015)
255. V. P. Snaith, *A stable decomposition of $\Omega^n S^n X$*, J. London Math. Soc. (2) **7** (1974), 577–583. MR0339155 (49 #3918)
256. E. Spanier, *Infinite symmetric products, function spaces, and duality*, Ann. of Math. (2) **69** (1959), 142–198[erratum, 733. MR0105106 (21 #3851)
257. James Stasheff, *A classification theorem for fibre spaces*, Topology **2** (1963), 239–246. MR0154286 (27 #4235)
258. N. E. Steenrod, *A convenient category of topological spaces*, Michigan Math. J. **14** (1967), 133–152. MR0210075 (35 #970)
259. Jeffrey Strom, *Miller spaces and spherical resolvability of finite complexes*, Fund. Math. **178** (2003), no. 2, 97–108. MR2029919 (2005b:55026)
260. Dennis Sullivan, *Geometric topology. Part I*, Massachusetts Institute of Technology, Cambridge, Mass., 1971, Localization, periodicity, and Galois symmetry, Revised version. MR0494074 (58 #13006a)
261. ———, *Infinitesimal computations in topology*, Inst. Hautes Études Sci. Publ. Math. (1977), no. 47, 269–331 (1978). MR0646078 (58 #31119)
262. W. A. Sutherland, *Path-components of function spaces*, Quart. J. Math. Oxford Ser. (2) **34** (1983), no. 134, 223–233. MR698208 (84g:55019)
263. ———, *Function spaces related to gauge groups*, Proc. Roy. Soc. Edinburgh Sect. A **121** (1992), no. 1-2, 185–190. MR1169902 (93i:55011)
264. Hirotaka Tamanoi, *Batalin-Vilkovisky Lie algebra structure on the loop homology of complex Stiefel manifolds*, Int. Math. Res. Not. (2006), Art. ID 97193, 23. MR2211159 (2006m:55026)
265. Daniel Tanré, *Homotopie rationnelle: modèles de Chen, Quillen, Sullivan*, Lecture Notes in Mathematics, vol. 1025, Springer-Verlag, Berlin, 1983. MR764769 (86b:55010)
266. Clifford Henry Taubes, *The stable topology of self-dual moduli spaces*, J. Differential Geom. **29** (1989), no. 1, 163–230. MR978084 (90f:58023)
267. Svjetlana Terzić, *The rational topology of gauge groups and of spaces of connections*, Compos. Math. **141** (2005), no. 1, 262–270. MR2099779 (2005h:55011)
268. R. Thom, *L'homologie des espaces fonctionnels*, Colloque de topologie algébrique, Louvain, 1956, Georges Thone, Liège, 1957, pp. 29–39. MR0089408 (19,669h)
269. Jean-Claude Thomas, *Rational homotopy of Serre fibrations*, Ann. Inst. Fourier (Grenoble) **31** (1981), no. 3, v, 71–90. MR638617 (83c:55016)
270. Kung-Kuen Tse, *Rational function spaces*, Bol. Soc. Mat. Mexicana (3) **9** (2003), no. 2, 323–337. MR2029280 (2005c:55014)
271. Shuichi Tsukuda, *A remark on the homotopy type of the classifying space of certain gauge groups*, J. Math. Kyoto Univ. **36** (1996), no. 1, 123–128. MR1381543 (97h:55017)
272. ———, *On the cohomology of the classifying space of a certain gauge group*, Proc. Roy. Soc. Edinburgh Sect. A **127** (1997), no. 2, 407–409. MR1447960 (98a:55014)
273. ———, *On isomorphism classes of gauge groups*, Topology Appl. **87** (1998), no. 3, 173–187. MR1624312 (99c:58022)
274. ———, *Comparing the homotopy types of the components of* $\mathrm{Map}(S^4, B\mathrm{SU}(2))$, J. Pure Appl. Algebra **161** (2001), no. 1-2, 235–243. MR1834088 (2002g:55013)
275. V. A. Vassiliev, *Topology of complements to discriminants and loop spaces*, Theory of singularities and its applications, Adv. Soviet Math., vol. 1, Amer. Math. Soc., Providence, RI, 1990, pp. 9–21. MR1089669 (92e:32019)
276. Micheline Vigué-Poirrier, *Dans le fibré de l'espace des lacets libres, la fibre n'est pas, en général, totalement non cohomologue à zéro*, Math. Z. **181** (1982), no. 4, 537–542. MR682673 (84e:55007)
277. ———, *Homotopie rationnelle et croissance du nombre de géodésiques fermées*, Ann. Sci. École Norm. Sup. (4) **17** (1984), no. 3, 413–431. MR777376 (86h:58027)
278. ———, *Sur l'homotopie rationnelle des espaces fonctionnels*, Manuscripta Math. **56** (1986), no. 2, 177–191. MR850369 (87h:55009)

279. ———, *Rational formality of function spaces*, J. Homotopy Relat. Struct. **2** (2007), no. 1, 99–108 (electronic). MR2326935 (2008e:55015)
280. Micheline Vigué-Poirrier and Dennis Sullivan, *The homology theory of the closed geodesic problem*, J. Differential Geometry **11** (1976), no. 4, 633–644. MR0455028 (56 #13269)
281. Craig Westerland, *Dyer-Lashof operations in the string topology of spheres and projective spaces*, Math. Z. **250** (2005), no. 3, 711–727. MR2179618 (2006h:55017)
282. ———, *Stable splittings of surface mapping spaces*, Topology Appl. **153** (2006), no. 15, 2834–2865. MR2248387 (2007g:55006)
283. ———, *String homology of spheres and projective spaces*, Algebr. Geom. Topol. **7** (2007), 309–325. MR2308947 (2008h:55010)
284. George W. Whitehead, *On products in homotopy groups*, Ann. of Math (2) **47** (1946), 460–475. MR0016672 (8,50b)
285. C. Wockel, *The Samelson product and rational homotopy for gauge groups*, Abh. Math. Sem. Univ. Hamburg **77** (2007), 219–228. MR2379340 (2008k:57064)
286. Moo Ha Woo and Yeon Soo Yoon, *T-spaces by the Gottlieb groups and duality*, J. Austral. Math. Soc. Ser. A **59** (1995), no. 2, 193–203. MR1346627 (96e:55009)
287. Kohhei Yamaguchi, *On the rational homotopy of* Map($\mathbf{HP}^m$, $\mathbf{HP}^n$), Kodai Math. J. **6** (1983), no. 3, 279–288. MR717319 (85j:55024)
288. ———, *Spaces of holomorphic maps with bounded multiplicity*, Q. J. Math. **52** (2001), no. 2, 249–259. MR1838367 (2002f:55024)
289. ———, *Spaces of polynomials with real roots of bounded multiplicity*, J. Math. Kyoto Univ. **42** (2002), no. 1, 105–115. MR1932739 (2003h:55013)
290. ———, *The topology of spaces of maps between real projective spaces*, J. Math. Kyoto Univ. **43** (2003), no. 3, 503–507. MR2028664 (2005h:55006)
291. ———, *Fundamental groups of spaces of holomorphic maps and group actions*, J. Math. Kyoto Univ. **44** (2004), no. 3, 479–492. MR2103778 (2005m:55011)
292. ———, *Spaces of free loops on real projective spaces*, Kyushu J. Math. **59** (2005), no. 1, 145–153. MR2134058 (2006a:55011)
293. ———, *The homotopy of spaces of maps between real projective spaces*, J. Math. Soc. Japan **58** (2006), no. 4, 1163–1184. MR2276186 (2007i:55007)
294. Toshihiro Yamaguchi, *Formality of the function space of free maps into an elliptic space*, Bull. Soc. Math. France **128** (2000), no. 2, 207–218. MR1772441 (2001d:55007)
295. ———, *On the genus of free loop fibrations over F_0-spaces*, Int. J. Math. Math. Sci. (2004), no. 65-68, 3617–3619. MR2128778 (2005k:55015)
296. ———, *An estimate in Gottlieb ranks of fibration*, Bull. Belg. Math. Soc. Simon Stevin **15** (2008), no. 4, 663–675. MR2475490 (2010a:55012)
297. Tsuneyo Yamanoshita, *On the spaces of self-homotopy equivalences of certain CW complexes*, Proc. Japan Acad. Ser. A Math. Sci. **60** (1984), no. 6, 229–231. MR758050 (86i:55010)
298. ———, *A note on the spaces of self-homotopy equivalences for fibre spaces*, Proc. Japan Acad. Ser. A Math. Sci. **61** (1985), no. 10, 308–310. MR834534 (87g:55012)
299. ———, *On the spaces of self-homotopy equivalences of certain CW complexes*, J. Math. Soc. Japan **37** (1985), no. 3, 455–470. MR792987 (86k:55004)
300. ———, *On the spaces of self-homotopy equivalences for fibre spaces. II*, Publ. Res. Inst. Math. Sci. **22** (1986), no. 1, 43–56. MR834347 (87g:55028)
301. ———, *On the spaces of self-homotopy equivalences for fibre spaces*, Kodai Math. J. **10** (1987), no. 1, 127–142. MR879390 (88f:55019)
302. ———, *On the space of self-homotopy equivalences of the projective plane*, J. Math. Soc. Japan **45** (1993), no. 3, 489–494. MR1219881 (94f:55007)
303. Yeon Soo Yoon, *Decomposability of evaluation fibrations*, Kyungpook Math. J. **35** (1995), no. 2, 361–369. MR1369453 (97a:55018)
304. Alex Zabrodsky, *On spaces of functions between classifying spaces*, Israel J. Math. **76** (1991), no. 1-2, 1–26. MR1177330 (93i:55020)
305. Wolfgang Ziller, *The free loop space of globally symmetric spaces*, Invent. Math. **41** (1977), no. 1, 1–22. MR0649625 (58 #31198)

Department of Mathematics, Saint Joseph's University, Philadelphia, PA 19131
E-mail address: smith@sju.edu

Contributed Articles

Contemporary Mathematics
Volume **519**, 2010

Upper bounds for the Whitehead-length of mapping spaces

Urtzi Buijs

ABSTRACT. In this brief paper, for any map between a simply connected and a nilpotent CW-complex of finite type (over $\mathbb{Q}$), we prove a known result which gives an upper bound for the rational Whitehead-length of the corresponding component of the pointed mapping space in terms of the rational cone-length of the source, using Quillen models of such spaces. Moreover we improve this bound in the case that the target space has rational Whitehead-length 1 using the same technique.

Introduction

A large variety of results has been published about the rational homotopy of mapping spaces since René Thom [**18**] proved that the space of continuous functions from X into the Eilenberg-MacLane space $K(G,n)$ has the homotopy type of the product $\Pi_{j=0}^{n} K_j$ where $K_j = K(H^{n-j}(X;G),j)$. From this, and arguing inductively over a Postnikov tower of Y, Thom concluded that the homotopy type of the mapping space is determined by a sequence of abelian group extensions arising from the cohomology of X with coefficients in the homotopy groups of Y. As example, taking $G = \mathbb{Q}$, he calculated the rational homotopy groups of certain function spaces.

Later on, and continuing with the original project of D. Sullivan [**17**], A. Haefliger [**12**] described an algebraic model of the sections of a nilpotent bundle, homotopic to a given one. This fundamental work, and in particular, the algebraic model developed on it, was later investigated by Y. Félix, S. Smith, D. Tanré, M. Vigué, to obtain very interesting results on the rational homotopy type of function spaces [**9, 11, 15, 16, 19**]. However, due mostly to the non functorial nature of the Haefliger model, it is not easy to use it to describe certain geometrical properties of the spaces.

2000 *Mathematics Subject Classification.* 55P62, 54C35.

Key words and phrases. Mapping space, Sullivan model, Quillen model, rational homotopy theory.

Partially supported by the *Ministerio de Educación y Ciencia* grant MTM2007-60016 and by the Junta de Andalucía grant FQM-213.

The author would like to thank the referee for very helpful comments which have improved the paper.

©2010 American Mathematical Society

A different treatment using Bousfield-Gugenheim [**1**] realization functor was employed by E. H. Brown and R. H. Szczarba [**2**] in which the realization of a certain commutative differential $\mathbb{Z}$ graded algebra (CDGA from now on) is homotopy equivalent to the (non connected) mapping space $\mathrm{map}(X,Y)$. It is also showed that, from this model, one can obtain a quotient algebra whose realization is a given path component of the mapping space. Indeed, this quotient turns out to be isomorphic to the Haefliger model.

This Brown-Szczarba approach was used by U. Buijs and A. Murillo to describe the rational homotopy Lie algebra of any component of the (free or pointed) mapping space in terms of derivations between CDGA models of X and Y [**4**], and subsequently Buijs, Félix and Murillo obtain Lie models for any component of the space of free and pointed sections of a nilpotent fibration [**5**].

G. Lupton and S. Smith described also in [**13, 14**] rational homotopy groups and Whitehead products in components of mapping spaces avoiding Brown-Szczarba models in terms of derivations between differential graded Lie algebras (DGL from now on) modeling X and Y. But this complexes of derivations turn out to carry more structure, they are DGL models in some cases and L_∞ models for the components of pointed mapping spaces in a general sense [**6, 7**].

Let $\mathrm{map}_f(X,Y)$ denote the path component of the space of free continuous functions from X to Y containing a given one $f\colon X \longrightarrow Y$, and $\mathrm{map}^*_f(X,Y)$ the component in the space of pointed functions.

By definition, the *rational Whitehead-length* of a space Z, denoted $\mathrm{Wh}_{\mathbb{Q}}(Z)$, is the length of longest non zero iterated Whitehead bracket in $\pi_{\geq 2}(Z)\otimes\mathbb{Q}$. In particular, $\mathrm{Wh}_{\mathbb{Q}}(Z)=1$ means that all Whitehead products vanish.

By definition the *rational cone-length* of a space Z, denoted $\mathrm{cl}_0(Z)$ is the least integer n such that Z has the rational homotopy type of an n-cone (see [**10**, p.359]). Here we will rely on the following characterization:

PROPOSITION 0.1. [**8, 10**] *The following conditions are equivalent for a simply connected topological space X with rational homology of finite type:*

(1) $\mathrm{cl}_0(X)\leq n$.
(2) $\mathrm{nil}\, A\leq n$ *for some commutative model* (A,d) *for* X *with* $A^0=\mathbb{Q}$.
(3) X *has a free Lie model* $(\mathbb{L}(V),\partial)$ *with an increasing filtration* $0=V(0)\subset V(1)\subset\cdots\subset V(n)=V$ *such that for each* i, $\partial\colon V(i)\longrightarrow\mathbb{L}(V(i-1))$.

In [**14**] the following topological application was proved:

THEOREM 0.2. [**14**] *Let X be a simply connected finite CW-complex and Y a simply connected complex of finite type. Then*

$$\mathrm{Wh}_{\mathbb{Q}}(\mathrm{map}^*_f(X,Y))\leq \mathrm{cl}_0(X)$$

for all maps $f\colon X\longrightarrow Y$.

In this paper following the models developed on [**6, 7**] we can prove the above Theorem in the more general setting of X non necessarily finite, and use the same technique to obtain the following improvement:

THEOREM 0.3. *Let X be a simply connected CW-complex and Y a nilpotent CW-complex both of finte type (over $\mathbb{Q}$). If* $\mathrm{Wh}_{\mathbb{Q}}(Y)=1$, *then*

$$\mathrm{Wh}_{\mathbb{Q}}(\mathrm{map}^*_f(X,Y))\leq \mathrm{cl}_0(X)-1$$

for all maps $f\colon X\longrightarrow Y$.

The paper is organized as follows. In Section 1 a brief exposition of CDGA and DGL models for the components of mapping spaces are given. Section 2 contains the proof of the main result of the paper in terms of DGL models in order to remove the finite condition on X. Section 3 is devoted to developing some geometrical examples.

I would like to thank the referee for its very helpful comments which have improved the paper.

1. Models for the components of mapping spaces

A basic background of rational homotopy theory which includes Sullivan models and Quillen models are needed. For an exhaustive exposition we refer to [**10**].

CDGA models for the components of mapping spaces

Here we explain briefly the rudiments of Brown-Szczarba approach to Haefliger model for the components of mapping space. For a complete description see [**2, 4**].

Let B be a finite dimensional CDGA model of the simply connected finite complex X, and let $(\Lambda V, d)$ be a Sullivan (non necessarily minimal) model of the nilpotent space Y. Denote by $B^\sharp$ the differential coalgebra dual of B with the grading $B^{\sharp^{-n}} = B^\sharp_n = Hom(B^n, \mathbb{Q})$ and consider $\Lambda(\Lambda V \otimes B^\sharp)$ the free commutative algebra generated by the $\mathbb{Z}$ graded vector space $\Lambda V \otimes B^\sharp$ endowed with the differential d induced by d and by the differential on $B^\sharp$. Let J be the differential ideal generated by $1 \otimes 1 - 1$, and the elements of the form

$$v_1 v_2 \otimes \beta - \sum_j (-1)^{|v_2||\beta'_j|}(v_1 \otimes \beta'_j)(v_2 \otimes \beta''_j), \quad v_1, v_2 \in V$$

where $\Delta\beta = \sum_j \beta'_j \otimes \beta''_j$. The map induced by the inclusion $V \otimes B^\sharp \hookrightarrow \Lambda V \otimes B^\sharp$,

$$\rho\colon \Lambda(V \otimes B^\sharp) \xrightarrow{\cong} \Lambda(\Lambda V \otimes B^\sharp)/J$$

is an isomorphism of graded algebras and thus $\widetilde{d} = \rho^{-1} d \rho$ defines a differential in $\Lambda(V \otimes B^\sharp)$ for which:

THEOREM 1.1. [**2, 12**] $(\Lambda(V \otimes B^\sharp), \widetilde{d})$ *(resp.* $(\Lambda(V \otimes \overline{B^\sharp}), \widetilde{d})$*) is a model of* $\mathrm{map}(X, Y)$ *(resp.* $\mathrm{map}^*(X, Y)$*).*

From now on $\overline{B^\sharp}$ denotes $ker\epsilon$ where $B^\sharp$ is augmented by $\epsilon\colon B^\sharp \longrightarrow \mathbb{Q}$.

For the models of the components of $\mathrm{map}(X, Y)$ and/or $\mathrm{map}^*(X, Y)$ we follow the approach and notation of [**3**]:

For any free CDGA $(\Lambda S, d)$, in which S is $\mathbb{Z}$ graded, and any algebra morphism $u\colon \Lambda S \longrightarrow \mathbb{Q}$ consider the differential ideal K_u generated by $A_1 \cup A_2 \cup A_3$, where

$$A_1 = S^{<0}, \ A_2 = dS^0, \ A_3 = \{\alpha - u(\alpha) : \alpha \in S^0\}.$$

$(\Lambda S, d)/K_u$ is again a free CDGA of the form $(\Lambda(\overline{S}^1 \oplus S^{\geq 2}), d_u)$ in which $\overline{S}^1$ is a complement in S^1 of $d(S^0)$ modulo identifications via A_1 and A_3 (see [**3**] for details). Note that, $\overline{S}^1$ depends also on u. Moreover, if $(\Lambda S, d)$ is a model of a (non connected) space X and u corresponds to a 0-simplex of X, as remarked in [**3**, 4.3], $(\Lambda(\overline{S}^1 \oplus S^{\geq 2}), d_u)$ is a Sullivan model of the path component of X containing the fixed 0-simplex.

We now apply this to the following: let $f\colon X \to Y$ be a map modeled by $\phi\colon (\Lambda V, d) \to (B, d)$, which determines an algebra morphism denoted also by $\phi\colon (\Lambda(V \otimes B^\sharp), \widetilde{d}) \to \mathbb{Q}$. Then:

THEOREM 1.2. [**2, 3**] *The projection*

$$(\Lambda(V \otimes B^\sharp), \widetilde{d}) \longrightarrow (\Lambda(V \otimes B^\sharp), \widetilde{d})/K_\phi \cong (\Lambda(\overline{V \otimes B^\sharp}^1 \oplus (V \otimes B^\sharp)^{\geq 2}), d_\phi)$$

is a model of the inclusion $\mathrm{map}_f(X,Y) \hookrightarrow \mathrm{map}(X,Y)$. *Moreover,*

$$(\Lambda(V \otimes \overline{B^\sharp}), \widetilde{d}) \longrightarrow (\Lambda(V \otimes \overline{B^\sharp}), \widetilde{d})/K_\phi \cong (\Lambda(\overline{V \otimes \overline{B^\sharp}}^1 \oplus (V \otimes \overline{B^\sharp})^{\geq 2}), d_\phi)$$

is a model of the inclusion $\mathrm{map}^*_f(X,Y) \hookrightarrow \mathrm{map}^*(X,Y)$.

In particular, $(\Lambda(\overline{V \otimes B^\sharp}^1 \oplus (V \otimes B^\sharp)^{\geq 2}), d_\phi)$ is a Sullivan model of $\mathrm{map}_f(X,Y)$ which shall be denoted by $(\Lambda S, d_\phi)$. If Y is 1-connected, $(\Lambda(S/V), d_\phi)$ is a Sullivan model of $\mathrm{map}^*_f(X,Y)$.

DGL models for the components of mapping spaces

Let $\gamma\colon L \longrightarrow L'$ be a DGL morphism and consider the differential graded vector space of Lie γ-derivations $(Der_\gamma(L,L'),\delta)$. Explicitly $Der_\gamma(L,L')_n$ is the space of linear maps of degree n, $\theta\colon L_* \longrightarrow L'_{*+n}$ for which $\theta\,[x,y] = [\theta(x),\gamma(y)] + (-1)^{n|x|}\,[\gamma(x),\theta(y)]$, $x,y \in L$. The differential is defined as usual $\delta\theta = \partial \circ \theta + (-1)^n \theta \circ \partial$. For our purposes we are interested in the space $\mathcal{D}er_\gamma(L,L')$ of positive γ-derivations,

$$\mathcal{D}er_\gamma(L,L')_i = \begin{cases} Der_\gamma(L,L')_i & \text{for } i > 1, \\ ZDer_\gamma(L,L')_1 & \text{for } i = 1, \end{cases}$$

in which Z denotes the space of cycles. We shall also denote by δ the differential of this complex.

Recall the *Quillen* functor, constructed for any CDGC C augmented by $\varepsilon\colon C \to \mathbb{Q}$ and coaugmented by $\mathbb{Q} \to C$. Denote by $\overline{C} = \ker \varepsilon$ and consider the reduced diagonal $\overline{\Delta}\colon \overline{C} \to \overline{C} \otimes \overline{C}$. Then $\mathcal{L}(C,d) = (\mathbb{L}(s^{-1}\overline{C}), d)$ in which:

(1) $\mathbb{L}(s^{-1}\overline{C})$ is the *free* Lie algebra generated by $s^{-1}\overline{C}$, i.e., the sub Lie algebra of the tensor Lie algebra (bracket is the commutator) $T(s^{-1}\overline{C})$ generated by $s^{-1}\overline{C}$.

(2) $d = d_1 + d_2$ where $d_1(s^{-1}c) = -s^{-1}dc$ and

$$d_2(s^{-1}c) = \frac{1}{2}\sum_i (-1)^{|a_i|}\left[s^{-1}a_i, s^{-1}b_i\right], \text{ being } \overline{\Delta}c = \sum a_i \otimes b_i.$$

Consider a map $f\colon X \longrightarrow Y$ between a finite simply connected CW-complex and a nilpotent CW-complex of finite type (over $\mathbb{Q}$), let L' be a DGL model of Y and choose a Quillen model of X of the form $\mathcal{L}(B^\sharp)$ for some CDGC $B^\sharp$. This is always possible by taking $B^\sharp$, for instance, the dual of B a CDGA model of X. Finally choose any DGL morphism $\gamma\colon \mathcal{L}(B^\sharp) \longrightarrow L'$ modeling the homotopy type of f.

The restriction of γ to $s^{-1}\overline{B^\sharp}$ gives a linear map $\gamma\colon s^{-1}\overline{B^\sharp} \longrightarrow L'$ which is also identified to a map $\gamma\colon \overline{B^\sharp} \longrightarrow sL'$. Composing γ with the degree -1 isomorphism $sL' \longrightarrow L'$ yields the map $\overline{\gamma}\colon \overline{B^\sharp} \longrightarrow L'$.

Consider the vector space $Hom(\overline{B^\sharp}, L')$ with the usual bracket $[f,g] = [,] \circ f \otimes g \circ \overline{\Delta}$, and the differential

$$D_\gamma f = \partial f + (-1)^{|f|} f\delta + [\overline{\gamma}, f].$$

Define $\mathcal{H}om(\overline{B^\sharp}, L')$ by:

$$\mathcal{H}om_i(\overline{B^\sharp}, L') = \begin{cases} Hom_i(\overline{B^\sharp}, L') & \text{for } i > 1, \\ Z(Hom_1(\overline{B^\sharp}, L')) & \text{for } i = 1. \end{cases}$$

Then the following holds:

THEOREM 1.3. [**5**] *$(\mathcal{H}om(\overline{B^\sharp}, L'), D_\gamma)$ is a Lie model of* $\mathrm{map}^*_f(X, Y)$.

The restriction of a given derivation in $\mathcal{D}er_\gamma(\mathcal{L}(B^\sharp), L')$ to $s^{-1}\overline{B^\sharp}$ produces an isomorphism of graded vector spaces

$$\Upsilon\colon s^{-1}\mathcal{D}er_\gamma(\mathcal{L}(B^\sharp), L') \xrightarrow{\cong} \mathcal{H}om(\overline{B^\sharp}, L'), \quad \Upsilon(s^{-1}\theta)(c) = (-1)^{|\theta|}\theta(s^{-1}c),$$

(which inverse is given by $s\Upsilon^{-1}(f)(s^{-1}c) = (-1)^{|f|+1}f(c)$).

This isomorphism commutes with the differentials $s^{-1}\delta$ and D_γ and define in $s^{-1}\mathcal{D}er_\gamma(\mathcal{L}(B^\sharp), L')$ a Lie bracket obtaining:

THEOREM 1.4. [**6**] *The DGL $s^{-1}\mathcal{D}er_\gamma(\mathcal{L}(B^\sharp), L')$, is a Lie model of* $\mathrm{map}^*_f(X, Y)$.

If X is a simply connected CW-complex of finite type over $\mathbb{Q}$ (non necessarily finite), denote by $f_n\colon X_n \longrightarrow Y$ the restriction of the n-skeleton of X and let $i_n\colon C_n \hookrightarrow C_{n+1}$ be an injective coalgebra model of the inclusion $j_n\colon X_n \hookrightarrow X_{n+1}$. Observe that $C = \lim_{\leftarrow n} C_n$ is a coalgebra model for X,

$$\gamma = \lim_{\leftarrow n} \gamma_n\colon \mathcal{L}(C) \longrightarrow L'$$

is a Lie model of f. Then we have:

THEOREM 1.5. [**6**] *$H_*(\mathcal{D}er_\gamma(\mathcal{L}(C), L')) \cong \pi_*\Omega(\mathrm{map}^*_f(X, Y))$ as graded Lie algebras.*

2. The main result

First of all some useful definition and Lemmas are described.

DEFINITION 2.1. Let $\varphi, \psi \in \mathcal{D}er_\gamma(\mathcal{L}(B^\sharp), L')$, we define the following operation on $\mathcal{D}er_\gamma(\mathcal{L}(B^\sharp), L')$:

$$\begin{aligned} \{\varphi, \psi\}(\Phi) &= 0, \text{if } \Phi \in s^{-1}\overline{B^\sharp}, \\ \{\varphi, \psi\}(\Phi) &= \{\varphi, \psi\}([\alpha, \beta]) \\ &= [\{\varphi, \psi\}\alpha, \gamma\beta] + (-1)^{|\alpha|(|\varphi|+|\psi|)}[\gamma\alpha, \{\varphi, \psi\}\beta] \\ &\quad +(-1)^{|\alpha||\psi|}[\varphi\alpha, \psi\beta] + (-1)^{|\varphi|(|\psi|+|\alpha|)}[\psi\alpha, \varphi\beta], \text{if } \Phi \in \mathbb{L}^{\geq 2}(s^{-1}\overline{B^\sharp}). \end{aligned}$$

LEMMA 2.2. *Let $\varphi, \psi \in \mathcal{D}er_\gamma(\mathcal{L}(B^\sharp), L')$, and $c \in \overline{B^\sharp}$, then*

$$\Upsilon([s^{-1}\varphi, s^{-1}\psi])(c) = (-1)^{|\varphi|}\{\varphi, \psi\}(ds^{-1}c).$$

PROOF. By the cocommutativity of $\overline{B^\sharp}$, we have

$$\overline{\Delta}(c) = \frac{1}{2}\sum(a_i \otimes b_i + (-1)^{|a_i||b_i|}b_i \otimes a_i).$$

Then by the one hand:

$$\begin{aligned}\Upsilon([s^{-1}\varphi, s^{-1}\psi])(c) &= \left[\Upsilon s^{-1}\varphi, \Upsilon s^{-1}\psi\right](c)\\ &= [,]\circ\Upsilon(s^{-1}\varphi)\otimes\Upsilon(s^{-1}\psi)\circ\overline{\Delta}(c)\\ &= (-1)^{|a_i||\psi|+|a_i|+|\varphi|+|\psi|}\frac{1}{2}\sum_i\left[\varphi(s^{-1}a_i),\psi(s^{-1}b_i)\right]\\ &+ (-1)^{|\varphi||\psi|+|\varphi||a_i|+|a_i|}\frac{1}{2}\sum_i\left[\psi(s^{-1}a_i),\varphi(s^{-1}b_i)\right].\end{aligned}$$

On the other hand:

$$\begin{aligned}(-1)^{|\varphi|}\{\varphi,\psi\}(ds^{-1}c) &= (-1)^{|\varphi|}\{\varphi,\psi\}(\frac{1}{2}\sum_i(-1)^{|a_i|}\left[s^{-1}a_i, s^{-1}b_i\right])\\ &= (-1)^{|\varphi|}\frac{1}{2}\sum_i(-1)^{|a_i|}\Big((-1)^{|\psi|(|a_i|-1)}\left[\varphi(s^{-1}a_i),\psi(s^{-1}b_i)\right]\\ &+ (-1)^{|\varphi|(|\psi|+|a_i|-1)}\left[\psi(s^{-1}a_i),\varphi(s^{-1}b_i)\right]\Big).\end{aligned}$$

A simple inspection shows that both terms agree. □

REMARK 2.3. We can define a bracket in $\mathcal{D}er_\gamma(\mathcal{L}(B^\sharp), L')$ by means of

$$[\varphi,\psi] = (-1)^{|\varphi|-1}s\left[s^{-1}\varphi, s^{-1}\psi\right], \text{where } \varphi,\psi\in\mathcal{D}er_\gamma(\mathcal{L}(B^\sharp), L').$$

Then with this notation, the previous yields that

$$\begin{aligned}[\varphi,\psi](s^{-1}c) &= (-1)^{|\varphi|-1}s\left[s^{-1}\varphi, s^{-1}\psi\right](s^{-1}c)\\ &= (-1)^{|\varphi|-1}s\Upsilon^{-1}\Upsilon\left[s^{-1}\varphi, s^{-1}\psi\right](s^{-1}c)\\ &= (-1)^{|\psi|}\Upsilon\left[s^{-1}\varphi, s^{-1}\psi\right](c)\\ &= (-1)^{|\varphi|+|\psi|}\{\varphi,\psi\}(ds^{-1}c).\end{aligned}$$

We can define in this way ($[\varphi,\psi](s^{-1}c) = (-1)^{|\varphi|+|\psi|}\{\varphi,\psi\}(ds^{-1}c)$) a bracket in $\mathcal{D}er_\gamma(\mathbb{L}(U), L')$, for an arbitrary DGL $(\mathbb{L}(U), d)$, and define recursively (as in [**7**]) higher order brackets as well.

Then n-ary brackets on $s^{-1}\mathcal{D}er_\gamma(\mathbb{L}(U), L')$ are defined as follows:

The first bracket is the differential of the complex, i.e., $\left[s^{-1}\theta\right] = s^{-1}\delta(s^{-1}\theta)$, $s^{-1}\theta\in s^{-1}\mathcal{D}er_\gamma(\mathbb{L}(U), L')$. For $n>1$,

$$\left[s^{-1}\theta_1,\ldots,s^{-1}\theta_n\right] = (-1)^\epsilon s^{-1}\left[\theta_1,\cdots,\theta_n\right],$$

where $s^{-1}\theta_1,\ldots,s^{-1}\theta_n\in s^{-1}\mathcal{D}er_\gamma(\mathbb{L}(U), L'), \epsilon = \sum_{j=1}^{n-1}(n-j)|\theta_j|$. With this bracket $s^{-1}\mathcal{D}er_\gamma(\mathbb{L}(U), L')$ is an L_∞ algebra (see [**6, 7**] for details). It turns out that when $(\mathbb{L}(U), d)$ is the minimal Lie model of X, then $s^{-1}\mathcal{D}er_\gamma(\mathbb{L}(U), L')$ is an L_∞ model of $\mathrm{map}^*_f(X,Y)$. If we choose $(\mathbb{L}(U), d) = (\mathcal{L}(B^\sharp), d = d_1+d_2)$ as model of X then the higher order brackets defined above vanish because the differential has only linear and quadratic parts, and the previous lemma shows that $s^{-1}\mathcal{D}er_\gamma(\mathbb{L}(U), L')$ is a DGL model of $\mathrm{map}^*_f(X,Y)$ (i.e. an L_∞ model with no brackets of order higher than 2).

Proof of theorem 0.3. Suppose that $\mathrm{cl}_0(X)\le n$, we may find [**10**, lemma 29.2] a commutative model (B,d) for X such that $B^0=\mathbb{Q}$, $B^1=0$, each B^i is finite dimensional, $d\colon B^+\longrightarrow B^+\cdot B^+$ and also $\mathrm{nil}B\le n$. The DGL $\mathcal{L}(B^\sharp) = \mathbb{L}(V)$

satisfy the condition that there is a filtration $0 = V(0) \subset \cdots \subset V(n) = V$ such that $\partial V(i) \subset \mathbb{L}(V(i-1))$, $0 < i \leq n$ [**10**, Theorem 29.1]. Consider the DGL model of $\mathrm{map}_f^*(X,Y)$ given by $s^{-1}\mathcal{D}er(\mathcal{L}(B^\sharp), L')$.

Let $s^{-1}\theta_1, \ldots, s^{-1}\theta_n$ be cycles in $(s^{-1}\mathcal{D}er(\mathcal{L}(B^\sharp), L'), \delta)$.

Observe that $\theta_k(s^{-1}c), k = 1,2$ is a cycle in L' if $s^{-1}c$ is a cycle in $\mathcal{L}(B^\sharp)$. For any $s^{-1}c \in V(2)$, $d(s^{-1}c) = d_1(s^{-1}c) + \frac{1}{2}(-1)^{|a_i|}\left[s^{-1}a_i, s^{-1}b_i\right] \in \mathbb{L}(V(1))$, and then $s^{-1}a_i, s^{-1}b_i$ are cycles in $\mathcal{L}(B^\sharp)$ and we can choose $x_c \in L'$ such that

$$\begin{aligned}\partial x_c &= \frac{1}{2}(-1)^{|\theta_1|}\sum_i((-1)^{|\theta_1|(|a_i|-1)}\left[\theta_2(s^{-1}a_i), \theta_1(s^{-1}b_i)\right] \\ &+ (-1)^{|\theta_2|(|\theta_1|+|a_i|-1)}\left[\theta_1(s^{-1}a_i), \theta_2(s^{-1}b_i)\right]),\end{aligned}$$

because we are assuming $\mathrm{Wh}_{\mathbb{Q}}(Y) = 1$.

Define a derivation $\theta_{1,2} \in Der(\mathcal{L}(B^\sharp), L')$ by $\theta_{1,2} = x_c$ if $s^{-1}c \in V(2)'$ and $\theta_{1,2|V(1)} = 0$ where $V(2) = V(1) \oplus V(2)'$. With this definition it's easy to check that $(\delta\theta_{1,2})_{|V(2)} = s\left[s^{-1}\theta_2, s^{-1}\theta_1\right]_{|V(2)}$.

Consider the border

$$\delta\left[s^{-1}\theta_n, \ldots \left[s^{-1}\theta_3, s^{-1}\theta_{1,2}\right]\ldots\right] = \pm\left[s^{-1}\theta_n, \ldots \left[s^{-1}\theta_3, s^{-1}\delta\theta_{1,2}\right]\ldots\right],$$

(we are not so rigid with signs at this point in order to clarity and write $\pm$ instead). Take $s^{-1}c_n \in V(n) = \overline{B^\sharp}$, then

$$\begin{aligned} & s\left[s^{-1}\theta_n, \ldots \left[s^{-1}\theta_3, s^{-1}\delta\theta_{1,2}\right]\ldots\right](s^{-1}c_n) = \\ &= \pm\Upsilon\left[s^{-1}\theta_n, \ldots \left[s^{-1}\theta_3, s^{-1}\delta\theta_{1,2}\right]\ldots\right](c_n) \\ &= \pm\{\theta_n, s\left[s^{-1}\theta_{n-1}, \left[\ldots, \left[s^{-1}\theta_3, s^{-1}\delta\theta_{1,2}\right]\right]\ldots\right]\}(ds^{-1}c_n) \\ &= \pm\sum\left[\theta_n(s^{-1}c'_{n-1}), s\left[s^{-1}\theta_{n-1}, \left[\ldots, \left[s^{-1}\theta_3, s^{-1}\delta\theta_{1,2}\right]\right]\ldots\right](s^{-1}c''_{n-1})\right],\end{aligned}$$

where $s^{-1}c'_{n-1}, s^{-1}c''_{n-1} \in V(n-1)$. Repeating this process we finally reach

$$\begin{aligned} s\left[s^{-1}\theta_3, s^{-1}\delta\theta_{1,2}\right](s^{-1}c_3) &= \pm\Upsilon\left[s^{-1}\theta_3, s^{-1}\delta\theta\right](c_3) \\ &= \pm\{\theta_3, \delta\theta_{1,2}\}(ds^{-1}c_3) \\ &= \pm\sum\pm\left[\theta_3(s^{-1}c'_2), \delta\theta_{1,2}(s^{-1}c''_2)\right] \\ &= \pm\sum\pm\left[\theta_3(s^{-1}c'_2), s\left[s^{-1}\theta_2, s^{-1}\theta_1\right](s^{-1}c''_2)\right],\end{aligned}$$

where $s^{-1}c'_k, s^{-1}c''_k \in V(k)$, $k = 2, \ldots, n-1$.

By the same argument as above, we see that the element

$$s\left[s^{-1}\theta_n, \ldots \left[s^{-1}\theta_3, \left[s^{-1}\theta_2, s^{-1}\theta_1\right]\right]\ldots\right](s^{-1}c_n),$$

can be expanded with the same form. This implies that every iterated Whitehead product of cycles in $s^{-1}\mathcal{D}er(\mathcal{L}(B^\sharp), L')$ is exact. We have

$$\mathrm{Wh}_{\mathbb{Q}}(\mathrm{map}_f^*(X,Y)) \leq n-1.$$

□

Proof of theorem 0.2. We can also use the same technique to prove Theorem 0.2 in the more general setting of X being non necessarily finite. The above process could be done for $s^{-1}\theta_1, \ldots, s^{-1}\theta_{n+1} \in (s^{-1}\mathcal{D}er(\mathcal{L}(B^\sharp), L'), \delta)$, and in the calculation of

$$s\left[s^{-1}\theta_{n+1}, \ldots \left[s^{-1}\theta_3, \left[s^{-1}\theta_2, s^{-1}\theta_1\right]\right]\ldots\right](s^{-1}c_n),$$

we end with

$$\begin{aligned}\pm\sum s\left[s^{-1}\theta_2,s^{-1}\theta_1\right](s^{-1}c_1'') &= \pm\sum \Upsilon\left[s^{-1}\theta_2,s^{-1}\theta_1\right](c_1'')\\ &= \pm\sum\{\theta_2,\theta_1\}(ds^{-1}c_1'')\\ &= 0,\end{aligned}$$

where $c_k', c_k'' \in V(k)$, $k=1,\dots,n$. Thus

$$\left[s^{-1}\theta_{n+1},\dots\left[s^{-1}\theta_3,\left[s^{-1}\theta_2,s^{-1}\theta_1\right]\right]\dots\right]=0,$$

instead of being exact. Note that in the above reasoning $s^{-1}\theta_1,\dots,s^{-1}\theta_{n+1}$ are not necessarily cycles. We have $\mathrm{Wh}_{\mathbb{Q}}(\mathrm{map}_f^*(X,Y))\le \mathrm{cl}_0(X)$. □

3. Some geometrical examples

In this section we offer three examples. First showing that theorem 0.2 is the finest bound possible for a general Y. With more detail, we can find spaces X and Y ($\mathrm{Wh}_{\mathbb{Q}}(Y)>1$ of course) satisfying $\mathrm{Wh}_{\mathbb{Q}}(\mathrm{map}_f^*(X,Y))=\mathrm{cl}_0(X)$. Second showing that theorem 0.3 is the finest bound possible for Y with $\mathrm{Wh}_{\mathbb{Q}}(Y)=1$. i.e., we find spaces X and Y ($\mathrm{Wh}_{\mathbb{Q}}(Y)=1$ in this example) satisfying $\mathrm{Wh}_{\mathbb{Q}}(\mathrm{map}_f^*(X,Y))=\mathrm{cl}_0(X)-1$. The last example is devoted to the general case in which X is not a finite complex.

Example 3.1. Consider $X=S^3\times S^3$ and $Y=S^4$. It's clear with the definition of cone-length that $\mathrm{cl}_0(X)=2$. Moreover considering the Sullivan model of Y, $(\Lambda a,b;da=0,db=a^2)$, $|a|=4$, $|b|=7$, it's also evident that $\mathrm{Wh}_{\mathbb{Q}}(Y)=2$. From a CDGA model of X, $B=(\Lambda x,y;dx=dy=0)$, $|x|=|y|=3$, we can consider the dual coalgebra $\overline{B^\sharp}=\mathbb{Q}\oplus\langle x^*,y^*\rangle\oplus\langle\gamma^*\rangle$, with differential $\delta=0$ and $\overline{\Delta}\gamma^*=x^*\otimes y^*-y^*\otimes x^*$. Then the Haefliger-Brown-Szczarba model of $\mathrm{map}_f(X,Y)$, for any $f\colon X\longrightarrow Y$ is of the form

$$(\Lambda(a\otimes x^*,a\otimes y^*,b\otimes\gamma^*,a\otimes 1,b\otimes x^*,b\otimes y^*,b\otimes 1);\tilde d),$$

where $|a\otimes x^*|=|a\otimes y^*|=|b\otimes\gamma^*|=1$, $|a\otimes 1|=|b\otimes x^*|=|b\otimes y^*|=4$, $|b\otimes 1|=7$, and the only generator with non-zero differential is $\tilde d(b\otimes\gamma^*)=2(a\otimes x^*)(a\otimes y^*)$. Recall that for any space Z, the Whitehead product in $\pi_*(Z)$ is dual to the quadratic part of the differential in the minimal Sullivan model $(\Lambda V_Z,d)$ of Z [**10**, Prop. 13.16]. More precisely:

$$\langle d_2v;\gamma_0,\gamma_1\rangle=(-1)^{k+n-1}\langle v;[\gamma_0,\gamma_1]_W\rangle,$$
$$v\in V_Z,\ \gamma_0\in\pi_k(Z),\ \gamma_1\in\pi_n(Z).$$

Then we have $\mathrm{Wh}_{\mathbb{Q}}(\mathrm{map}_f^*(X,Y))=2=\mathrm{cl}_0(X)$.

Example 3.2. Let X be the space with CDGA model $B=(\Lambda x,y,z;d)$, where $|x|=|y|=3$, $|z|=5$ and $dx=dy=0$, $dz=xy$. Then $\mathrm{cl}_0(X)=3$ (see [**10**, p.387]).

Let $\overline{B^\sharp}=\mathbb{Q}\oplus\langle x^*,y^*\rangle\oplus\langle z^*\rangle\oplus\langle\gamma^*\rangle\oplus\langle\alpha^*,\beta^*\rangle\oplus\langle\epsilon^*\rangle$, be the reduced dual coalgebra, where

$$\overline{\Delta}\gamma^*=x^*\otimes y^*-y^*\otimes x^*,\overline{\Delta}\alpha^*=x^*\otimes z^*-z^*\otimes x^*,\overline{\Delta}\beta^*=y^*\otimes z^*-z^*\otimes y^*,$$
$$\overline{\Delta}\epsilon^*=\gamma^*\otimes z^*-\alpha^*\otimes y^*+\beta^*\otimes x^*+x^*\otimes\beta^*-y^*\otimes\alpha^*+z^*\otimes\gamma^*,$$

and the only non zero differential is $\delta\gamma^*=z^*$.

Now consider Y the space whose minimal Sullivan model is $(\Lambda(a,b,c,v);d)$, where $|a|=|b|=|c|=5$, $|v|=14$ and $dv=abc$ is the only non trivial differential.

Finally take the map $f\colon X \longrightarrow Y$ whose model is given by $\phi\colon \Lambda V \longrightarrow B$, $\phi a = z$ and $\phi = 0$ for the rest of generators. Then the Haefliger-Brown-Szczarba model for the component of that map is of the form

$$(\Lambda S/V, d_\phi) = (\Lambda(V \otimes \overline{B^\sharp})^{\geq 2}, d_\phi),$$

where the generators are

degree	generators
2	$a\otimes x^*, a\otimes y^*, b\otimes x^*, b\otimes y^*, c\otimes x^*, c\otimes y^*$
3	$v\otimes \epsilon^*$
5	$a\otimes 1, b\otimes 1, c\otimes 1$
6	$v\otimes\alpha^*, v\otimes\beta^*$
8	$v\otimes\gamma^*$
9	$v\otimes z^*$
11	$v\otimes x^*, v\otimes y^*$
14	$v\otimes 1,$

and $d_\phi(v\otimes\epsilon^*) = (b\otimes x^*)(c\otimes y^*) + (b\otimes y^*)(c\otimes x^*)$, $d_\phi(v\otimes\gamma^*) = v\otimes z^*$ are the only non trivial differentials. Then of course

$$\mathrm{Wh}_{\mathbb{Q}}(\mathrm{map}^*_f(X,Y)) = 2 = \mathrm{cl}_0(X) - 1.$$

EXAMPLE 3.3. Consider $X = S^2_e \vee (\vee_{i\geq 1} S^{2i}_{a_i})$. Let $\epsilon\in\pi_2(X)$, $\alpha_i\in\pi_{2i}(X)$ be the elements represented by S^2_e, and $S^{2i}_{a_i}$ respectively. Then a Lie model for X is just $(\mathbb{L}(e,a_i),0)$ with $i\geq 1$, $|e| = 1$, $|a_i| = 2i-1$ and e, a_i corresponding to ϵ, α_i. Moreover, the isomorphism $\pi_*(X)\otimes\mathbb{Q}\cong s\mathbb{L}(e,a_i)$ identifies $[\epsilon,\alpha_i]_W$ with $s\,[e,a_i]$. Hence

$$(\mathbb{L}(e,a_i,b_i), \partial b_i = [e,a_i]) \text{ is a Lie model for } X\cup_{i;[\epsilon,\alpha_i]_W} D^{2i+2}_i.$$

By construction $X\cup_{i;[\epsilon,\alpha_i]_W} D^{2i+2}_i$ is an infinite CW-complex. Thus the above model is of the form $\mathcal{L}(C)$ for some CDGC C, because it has only quadratic part in the differential, and $\mathrm{cl}_0(X\cup_{i;[\epsilon,\alpha_i]_W} D^{2i+2}_i) = 2$.

Consider the space $\mathbb{C}P^\infty$, which has $\mathrm{Wh}_{\mathbb{Q}}(\mathbb{C}P^\infty) = 1$ and a Lie model of the form

$$(\mathbb{L}(v_i), \partial v_k = \frac{1}{2}\sum_{i+j=k}[v_i,v_j]), \text{ where } i\geq 1.$$

Then we can consider the map $f\colon X\cup_{i;[\epsilon,\alpha_i]_W} D^{2i+2}_i \longrightarrow \mathbb{C}P^\infty$ whose Lie model $\gamma\colon \mathbb{L}(e,a_i,b_i)\longrightarrow\mathbb{L}(v_i)$ is given by $\gamma(e) = v_1$, $\gamma(a_i) = \gamma(b_i) = 0$.

Then

$$s^{-1}\mathcal{D}er_\gamma(\mathbb{L}(e,a_i,b_i),\mathbb{L}(v_i)) \text{ is a Lie model of } \mathrm{map}^*_f(X\cup_{i;[\epsilon,\alpha_i]_W} D^{2i+2}_i, \mathbb{C}P^\infty).$$

Theorem 0.3 states that

$$\mathrm{Wh}_{\mathbb{Q}}(\mathrm{map}^*_f(X\cup_{i;[\epsilon,\alpha_i]_W} D^{2i+2}_i,\mathbb{C}P^\infty)) \leq \mathrm{cl}_0(X\cup_{i;[\epsilon,\alpha_i]_W} D^{2i+2}_i) - 1 = 1,$$

and then all Whitehead products vanish.

Let see how the idea of the Theorem 0.3 works in this particular example:

Consider $\theta \in \mathcal{D}er_\gamma(\mathbb{L}(e,a_i,b_i),\mathbb{L}(v_i))$ given by $\theta e = [v_1,v_2]$, $\theta a_i = \partial v_{i+2}$, $\theta b_i = [v_1, v_{i+2}]$, where $|\theta| = 3$. A simple inspection shows that θ is a cycle:

$$\begin{aligned}
\delta\theta e &= \partial\theta e - \theta\partial e = \partial\,[v_1,v_2] = 0,\\
\delta\theta a_i &= \partial\theta a_i - \theta\partial a_i = \partial\partial v_{i+2} = 0,\\
\delta\theta b_i &= \partial\theta b_i - \theta\partial b_i = \partial\,[v_1,v_{i+2}] - \theta\,[e,a_i]\\
&= -\,[v_1,\partial v_{i+2}] + [\gamma e,\theta a_i] = -\,[v_1,\partial v_{i+2}] + [v_1,\partial v_{i+2}] = 0.
\end{aligned}$$

Then $[\theta,\theta]$ is also a cycle that must be exact.

In $\mathbb{L}(V) = \mathbb{L}(e,a_i,b_i)$, V can be filtered according to cone-length characterization as

$$0 = V(0) \subset V(1) \subset V(2) = V(1) \oplus V(2)' = V,$$

taking $V(1) = \langle e, a_i\rangle$, $V(2)' = \langle b_i\rangle$. Then for any $i \geq 1$, $[\theta e, \theta a_i] = [[v_1,v_2]\,,\partial v_{i+2}]$ must be exact. Indeed for $\Psi_i = [[v_1,v_2]\,,v_{i+2}]$, we have $\partial\Psi_i = [\theta e,\theta a_i]$. Then define $\eta \in \mathcal{D}er_\gamma(\mathbb{L}(e,a_i,b_i),\mathbb{L}(v_i))$, where $|\eta| = 6$ by $\eta e = \eta a_i = 0$, and $\eta b_i = \Psi_i$. A straightforward computation shows that $\delta\eta e = 0 = [\theta,\theta]\,e$, $\delta\eta a_i = 0 = [\theta,\theta]\,a_i$, and

$$\begin{aligned}
\delta\eta b_i &= \partial\eta b_i + (-1)^{|\eta|}\eta\partial b_i\\
&= \partial\Psi_i + \eta\,[e,a_i] = \partial\Psi_i,\\
[\theta,\theta]\,b_i &= (-1)^{2|\theta|}\{\theta,\theta\}(\partial b_i)\\
&= \{\theta,\theta\}\,[e,a_i] = [\theta e,\theta a_i]\,.
\end{aligned}$$

We conclude that $\delta\eta = [\theta,\theta]$.

References

[1] A. BOUSFIELD AND V. GUGENHEIM, On PL de Rham theory and rational homotopy type. *Mem. Amer. Math. Soc.* **179** (1978).

[2] E. H. BROWN AND R. H. SZCZARBA, On the rational homotopy type of function spaces, *Trans. Amer. Math. Soc.* , **349** (1997), 4931–4951.

[3] U. BUIJS AND A. MURILLO, Basic constructions in rational homotopy theory of function spaces, *Annales de l'Inst. Fourier*, **56**(3) (2006), 815–838.

[4] ———, The rational homotopy Lie algebra of function spaces, *Comment. Math. Helv.* **83** (2008) 723–739.

[5] U. BUIJS, Y. FÉLIX AND A. MURILLO, Lie models for the components of sections of a nilpotent fibration, *Trans. Amer. Math. Soc.* , **361**(10) (2009), 5601–5614.

[6] ———, L_∞ models of mapping spaces, Preprint (2009).

[7] U. BUIJS, An explicit L_∞ structure for the rational homotopy type of the components of mapping spaces, Preprint (2009).

[8] O. CORNEA, Cone length and LS category, *Topology* , **33** (1994), 95–111.

[9] Y. FÉLIX, Rational category of the space of sections of a nilpotent bundle, *Comment. Math. Helv.* **65** (1990) 1–37.

[10] Y. FÉLIX, S. HALPERIN, J. THOMAS, Rational homotopy theory, *Springer GTM* , **205** (2000).

[11] Y. FÉLIX, D. TANRÉ, H-space structure on pointed mapping spaces. *Algebraic and Geometric Topology*, **29** (2005), 713–724.

[12] A. HAEFLIGER, Rational Homotopy of the space of sections of a nilpotent bundle, *Trans. Amer. Math. Soc.*, **273** (1982), 609–620.

[13] G. LUPTON AND S. SMITH, Rationalized evaluation subgroups of a map II: Quillen models and adjoint maps, *J. Pure Appl. Algebra*, **209** (2007), no 1, 173–188.

[14] ———, Whitehead products in function spaces: Quillen model formulae, *J. of the Math. Soc. of Japan* , **62** (2010), 1–33.

[15] S. SMITH, Rational homotopy of the space of self-maps of complexes with finitely many homotopy groups, *Trans. Amer. Math. Soc.* **342**(2) (1994), 895–915.

[16] ———, Rational evaluation subgroups, Math. Zeit. **221** (1996), 387–400.

[17] D. SULLIVAN, Infinitesimal computations in Topology, *Publ. Math. de l'I.H.E.S.*, **47** (1978), 269–331.

[18] R. THOM, L'homologie des espaces fonctionnels, *Colloque de topologie algébrique*, Louvain, (1957), 299–39.

[19] M. VIGUÉ-POIRRIER, Sur l'homotopie rationelle des espaces fonctionels, *Manuscripta Math.* **56** (1986), 177–191.

Departamento de Matemática Aplicada, Universidad de Málaga, 29000 Málaga, Spain

Current address: Facultad de Matemáticas, Edifici Històric, Gran Via de les Corts Catalanes, 585, 08007 Barcelona, Spain

E-mail address: urtzi@agt.cie.uma.es

Contemporary Mathematics
Volume **519**, 2010

String topology of classifying spaces and gravity algebras

David Chataur

ABSTRACT. In this short note we prove that the S^1-equivariant singular homology of the free loop space of the classifying space of a connected compact Lie group when taken with coefficients into a field is a full gravity algebra.

1. Introduction

Among functional spaces the free loop space $\mathcal{L}X = map(S^1, X)$ has been intensively studied by algebraic topologists. This is due to the fundamental role played by this space in riemannian geometry and in the study of the space of automorphisms of manifolds.
A crucial step in our understanding of the homology of these spaces is the recent development of the string topology operations. In [**CS1**] Chas and Sullivan introduced on the shifted homology of the space of free loops of a closed oriented manifold a very rich algebraic structure. This structure is encoded by the action of a prop, the prop of moduli spaces of curves with boundary components, we refer the reader to Cohen and Godin's paper [**CG**] and Godin's preprint [**Go**]. This structure is highly non-trivial and is related to the world of topological field theories and symplectic topology.
One can also endow the homology of the space of free loops of the classifying space of a Lie group G with such an algebraic structure see [**CM**].
In this paper we study the S^1-equivariant side of the theory and we prove the S^1-equivariant homology of $\mathcal{L}BG$ is acted upon by a prop. This prop is related to moduli spaces of curves with marked points. In the case of closed manifolds part of this structure was studied by Chas and Sullivan in [**CS1**] and by Westerland in [**W**].

2. Props

The aim of this section is to introduce the algebraic notions that encompass the "stringy" operations acting on $\mathcal{L}BG$. This section is mainly expository.

Let us recall that a *prop* is a symmetric (strict) monoidal category $\mathcal{P}$ whose set

Key words and phrases. equivariant topology, free loop space, Properads, string topology, topological field theories.
The author is supported in part by ANR grant JCJC06 OBTH.

©2010 American Mathematical Society

of objects is identified with the set $\mathbb{Z}_+$ of nonnegative integers. The tensor law on objects should be given by addition of integers $p \otimes q = p + q$. Strict monoidal means that the associativity and neutral conditions are the identity.
We thus have two composition products on morphisms a horizontal one given by the tensor law :

$$- \otimes - : \mathcal{P}(p,q) \otimes \mathcal{P}(p',q') \longrightarrow \mathcal{P}(p+p',q+q'),$$

and a vertical one given by composition of morphisms :

$$- \circ - : \mathcal{P}(q,r) \otimes \mathcal{P}(p,q) \longrightarrow \mathcal{P}(p,r).$$

Let X be an object of a strict symmetric monoidal category $\mathbf{C}$. A fundamental example of prop is given by the *endomorphims prop of* X denoted $\mathcal{E}nd_X$. The set of morphisms is defined as

$$\mathcal{E}nd_X(p,q) = Hom(X^{\otimes p}, X^{\otimes q}).$$

The horizontal composition product is just given by the tensor of $\mathbf{C}$ while the vertical is the composition of morphisms.

A morphism of props is a symmetric (strict) monoidal F with values in $\mathbf{C}$ such that $F(1) = 1_{\mathbf{C}}$.

DEFINITION 2.1. An object X of a strict symmetric monoidal category is said to be a $\mathcal{P}$*-algebra* if there is a morphism of linear props

$$F : \mathcal{P} \longrightarrow \mathcal{E}nd_X.$$

This means that we have a a family of morphisms

$$F : \mathcal{P}(m,n) \to \mathrm{Hom}(X^{\otimes m}, X^{\otimes n})$$

such that

(monoidal) $F(f \otimes g) = F(f) \otimes F(g)$ for $f \in \mathcal{P}(m,n)$ and $g \in \mathcal{P}(m',n')$.

(identity) The image $F(id_n)$ of the identity morphism $id_n \in \mathcal{P}(n,n)$, is equal to the identity morphism of $X^{\otimes n}$.

(symmetry) $F(\tau_{m,n}) = \tau_{X^{\otimes m},X^{\otimes n}}$. Here $\tau_{m,n} : m \otimes n \to n \otimes m$ denotes the natural twist isomorphism of $\mathcal{P}$.

(composition) $F(g \circ f) = F(g) \circ F(f)$ for $f \in \mathcal{P}(p,q)$ and $g \in \mathcal{P}(q,r)$.

By adjunction, this morphism determines evaluation products

$$\mu : \mathcal{P}(p,q) \otimes X^{\otimes p} \longrightarrow X^{\otimes q}.$$

3. Wrong way maps

An *Umkehr map* or wrong way map is a map $f_!$ in homology related to an original continuous map $f : X \to Y$ which reverses the arrow. Umkehr maps can also be considered in cohomology and some of them are refined to stable maps.
A typical example is given when one considers a continuous map $f : M^m \to N^n$ between two oriented closed manifolds. Then using Poincaré duality one defines the associated Umkehr, Gysin, wrong way, surprise or transfer map (depending on your prefered name)

$$f_! : H_*(N^n) \to H_{*+m-n}(M^m).$$

In this paper we will deal with integration along the fiber, this type of transfer map is associated to a fibration.

3.1. Integration along the fiber.

3.1.1. *A spectral sequence version.* Let $F \hookrightarrow E \stackrel{p}{\twoheadrightarrow} B$ be a fibration over a path-connected base B. We suppose that the homology of the fiber $H_*(F,\mathbb{F})$ is concentrated in degree less than or equal to n and has a top non-zero homology group $H_n(F,\mathbb{F}) \cong \mathbb{F}$. Let us assume that the action of the fundamental group $\pi_1(B)$ on $H_n(F,\mathbb{F})$ induced by the holonomy is trivial. Let ω be a generator of $H_n(F,\mathbb{F})$ i.e. an orientation class. We shall refer to such data as an *oriented fibration.*
Using the Serre spectral sequence, one can define the integration along the fiber as a map

$$p_! : H_*(B) \to H_{*+n}(E).$$

Let us recall the construction, we consider the spectral sequence with local coefficients.

$$E^2_{l,m} = H_l(B, \mathcal{H}_m(F_b,\mathbb{F})) \Rightarrow H_*(E,\mathbb{F})$$

As the Serre spectral sequence is concentrated under the n-th line, the filtration on the abutment $H_{l+n}(E)$ is of the form

$$0 = F^{-1} = F^0 = \cdots = F^{l-1} \subset F^l \subset F^{l+1} \subset \cdots \subset F^{l+n} = H_{l+n}(E).$$

As the local coefficients are trivial by hypothesis, the orientation class ω defines an isomorphism of local coefficients $\tau : \mathbb{F} \to \mathcal{H}_n(F_b,\mathbb{F})$. By definition $p_!$ is the composite

$$p_! : H_l(B,\mathbb{F}) \stackrel{H_l(B;\tau)}{\to} H_l(B, \mathcal{H}_n(F_b,\mathbb{F})) = E^2_{l,n} \twoheadrightarrow E^\infty_{l,n} = \frac{F^l}{F^{l-1}} = F^l \subset H_{l+n}(E,\mathbb{F}).$$

3.1.2. *A stable version.* Let us review a more geometric construction of integration along the fiber, let us suppose that we have a smooth fiber bundle $F \hookrightarrow P^{m+n} \stackrel{p}{\to} M^m$ with a compact fiber and base. The fibration is oriented if and only if the tangent bundle along the fiber

$$\mathbb{R}^n \hookrightarrow T^v P^{m+n} \twoheadrightarrow P^{m+n}$$

is oriented. Choose an embedding of P^{m+n} in some Euclidean space $\mathbb{R}^k$ and consider the normal bundle $N^v P^{m+n}$ of $i : P^{m+n} \subset M^m \times \mathbb{R}^k$ we have

$$T^v P^{m+n} \oplus N^v P^{m+n} \simeq E \times \mathbb{R}^k$$

applying the Thom-Pontryagin collapse with respect to i we get a map

$$M^m_+ \wedge S^k \to Th(N^v P^{m+n}).$$

In homology if we compose this map with the Thom isomorphism we get integration along the fiber

$$p_! : H_*(M^m,\mathbb{F}) \stackrel{i_*}{\to} H_*(Th(N^v P^{m+n}),\mathbb{F}) \to H_{*+n}(P^{m+n},\mathbb{F}).$$

Here it may be useful to notice that the composition

$$B_+ \wedge S^k \to Th(N^v P^{m+n}) \to Th(T^v P^{m+n} \oplus N^v P^{m+n}) \simeq P^{m+n}_+ \wedge S^k$$

is the transfer map when $k \to +\infty$.

Let us review the equivariant version. Fix a compact Lie Group G and suppose that it acts smoothly on P^{m+n} and M^m and that p is G-equivariant. We also suppose that the action preserves the orientation. Now we choose a G-equivariant embedding of P^{m+n} into a finite dimensional real G-module V equipped with a G-invariant metric, applying the construction described above we end the day with a map :

$$p_! : H_*(M^m_{hG}, \mathbb{F}) \to H_{*+n}(P^{m+n}_{hG}, \mathbb{F}).$$

This map is equal to the integration along the fiber built thanks to the Serre spectral sequence associated to the fibration

$$F \to P^{m+n}_{hG} \to M^m_{hG}.$$

3.2. Properties of Umkehr maps. Let us give a list of properties that are satisfied by transfer maps and integration along the fibers. In fact all reasonable notion of Umkehr map must satisfy this Yoga. We write these properties for integration along the fiber taking into account the degree shifting, we let the reader do the easy translation for transfer maps.

Naturality: Consider a commutative diagram

$$\begin{array}{ccc} E_1 & \xrightarrow{g} & E_2 \\ \downarrow{p_1} & & \downarrow{p_2} \\ B_1 & \xrightarrow{h} & B_2 \end{array}$$

where p_1 is a fibration over a path-connected base and p_2 equipped with the orientation class $w_2 \in H_n(F_2)$ is an oriented fibration. Let $f : F_1 \to F_2$ the map induced between the fibers. Suppose that $H_*(f)$ is an isomorphism. Then the fibration p_1 equipped with the orientation class $w_1 := H_n(f)^{-1}(w_2)$ is an oriented fibration and the following diagram commutes

$$\begin{array}{ccc} H_{*+n}(E_1) & \xrightarrow{H_*(g)} & H_{*+n}(E_2) \\ \uparrow{p_1!} & & \uparrow{p_2!} \\ H_*(B_1) & \xrightarrow{H_*(h)} & H_*(B_2) \end{array}$$

Composition: Let $f : X \twoheadrightarrow Y$ be an oriented fibration with path-connected fiber F_f and orientation class $w_f \in H_m(F_f)$. Let $g : Y \twoheadrightarrow Z$ be a second oriented fibration with path-connected fiber F_g and orientation class $w_g \in H_n(F_g)$. Then the composite $g \circ f : X \to Z$ is an oriented fibration with path-connected fiber $F_{g\circ f}$. By naturality with respect to pull-back, we obtain an oriented fibration $f' : F_{g\circ f} \twoheadrightarrow F_g$ with orientation class $w_f \in H_m(F_f)$. By definition, the orientation class of $g \circ f$ is $w_{g\circ f} := f'_!(w_g) \in H_{n+m}(F_{g\circ f})$. Then we have the commutative diagram

$$\begin{array}{ccccc} & & H_{*+n}(Y) & & \\ & \nearrow{g!} & & \searrow{f!} & \\ H_*(Z) & & \xrightarrow{(g\circ f)!} & & H_{*+m+n}(X). \end{array}$$

Product: Let $p : E \twoheadrightarrow B$ be an oriented fibration with fiber F and orientation class $w \in H_m(F)$. Let $p' : E' \twoheadrightarrow B'$ be a second oriented fibration with fiber F' and orientation class $w' \in H_n(F')$. Then if one work with homology with field coefficients, $p \times p' : E \times E' \twoheadrightarrow B \times B'$ is a third oriented fibration with fiber $F \times F'$

and orientation class $w \times w' \in H_{m+n}(F \times F')$ and one has for $a \in H_*(B)$ and $b \in H_*(B')$,

$$(p \otimes p')_!(a \otimes b) = (-1)^{|a|n} p_!(a) \otimes p'_!(b).$$

Notice that since $p'_!$ is of degree n, the sign $(-1)^{|a|n}$ agrees with the Koszul rule.
Borel construction: Let G be a topological group acting continuously on two topological spaces E and B, we also suppose that we have a continuous G-equivariant map

$$p : E \to B$$

the induced map on homotopy G-quotients (we apply the Borel functor $EG \times_G -$ to p) is denoted by

$$p_{hG} : E_{hG} \to B_{hG}.$$

We suppose that the action of G on B has a fixed point b and that $p : E \twoheadrightarrow B$ is an oriented fibration with fiber $F := p^{-1}(b)$ and orientation class $w \in H_n(F)$. This fiber F is a sub G-space of E. Then we suppose that the action of G preserves the orientation, to be more precise we suppose that the action of $\pi_0(G)$ on $H_n(F)$ is trivial. Then

$$p_{hG} : E_{hG} \twoheadrightarrow B_{hG}$$

is locally an oriented (Serre) fibration and therefore is an oriented (Serre) fibration with fiber F and orientation class $w \in H_n(F)$. Note that under the same hypothesis, $p_!$ is $H_*(G)$-linear.

4. Topological models for moduli spaces

Let Σ be a genus g compact connected oriented Riemann surface with boundary, we assume that this surface comes equipped with p incoming boundary components and q outgoing components. Each boundary component ∂_i has a collar C_i. A collar is given by a diffeomorphism from $S^1 \times [0, \epsilon)$ to an open neighbourhood of ∂_i if it is an ingoing component or by a diffeomorphism from $S^1 \times (\epsilon, 0]$to an open neighborhood of ∂_i if it is an outgoing component.

Let $D^+(\Sigma)$ be the group of orientation preserving diffeomorphisms of Σ whose restriction to each collar is the identity and let $D^+_{rot}(\Sigma)$ be the group of orientation preserving diffeomorphisms of the surface Σ whose restriction to each collar is a rotation. To be more precise a diffeomorphism $r_{\alpha_i} : C_i \to C_i$ is in $D^+_{rot}(\Sigma)$ if its restriction to each collar is given by

$$r_{\alpha_i}(\theta, l) = (\theta + \alpha_i, l).$$

We also define their associated mapping class groups by

$$\Gamma_{g,p,q} := \pi_0(D^+(\Sigma)) \; and \; \Gamma^{rot}_{p,q} := \pi_0(D^+_{rot}(\Sigma)).$$

We have a fibration (it can be made into a locally trivial $(S^1)^{p+q}$ fiber bundle)

$$(S^1)^{p+q} \rightarrowtail B\Gamma_{g,p,q} \twoheadrightarrow B\Gamma^{rot}_{g,p,q}$$

which comes from the exact sequence of topological groups

$$D^+(\Sigma) \rightarrowtail D^+_{rot}(\Sigma) \overset{res}{\twoheadrightarrow} (S^1)^{p+q}$$

where $res(\phi) := (\alpha_1, \dots, , \alpha_{p+q})$.

Remarks. We notice that we have a homotopy equivalence

$$B\Gamma^{rot}_{g,p,q} \simeq (B\Gamma_{g,p,q})_{h(S^1)^{p+q}}.$$

Moreover we know that the space $B\Gamma_{g,p,q}$ is a homotopical model of the moduli space of curves associated to the surface Σ, whereas $B\Gamma^{rot}_{g,p,q}$ is rationnaly homotopy equivalent to the moduli space associated to connected curves of genus g with $p+q$ marked points.

Given another surface Σ' of genus g' with q ingoing components and r outgoing components. One can glue Σ together with Σ' in order to get a new surface $\Sigma"$ of genus $g" = g + g' + q - 1$ with p ingoing components and r outgoing components. Thanks to the collars, one get at the level of diffeomorphisms and of their classifying a map

$$glue : B\Gamma_{g,p,q} \times B\Gamma_{g',q,r} \to B\Gamma_{g",p,r}.$$

All these operations extend to non-connected surfaces. Let us consider a surface $F = \Sigma \coprod \Sigma'$, then we have an isomorphism

$$D^+(F) \cong D^+(\Sigma) \times D^+(\Sigma').$$

If F is an oriented surface with p ingoing components and q outgoing components then we set

$$\Gamma_F := \pi_0(D^+(F))$$

and

$$B\Gamma_{p,q} = \coprod_F B\Gamma_F$$

where the disjoint union is taken over all diffeomorphism types of surfaces with p ingoing and q outgoing components. Moreoer we suppose that each connected component has at least one outgoing component.
Together with the morphisms *glue* the collection of topological spaces

$$B\Gamma := \{B\Gamma_{p,q}\}_{p,q}$$

is a prop in the category of topological spaces.

DEFINITION 4.1. *The collection $\{H_*(B\Gamma_{p,q}, R)\}_{p,q>0}$ of graded R-modules is a prop called the homological Segal prop, an algebra over this prop is called an homological Segal algebra.*

Unlike $B\Gamma$ the spaces $\left\{B\Gamma^{rot}_{p,q}\right\}_{p,q}$ do not form a prop in the category of topological spaces, there is no nice gluing morphism. But rather we will make it into a prop in the stable homotopy category of spaces (up to a shift). Let us consider the following commutative diagram

$$\begin{array}{ccccc} D^+_{rot}(\Sigma") & \overset{glue}{\longleftarrow} & D^+_{p,q,r} & \overset{\Delta_D}{\longrightarrow} & D^+_{rot}(\Sigma) \times D^+_{rot}(\Sigma') \\ & & \downarrow & & \downarrow out \times in \\ & & (S^1)^{\times q} & \overset{\Delta}{\longrightarrow} & (S^1)^{\times q} \times (S^1)^{\times q} \end{array}$$

where $in(\phi) = (\alpha_1, \ldots, \alpha_q)$, $out(\psi) = (\alpha_{p+1}, \ldots, \alpha_{p+q})$, we also set

$$D^+_{p,q,r} = \{(\psi, \phi) \in D^+_{rot}(\Sigma) \times D^+_{rot}(\Sigma')/out(\psi) = in(\phi)\}$$

and the morphism *glue* is just the gluing morphism of ϕ and ψ. The preceding diagram is a diagram of topological groups, since at the level of classifying spaces one has the correspondence

$$B\Gamma^{rot}_{p,r} \stackrel{glue}{\longleftarrow} B\Gamma^{rot}_{p,q,r} \stackrel{\Delta_B}{\longrightarrow} B\Gamma^{rot}_{p,q} \times B\Gamma^{rot}_{q,r}.$$

the morphism Δ_B can be made into a $(S^1)^{\times q}$-fiber bundle. Thus one can consider the S^1-transfer map and one gets a map in the homotopy category of spectra

$$(\Delta_B)_! : \Sigma^\infty(B\Gamma^{rot}_{p,q})_+ \wedge \Sigma^\infty(B\Gamma^{rot}_{q,r})_+ \to \Sigma^q\Sigma^\infty(B\Gamma^{rot}_{p,q,r})_+$$

composing by the map $\Sigma^q\Sigma^\infty glue$ one has

$$\mu_{p,q,r} : \Sigma^\infty(B\Gamma^{rot}_{p,q})_+ \wedge \Sigma^\infty(B\Gamma^{rot}_{q,r})_+ \to \Sigma^q\Sigma^\infty(B\Gamma^{rot}_{p,r})_+$$

Let us define $\Sigma\mathcal{B}\Gamma_{p,q} := \Sigma^{-q}\Sigma^\infty(B\Gamma^{rot}_{p,q})_+$, then one has composition products $m_{p;q;r} = \Sigma^{-q-r}\mu_{p,q,r}$:

$$m_{p,q,r} : \Sigma\mathcal{B}\Gamma_{p,q} \wedge \Sigma\mathcal{B}\Gamma_{q,r} \to \Sigma\mathcal{B}\Gamma_{p,r}.$$

This composition products satisfy the following associativity condition

$$m_{p,r,s}\circ(m_{p,q,r}\wedge Id_{r,s}) = m_{p,q,s}\circ(Id_{p,q}\wedge m_{q,r,s}) : \Sigma\mathcal{B}\Gamma_{p,q}\wedge\Sigma\mathcal{B}\Gamma_{q,r}\wedge\Sigma\mathcal{B}\Gamma_{r,s} \to \Sigma\mathcal{B}\Gamma_{p,s}.$$

DEFINITION 4.2. *The collection of spectra $\{\Sigma\mathcal{B}\Gamma_{p,q}\}_{p,q>0}$ is a prop in the stable homotopy category of spaces, this prop is called the stable full gravity prop. The collection $\{H_*(\Sigma\mathcal{B}\Gamma_{p,q}, R)\}_{p,q>0}$ of graded R-modules is a linear prop called the full gravity properad. An algebra over the full gravity prop is called a full gravity algebra.*

We notice that the operadic parts of these props have been studied by E. Getzler on the algebraic side in [**Ge**] and by C. Westerland on the topological side [**W**]. In fact using C. Westerland's techniques one can produce a prop in the stable category of spectra (one has to work with a category of spectra equipped with an associative smash product).

5. Sigma models and String topology

In this section we will always consider singular homology with coefficients into a field $\mathbb{F}$ and denote it by $H_*(-)$. We restrict ourself to field coefficients because one considers products and coproducts, thus one wants to use the Kunneth isomorphism.

5.1. String topology and moduli spaces. Let us consider a topological space X and the following correspondence

$$(\mathcal{L}X)^{\times q} \xleftarrow{r_{out}} map(\Sigma, X) \xrightarrow{r_{in}} (\mathcal{L}X)^{\times p}$$

one would like to define an Umkher map

$$(r_{in})_! : H_*((\mathcal{L}X)^{\times p}) \to H_{*+?}(map(\Sigma, X))$$

in order to get a "stringy" homological operation

$$H_*(\mathcal{L}X)^{\otimes p} \cong H_*((\mathcal{L}X)^{\times p}) \xrightarrow{(r_{in})_!} H_{*+?}(map(\Sigma, X)) \xrightarrow{(r_{out})_*} H_{*+?}((\mathcal{L}X)^{\times q}) \cong H_{*+?}(\mathcal{L}X)^{\otimes q}.$$

When $X = M$ is a closed compact oriented d-dimensional manifold this can be achieved by replacing the surface Σ by a metric chord diagram c, the map r_{in} becomes homotopic to a finite codimensional embedding

$$\rho_{in} : map(\Sigma, M) \to (\mathcal{L}M)^{\times p}$$

of codimension $-\chi(\Sigma).d = (2g + p + q - 2).d$ where $\chi(\Sigma)$ is the Euler-Poincar characteristic of the surface. In that particular and fundamental example one can play with Gysin map associated to finite codimensional embedding this the clasical topological apparatus of string topology as defined and studied by M. Chas, D. Sullivan [**CS1**], [**CS2**], R. Cohen and J. D. S. Jones [**CJ**] and R. Cohen and V. Godin [**CG**].

As focused in the preceding cited papers one wants to study family of operations parametrized by moduli spaces of surfaces or graphs. To be more precise the restriction maps r_{in} and r_{out} are equivariant with respect to the action of the groups $D^+(\Sigma)$ and $D^+_{rot}(\Sigma)$. By applying the Borel construction to our initial correspondence one gets

$$(\mathcal{L}X)^{\times q} \xleftarrow{r_{out}} ED^+(\Sigma) \times_{D^+(\Sigma)} map(\Sigma, X) \xrightarrow{r_{in}} BD^+(\Sigma) \times (\mathcal{L}X)^{\times p}$$

defining an Umkher map for r_{in} in this equivariant setting is much more delicate, this was done for the genus zero operadic part of the properad $B\Gamma$ i.e. the cacti part by R. Cohen and J. D. S. Jones [**CJ**] and for the TQFT part by R. Cohen and V. Godin [**CG**] and recently for the full properadic part by V. Godin [**Go**].

5.2. String topology of classifying spaces. When $X = BG$ where G is a compact connected Lie group, the homotopy fiber of the map

$$r_{in} : ED^+(\Sigma) \times_{D^+(\Sigma)} map(\Sigma, BG) \longrightarrow BD^+(\Sigma) \times (\mathcal{L}BG)^{\times p}$$

is homotopy equivalent to the based loop space $\Omega(BG)^{\times -\chi(\Sigma).dim(G)}$ therefore to a finite space and we can apply integration along the fiber. We are led to the following theorem [**CM**]

If G a connected compact Lie group then the singular homology $H_*(\mathcal{L}BG)$ is an homological Segal algebra.

5.3. The action of the gravity properad. One can also notice that the diffeomorphism group $D^+_{rot}(\Sigma)$ acts on the preceding correspondence, by applying the Borel construction one gets

$$\begin{array}{ccc} ED^+_{rot}(\Sigma)\times_{D^+_{rot}(\Sigma)}(\mathcal{L}BG)^{\times q} & \stackrel{r_{out}}{\longleftarrow} & ED^+_{rot}(\Sigma)\times_{D^+_{rot}(\Sigma)} map(\Sigma,BG) \\ & & \downarrow r_{in} \\ & & ED^+_{rot}(\Sigma)\times_{D^+_{rot}(\Sigma)}(\mathcal{L}BG)^{\times p} \end{array}$$

in order to define stringy operations let us modify slighty the equivariant restriction maps r_{in} and r_{out}. For the sake of simplicity, let us consider a Riemann surface S_n with n boundary components, for the moment we do not distinguish between incoming and outgoing components. We have a restriction map

$$Res: map(S_n,BG)\to(\mathcal{L}BG)^{\times n},$$

let us also consider the morphism of topological groups

$$\begin{array}{cccc} Rot: & D^+_{rot}(S_n) & \longrightarrow & D^+_{rot}(S_n)\times(S^1)^{\times n} \\ & \phi & \mapsto & (\phi,\alpha_1,\dots,\alpha_n) \end{array}$$

the group $D^+_{rot}(S_n)\times(S^1)^{\times n}$ acts on $(\mathcal{L}BG)^{\times n}$ via the formula

$$(\phi,\alpha_1,\dots,\alpha_n).(\gamma_1,\dots,\gamma_n):=(\gamma_1(-+\alpha_1),\dots,\gamma_n(-+\alpha_n))$$

and the following diagram is commutative

$$\begin{array}{ccc} D^+_{rot}(S_n)\times map(S_n,BG) & \stackrel{act}{\longrightarrow} & map(S_n,BG) \\ Rot\times Res\downarrow & & \downarrow Res \\ D^+_{rot}(S_n)\times(S^1)^{\times n}\times(\mathcal{L}BG)^{\times n} & \stackrel{act}{\longrightarrow} & (\mathcal{L}BG)^{\times n} \end{array}$$

therefore as $D^+_{rot}(S_n)$ acts trivially on $(\mathcal{L}BG)^{\times n}$ we end the day with a map

$$\mathcal{R}es: ED^+_{rot}(S_n)\times_{D^+_{rot}(S_n)} map(S_n,BG)\longrightarrow BD^+_{rot}(S_n)\times(\mathcal{L}BG_{hS^1})^{\times n}$$

where $\mathcal{L}BG_{hS^1}:=ES^1\times_{S^1}\mathcal{L}BG$. Finally we consider the correspondence

$$(\mathcal{L}BG)^{\times q}\stackrel{\mathcal{R}_{out}}{\longleftarrow} ED^+_{rot}(\Sigma)\times_{D^+_{rot}(\Sigma)} map(\Sigma,BG)\stackrel{\mathcal{R}_{in}}{\longrightarrow} BD^+_{rot}(\Sigma_{p,q})\times(\mathcal{L}BG_{hS^1})^{\times p}.$$

LEMMA 5.1. *The homotopy fiber of the map $\mathcal{R}_{in}$ is homotopy equivalent to the space $(S^1)^{\times q}\times G^{\times-\chi(\Sigma)}$.*

PROOF. The morphism

$$Rot : D^+_{rot}(\Sigma) \to D^+_{rot}(\Sigma) \times (S^1)^{\times q}$$

is an injection, the group $Rot(D^+_{rot}(\Sigma))$ is a normal subgroup and we have an isomorphism

$$\Phi : D^+_{rot}(\Sigma) \times (S^1)^{\times q}/Rot(D^+_{rot}(\Sigma)) \to (S^1)^{\times q}.$$

The result follows from the classical properties of the Borel Construction. □

We recall that if we want to define an Umkher map one needs an orientation condition. In general if one has a fibration

$$F \to E \to B$$

where F is a d-dimensional oriented manifold and if B is connected, we say that the fibration is oriented if the action of the fundamental group $\pi_1(B,b)$ on the homology of F preserves the fundamental class $[F] \in H_d(F)$. In that case one can define a wrong way map

$$H_i(B) \to H_{i+d}(E).$$

Oriented fibrations are stable by pull-back and composition.

LEMMA 5.2. *The map $\mathcal{R}_{in}$ is oriented.*

PROOF. In [**CM**, prop 15] we proved that the restriction map

$$map(\Sigma, BG) \to (\mathcal{L}BG)^{\times q}$$

is an oriented fibration and the Borel construction respects this orientation therefore the map

$$r_{in} : ED^+_{rot}(\Sigma) \times_{D^+_{rot}(\Sigma)} map(\Sigma, BG) \to ED^+_{rot}(\Sigma) \times_{D^+_{rot}(\Sigma)} (\mathcal{L}BG)^{\times q}$$

is an oriented fibration. Moreover the map

$$BRot : ED^+_{rot}(\Sigma) \times_{D^+_{rot}(\Sigma)} (\mathcal{L}BG)^{\times q} \to BD^+_{rot}(\Sigma) \times (\mathcal{L}BG_{hS^1})^{\times q}$$

is also an oriented fibration, this follows from the fact that for any space the fibration $\pi : ES^1 \times \mathcal{L}X \to \mathcal{L}X_{hS^1}$ is an oriented fibration as pull-back of the universal fibration

$$S^1 \hookrightarrow ES^1 \to BS^1$$

which is oriented. Moreover as we have $\mathcal{R}_{in} = BRot \circ r_{in}$ it follows that $\mathcal{R}_{in}$ is oriented as composition of two oriented maps. □

DEFINITION 5.3. *Let G be a compact Lie group, we define stringy evaluation products*

$$ev_{p,q} : H_i(B\Gamma^{rot}_{p,q}) \otimes H^{S^1}_j(\mathcal{L}BG^{\times p}) \longrightarrow H^{S^1}_{i+j+sh}(\mathcal{L}BG^{\times q})$$

where $sh := q - \chi(\Sigma_{p,q}).dim(G)$ as the composition

$$ev_{p,q} := (\mathcal{R}_{out})_* \circ (\mathcal{R}_{in})_! = (\mathcal{R}_{out})_* \circ (r_{in})_! \circ (BRot)_!.$$

THEOREM 5.4. *Let G be a connected compact Lie group, then the S^1-singular homology $H_*^{S^1}(\mathcal{L}BG)$ of the free loop space of the classifying space BG of G is a full gravity algebra.*

PROOF. The proof is just a routine adaption of the techniques used in [**CM**] which follows from the basic properties of the umkher maps. Moreover C. Westerland in [**W**] has proved a similar result for closed manifolds rather than classifying spaces and for the operadic part of the action. □

6. Two colours are better than one

Let us fix a connected surface σ of genus g with p ingoing an q outgoing components. We choose to decorate the boundary components with two colours fix and rot. We thus have $p = p_f + p_r$ and $q = q_f + q_r$. We define the group of diffeomorphisms $D^+(\Sigma)_{p_f,p_r,q_f,q_r}$ as the group of diffeomorphisms that preserve the orientation and that fix the boundary components colored in fix and that restrict to a rotation on boundary components colored in rot.
It is not hard to extend the preceding constructions and to define a stable colored prop $B\Gamma_{fix,rot}$ and a linear colored prop

$$Seg - Grav := H_*(B\Gamma_{fix,rot}).$$

For the definition of colored props the reader is refered to Johnson and Yau's preprint [**JY**]. We thus have the theorem

THEOREM 6.1. *Let G be a connected compact Lie group, then the couple $(H_*(\mathcal{L}BG), H_*^{S^1}(\mathcal{L}BG))$ is a $Seg - Grav$-algebra.*

References

[CS1] *Chas, Moira and Sullivan, Dennis*; String Topology, preprint math.GT/9911159

[CS2] *Chas, Moira and Sullivan, Dennis*; Closed string operators in topology leading to Lie bialgebras and higher string algebra. The legacy of Niels Henrik Abel, 771-784, Springer, Berlin, 2004.

[CM] *Chataur, David and Menichi, Luc*; String topology of classifying spaces, arXiv : 0801.0174v3

[CG] *Cohen, Ralph and Godin, Vronique*; A polarized view of string topology. Topology, geometry and quantum field theory, 127-154,London Math. Soc. Lecture Note Ser., 308, Cambridge Univ. Press, 2004.

[CJ] *Cohen, Ralph and Jones, John*; A homotopy theoretic realization of string topology, Math. Ann. 324 (2002), no. 4, 773-798.

[Cos] *Costello, Kevin*; Topological conformal Field theories and Calabi-Yau categories. Advances in Mathematics, Volume 210, Issue 1, March 2007

[Ge] *Getzler, Ezra*; Operads and moduli spaces of genus 0 Riemann surfaces, The moduli space of curves, Progr. Math., vol. 129, 1995.

[Go] *Godin, Veronique*; Higher string topology operations; arXiv:0711.4859

[JY] *Mark W. Johnson, Donald Yau*; On homotopy invariance for algebras over colored PROPs; arXiv:0906.0015

[Sul] *Sullivan, Dennis*; Sigma models and string topology. Graphs and patterns in mathematics and theoretical physics, 1-11, Proc. Sympos. Pure Math., 73, Amer. Math. Soc., Providence, RI, 2005.

[W] *Westerland, Craig* Equivariant operads, string topology, and Tate cohomology; Mathematische Annalen, 340 (2008), no. 1, 97-142.

E-mail address: `david.chataur@math.univ-lille1.fr`

Current address: David Chataur, Laboratoire Paul Painlev, USTL, 59655 Villeneuve d'Ascq Cdex, France

Contemporary Mathematics
Volume **519**, 2010

A FIBREWISE STABLE SPLITTING AND FREE LOOPS ON PROJECTIVE SPACES

M. C. Crabb

ABSTRACT. Stable decompositions of the spaces of free loops on real, complex and quaternionic projective spaces (due to Bauer, Crabb and Spreafico [**3**] and Bökstedt and Ottosen [**4**]) are obtained as consequences of a fibrewise stable splitting theorem depending only on elementary homology calculations.

1. Introduction

In [**3**] we established the following stable decomposition of the space of free loops on a real projective space. (See also [**1, 7**].) Some points of notation are explained below the statement of the theorem.

THEOREM 1.1. (Bauer, Crabb, Spreafico [**2, 3**].) *Let V be a finite-dimensional real vector space. Then there is a natural stable decomposition of the space $\mathcal{L}\mathbb{R}\mathrm{P}(V)$ of free loops on the real projective space $\mathbb{R}\mathrm{P}(V)$ of V as*

$$(\mathcal{L}\mathbb{R}\mathrm{P}(V))_+ \simeq \mathbb{R}\mathrm{P}(V)_+ \vee \bigvee_{l\geq 1} S(\eta)^{(l-1)\zeta},$$

where η is the tangent bundle of $\mathbb{R}\mathrm{P}(V)$ and ζ is the fibrewise tangent bundle of $S(\eta)$.

For a space B, we write B_+ for the pointed space obtained by adjoining a disjoint basepoint to B. The sphere bundle of a finite-dimensional real vector bundle ξ over B is written as $S(\xi)$ and the Thom space as B^{ξ}.

A little later, and independently, Bökstedt and Ottosen [**4**] obtained similar splittings of the spaces of free loops on complex and quaternionic projective spaces.

THEOREM 1.2. (Bökstedt, Ottosen [**4**]). *Let V be a finite-dimensional complex vector space. Then there is a natural stable decomposition*

$$(\mathcal{L}\mathbb{C}\mathrm{P}(V))_+ \simeq \mathbb{C}\mathrm{P}(V)_+ \vee \bigvee_{l\geq 1} S(\eta)^{l\mathbb{R}\oplus(l-1)\zeta},$$

where η is the tangent bundle of the complex projective space $\mathbb{C}\mathrm{P}(V)$, $\mathbb{R}$ is written for the trivial real line bundle and ζ is the fibrewise tangent bundle of $S(\eta)$.

2000 *Mathematics Subject Classification.* Primary 55P35; Secondary 55P91.

©2010 American Mathematical Society

THEOREM 1.3. (Bökstedt, Ottosen [**4**]). *Let V be a finite-dimensional quaternionic vector space. Then there is a natural stable decomposition*

$$(\mathcal{L}\mathbb{HP}(V))_+ \simeq \mathbb{HP}(V)_+ \vee \bigvee_{l\geq 1} S(\eta)^{l\xi\oplus(l-1)\zeta},$$

where η is the tangent bundle of the quaternionic projective space $\mathbb{HP}(V)$, ξ is the 3-dimensional Lie algebra bundle over $\mathbb{HP}(V)$ and ζ is the fibrewise tangent bundle of $S(\eta)$.

From a geometric point of view one can regard the tangent sphere bundle $S(\eta)$ appearing in each splitting as a projective Stiefel manifold. In the real case, $S(\eta)$ can be identified with the quotient $\mathrm{PO}(\mathbb{R}^2, V)$ of the real Stiefel manifold $\mathrm{O}(\mathbb{R}^2, V)$ of isometric linear maps $\mathbb{R}^2 \to V$ by the action of $\{\pm 1\}$. In the complex case, we can express $S(\eta)$ as the space of pairs

$$\{(x, y) \in S(V) \times S(V) \mid \langle x, y\rangle = 0\}$$

modulo the action of the group $\mathbb{T}$ of complex numbers of modulus 1 by $z \cdot (x, y) = (zx, zy)$ for $z \in \mathbb{T}$. This is the quotient $\mathrm{PU}(\mathbb{C}^2, V) = \mathrm{U}(\mathbb{C}^2, V)/\mathbb{T}$ of the complex Stiefel manifold of isometric linear maps $\mathbb{C}^2 \to V$. In the quaternionic case, we have the quotient of the quaternionic Stiefel manifold $\mathrm{Sp}(\mathbb{H}^2, V)$ by the action of the group of unit quaternions.

In this note we shall give a uniform proof of these three splitting theorems using the methods developed in [**2**] and [**3**]. The main result is the fibrewise splitting theorem stated as Theorem 3.1. In Section 2 we establish, by elementary methods, a natural splitting of the fibre, from which the global decomposition readily follows. The application to the splitting of the free loop spaces is contained in Section 4.

For fibre bundles over a compact ENR base B, we shall use a subscript 'B' to indicate fibrewise constructions. Thus if $X \to B$ is a fibre bundle, with fibre at $b \in B$ written as X_b, the bundle $X \sqcup B \to B$ with fibre $(X_b)_+$, obtained by adjoining a basepoint to each fibre, is denoted by X_{+B}. If ξ is a real vector bundle over X, the fibrewise Thom space X_B^{ξ} is the pointed fibre bundle with fibre at $b \in B$ the Thom space of ξ restricted to X_b.

This work continues a fruitful collaboration [**1, 2, 3**] with Sven Bauer and Mauro Spreafico, and it is a pleasure to record my gratitude to them.

2. A stable splitting

Let E and F be real vector spaces of dimensions m and n respectively. We assume that both are equipped with a Euclidean inner product. The unit sphere $S(\mathbb{R} \oplus E)$ has a basepoint $(1, 0)$, which we write simply as 1. We define $\Phi(E, F)$ to be the homotopy fibre of the inclusion $S(\mathbb{R} \oplus E) \to S(\mathbb{R} \oplus E \oplus F)$, that is, the space of continuous paths

$$\Phi(E, F) = \{\omega : [0, 1] \to S(\mathbb{R} \oplus E \oplus F) \mid \omega(0) = 1,\ \omega(1) \in S(\mathbb{R} \oplus E)\}.$$

Our goal in this section is to construct in a natural way a stable splitting of $\Phi(E, F)_+$. When $F = 0$, the space $\Phi(E, F)$ is contractible. We suppose now that F is non-zero.

There is a natural geodesic path in the sphere $S(\mathbb{R} \oplus E \oplus F)$ joining a given point $x \in S(\mathbb{R} \oplus E)$ along a great circle to a given point $y \in S(F)$:

$$\alpha_{x,y} : [0, 1] \to S(\mathbb{R} \oplus E \oplus F), \qquad t \mapsto (\cos(\pi t/2)x, \sin(\pi t/2)y).$$

For an integer $l \geq 1$, we define a map

$$\psi_l : S(\mathbb{R} \oplus E)^l \times S(F)^l \to \Phi(E, F),$$

by mapping $((x_1, \dots, x_l), (y_1, \dots, y_l))$ to the piecewise geodesic path α alternating through the $2l + 1$ points of $S(\mathbb{R} \oplus E)$ and $S(F)$ in $S(\mathbb{R} \oplus E \oplus F)$:

$$1,\ y_1,\ x_1,\ y_2,\ x_2,\ \dots,\ y_l,\ x_l.$$

Precisely, with the convention that $x_0 = 1$, we define

$$\alpha(t) = \begin{cases} \alpha_{x_{i-1}, y_i}(2lt - 2(i-1)) & \text{if } 2i - 2 \leq 2lt \leq 2i - 1, \\ \alpha_{x_i, y_i}(2i - 2lt) & \text{if } 2i - 1 \leq 2lt \leq 2i, \end{cases}$$

for $i = 1, \dots, l$.

We need an explicit stable splitting of the space $S(\mathbb{R} \oplus E)_+$. The Pontrjagin-Thom construction applied to the embedding $S(\mathbb{R} \oplus E) \subseteq \mathbb{R} \oplus E$ of the sphere in Euclidean space with trivial normal bundle $S(\mathbb{R} \oplus E) \times \mathbb{R}$ gives a map

$$r : \Sigma E^+ = (\mathbb{R} \oplus E)^+ \to \big(S(\mathbb{R} \oplus E) \times \mathbb{R}\big)^+ = \Sigma S(\mathbb{R} \oplus E)_+,$$

where a superscript $+$ is used for one-point compactification (with basepoint at infinity). The stable map $\rho : E^+ \to S(\mathbb{R} \oplus E)_+$ defined by r is a splitting of the projection $S(\mathbb{R} \oplus E)_+ \to E^+$ onto the 'top cell' $S(\mathbb{R} \oplus E) - \{1\}$ of the framed manifold. For $l \geq 1$, we define

$$\rho_l : (lE)^+ = E^+ \wedge \cdots \wedge E^+ \to (S(\mathbb{R} \oplus E)^l)_+ = S(\mathbb{R} \oplus E)_+ \wedge \cdots \wedge S(\mathbb{R} \oplus E)_+$$

to be the l-fold smash product of ρ.

We now apply this construction on a vector space E fibrewise to the tangent bundle $\zeta = \tau S(F)$ of $S(F)$. The Pontrjagin-Thom construction gives a fibrewise stable map

$$\zeta^+_{S(F)} \to S(\mathbb{R} \oplus \zeta)_{+S(F)},$$

over $S(F)$, where $\mathbb{R}$ stands, also, for the one-dimensional real trivial bundle and the subscript $S(F)$ indicates the appropriate fibrewise construction. The $(l-1)$-fold smash product over $S(F)$ is a fibrewise stable map

$$\tilde{\sigma}_l : ((l-1)\zeta)^+_{S(F)} \to (S(\mathbb{R} \oplus \zeta)^{(l-1)}_{S(F)})_{+S(F)}.$$

Now $\mathbb{R} \oplus \zeta$ is the trivial bundle $S(F) \times F$ over $S(F)$, so that $S(\mathbb{R} \oplus \zeta) = S(F) \times S(F)$ and $S(\mathbb{R} \oplus \zeta)^{l-1}_{S(F)} = S(F) \times S(F)^{l-1}$. Making these identifications and collapsing the basepoint sections $S(F)$ to a point, we obtain from $\tilde{\sigma}_l$ a stable map

$$\sigma_l : S(F)^{(l-1)\tau S(F)} \to (S(F)^l)_+$$

for $l \geq 1$. (The map σ_1 is the identity.)

In this way we obtain, for $l \geq 1$, more-or-less explicit stable maps

$$\phi_l = (\psi_l)_+ \circ (\rho_l \wedge \sigma_l) : (lE)^+ \wedge S(F)^{(l-1)\tau S(F)} \to \Phi(E, F)_+.$$

We also define $\phi_0 : S^0 \to \Phi(E, F)_+$ to be the map given by the inclusion of the constant loop 1 in $\Phi(E, F)$.

Proposition 2.1. *The stable map*

$$\phi_0 \vee \bigvee_{l \geq 1} \phi_l : S^0 \vee \bigvee_{l \geq 1} (lE)^+ \wedge S(F)^{(l-1)\tau S(F)} \to \Phi(E, F)_+$$

constructed above is a stable homotopy equivalence.

PROOF. It is enough to check that the induced map in integral homology is an isomorphism. This is an essentially elementary computation achieved by the sequence of lemmas below in which 'H' always denotes homology with $\mathbb{Z}$-coefficients. □

To carry out the calculations we may choose a point $+ \in S(F)$. This allows us to write down a homotopy equivalence

$$\Phi(E, F) \to S(\mathbb{R} \oplus E) \times \Omega S(\mathbb{R} \oplus E \oplus F)$$

mapping ω to $\omega(1) \in S(\mathbb{R} \oplus E)$ and the loop in $S(\mathbb{R} \oplus E \oplus F)$ obtained by concatenation of ω with the piecewise geodesic path from $\omega(1)$ to $+$ to 1.

Using the point $+$ we can also define a map

$$\beta : S(\mathbb{R} \oplus E) \times S(F) \to \Omega S(\mathbb{R} \oplus E \oplus F)$$

sending (x, y) to the piecewise geodesic $\beta_{x,y}$ through 1, y, x, $+$, 1. Then ψ_l is homotopic to the map

$$S(\mathbb{R} \oplus E)^l \times S(F)^l \to S(\mathbb{R} \oplus E) \times \Omega S(\mathbb{R} \oplus E \oplus F)$$

taking $((x_1, \ldots, x_l), (y_1, \ldots, y_l))$ to the pair

$$(x_l,\ \beta_{x_1,y_1} \cdot \beta_{x_1,y_2}^{-1} \cdot \beta_{x_2,y_2} \cdot \beta_{x_2,y_3}^{-1} \cdot \ldots \cdot \beta_{x_l,y_l}),$$

in which the loop is written in terms of the loop product and inverse.

For $m > 0$, let us write e for the standard generator of $H_0(S(\mathbb{R} \oplus E)) = \mathbb{Z}$ and u for a chosen generator of $H_m(S(\mathbb{R} \oplus E)) \cong \mathbb{Z}$. For $m = 0$, we have $H_0(S(\mathbb{R} \oplus E)) = \mathbb{Z}e_+ \oplus \mathbb{Z}e_-$, where e_+ and e_- are the canonical generators corresponding to the components of 1 and -1; we set $e = e_+$ and $u = e_+ - e_-$. Similarly, for $n > 1$, we write $H_0(S(F)) = \mathbb{Z}f$ and $H_{n-1}(S(F)) = \mathbb{Z}v$. When $n = 1$, we write $H_0(S(F)) = \mathbb{Z}f_+ \oplus \mathbb{Z}f_-$, where f_+ is the standard generator for the component of the chosen basepoint $+ \in S(F)$ and f_- is the generator for the other component; we set $f = f_+$ and $v = f_+ - f_-$.

LEMMA 2.2. *Suppose that $m + n - 1 > 0$ and put*

$$w = \beta_*(u \otimes v) \in H_{m+n-1}(\Omega S(\mathbb{R} \oplus E \oplus F)).$$

Then

$$\beta_*(e \otimes f) = 1, \quad \beta_*(u \otimes f) = 0, \quad \beta_*(e \otimes v) = 0,$$

and the Pontrjagin ring $H_*(\Omega S(\mathbb{R} \oplus E \oplus F)) = \mathbb{Z}[w]$ is polynomial on w.

PROOF. Suppose first that $m = 0$ (and $n > 1$). Then $y \mapsto \beta_{1,y}$ is homotopic to the constant map and $y \mapsto \beta_{-1,y}$ is the inclusion of the classical generating variety $S(F) \to \Omega S(\mathbb{R} \oplus F)$ of the loop space of the suspension of the sphere $S(F)$ with basepoint $+$. Hence $\beta_*(e_+ \otimes v) = 0$ and the Pontrjagin ring is polynomial on $\beta_*(e_- \otimes v)$. Of course, $\beta_*(e_+ \otimes f) = 1 = \beta_*(e_- \otimes f)$. The results involving u clearly follow.

The case in which $n = 1$ (and $m > 0$) is treated in the same way.

Suppose now that $m > 0$ and $n > 1$. Then $\beta_*(u \otimes f)$ and $\beta_*(e \otimes v)$ vanish for dimensional reasons. We only have to show that $w = \beta_*(u \otimes f)$ generates $H_{m+n-1}(\Omega S(\mathbb{R} \oplus E \oplus F))$. This is easily verified by checking that the adjoint of β

$$S^1 \times S(\mathbb{R} \oplus E) \times S(F) \to S(\mathbb{R} \oplus E \oplus F)$$

has degree ± 1. (Indeed, this map restricts to a homeomorphism from an open subspace onto the complement of $S(\mathbb{R} \oplus E) \sqcup S(F)$ in $S(\mathbb{R} \oplus E \oplus F)$.) □

We now use the loop multiplication and the description of ψ_l in terms of β to compute

$$(\psi_l)_* : H_*(S(\mathbb{R}\oplus E)^l \otimes S(F)^l) \to H_*(S(\mathbb{R}\oplus E)\otimes \Omega S(\mathbb{R}\oplus E\oplus F)).$$

LEMMA 2.3. *For $m>0$, $n>1$ and $l\geq 1$, we have*

$$(\psi_l)_*((u\otimes\cdots\otimes u)\otimes(v\otimes\cdots\otimes v)) = e\otimes w^l$$

and

$$(\psi_l)_*((u\otimes\cdots\otimes u)\otimes(f\otimes v\otimes\cdots\otimes v)) = (-1)^{l-1}u\otimes w^{l-1}.$$

PROOF. We make the abbreviations $X = S(\mathbb{R}\oplus E)$ and $Y = S(F)$. The diagonal maps

$$\Delta_X : X\to X\times X,\quad \Delta_Y : Y\to Y\times Y$$

are given in homology by

$$(\Delta_X)_*(e) = e\otimes e,\quad (\Delta_X)_*(u) = e\otimes u + u\otimes e$$

and

$$(\Delta_Y)_*(f) = f\otimes f,\quad (\Delta_Y)_*(v) = f\otimes v + v\otimes f.$$

The map ψ_l factors through $\Delta_X^l\times(\Delta_Y^{l-1}\times 1)$ as a composition:

$$X^l\times Y^l\to X^{2l}\times Y^{2l-1} = X\times(X\times Y)^{2l-1}\to X\times\Omega S(\mathbb{R}\oplus E\oplus F),$$

where the identification '$=$' involves a permutation of the factors and the second map is described by β and the product and antipode in the loop space.

The image of $(u\otimes\cdots\otimes u)\otimes(v\otimes\cdots\otimes v)$ in the homology of $X\times(X\times Y)^{2l-1}$ is the sum of $2^l\cdot 2^{l-1}$ terms of the form:

$$a_l'\otimes((a_1\otimes v)\otimes(a_1'\otimes b_2')\otimes(a_2\otimes b_2)\otimes(a_2'\otimes b_3')\otimes\cdots\otimes(a_l\otimes b_l)),$$

where $\{a_i, a_i'\} = \{e, u\}$ and $\{b_i, b_i'\} = \{f, v\}$. The image in the homology of $X\times\Omega S(\mathbb{R}\oplus E\oplus F)$ will, according to Lemma 2.2, be non-zero only if: $a_1 = u$, $a_1' = e$, $b_2' = f$, $b_2 = v$, $a_2 = u$, $a_2' = e$, $b_3' = f$, $b_3 = v$, $\ldots$, $b_l = v$, $a_l = u$ and $a_l' = e$. So the image is $e\otimes w^l$ as asserted.

The image of $(u\otimes\cdots\otimes u)\otimes(f\otimes v\otimes\cdots\otimes v)$ is, similarly, a sum of terms

$$a_l'\otimes((a_1\otimes f)\otimes(a_1'\otimes b_2')\otimes(a_2\otimes b_2)\otimes(a_2'\otimes b_3')\otimes\cdots\otimes(a_l\otimes b_l)).$$

For a non-zero image we must have $a_1 = e$, $a_1' = u$, $b_2' = v$, $b_2 = f$, $\ldots$, $b_l = f$, $a_l = e$ and $a_l' = u$. Noting that reversing loops multiplies w by -1, we deduce that the image is $(-1)^{l-1}u\otimes w^{l-1}$. □

LEMMA 2.4. *For $m>0$, $n=1$ and $l\geq 1$, we have*

$$(\psi_l)_*((u\otimes\cdots\otimes u)\otimes(v\otimes\cdots\otimes v)) = e\otimes w^l$$

and

$$(\psi_l)_*((u\otimes\cdots\otimes u)\otimes(f\otimes v\otimes\cdots\otimes v)) = (-1)^{l-1}u\otimes w^{l-1} + \tfrac{(1+(-1)^l)}{2}e\otimes w^l.$$

PROOF. We follow the proof of Lemma 2.3, but must take account of the fact that $(\Delta_Y)_*(v) = f\otimes v + v\otimes f - v\otimes v$ and so consider a sum of $2^l\cdot 3^{l-1}$ terms with $(b_i, b_i') = (f, v)$, (v, f) or (v, v). The completion of the proof is a routine exercise. □

LEMMA 2.5. *For $m = 0$, $n > 1$ and $l \geq 1$, we have*

$$(\psi_l)_*((u \otimes \cdots \otimes u) \otimes (v \otimes \cdots \otimes v)) = e_l \otimes w^l$$

and

$$(\psi_l)_*((u \otimes \cdots \otimes u) \otimes (f \otimes v \otimes \cdots \otimes v)) = (-1)^{l-1} u \otimes w^{l-1},$$

where e_l is equal to e_+ if l is even, e_- if l is odd.

PROOF. The proof again follows the pattern of Lemma 2.3. This time $\Delta_X(u) = e \otimes u + u \otimes e - u \otimes u$, and one has to consider a sum of $3^l \cdot 2^{l-1}$ terms with $(a_i, a_i') = (e, u)$, (u, e) or (u, u). In the first case, there are $l+1$ non-zero terms summing to $(e + (-1 + 1 - 1 + \ldots + (-1)^l)u) \otimes w^l$. In the second case, there is a single non-zero term. □

We turn to the homological description of the stable splitting maps ρ_l and σ_l. The first is classical; the second was embedded in [**2**].

LEMMA 2.6. *For $m \geq 0$ and $l \geq 1$, the homomorphism induced by ρ_l*

$$\tilde{H}_*((lE)^+) \to H_*(S(\mathbb{R} \oplus E)^l)$$

is the inclusion of the summand $\mathbb{Z}(u \otimes \cdots \otimes u)$. □

LEMMA 2.7. *For $n \geq 1$ and $l \geq 1$, the homomorphism induced by σ_l*

$$\tilde{H}_*(S(F)^{(l-1)\tau S(F)}) \to H_*(S(F)^l)$$

is the inclusion of the summand $\mathbb{Z}(v \otimes \cdots \otimes v) \oplus \mathbb{Z}(f \otimes v \otimes \cdots \otimes v)$.

PROOF. By construction, $(\sigma_l)_*$ is the inclusion of a direct summand. And this summand is annihilated by each of the projection maps

$$S(F)^l \to S(F)^{l-1}$$

which omits the ith factor for $i = 2, \ldots, l$. (This follows from the fact that the composition $E^+ \to S(\mathbb{R} \oplus E)_+ \to S^0$ of the Pontrjagin-Thom map ρ with the projection of the sphere to a point is zero.) □

The proof of Proposition 2.1 is easily completed, if $m + n - 1 > 0$, by using the Lemmas 2.3–2.7 and the observation that ϕ_0 picks out $e \otimes 1$. This leaves the special case in which $m = 0$ and $n = 1$. The space $\Phi(0, \mathrm{i}\mathbb{R})$ (where $\mathrm{i}\mathbb{R} \subseteq \mathbb{C}$) is homotopy equivalent to the discrete space $\mathbb{Z}$ with components represented by paths $[0,1] \to \mathbb{R} \oplus \mathrm{i}\mathbb{R} = \mathbb{C} : t \mapsto \mathrm{e}^{d\pi \mathrm{i}t}$ for $d \in \mathbb{Z}$. The image of ψ_l for $l \geq 1$ is $\{d \in \mathbb{Z} \mid |d| \leq l\}$. We omit the details of the proof of Proposition 2.1 in this case.

3. The main theorem

Let ξ and η be real vector bundles, of dimension m and n respectively, over a compact ENR B. It is again convenient to write simply $\mathbb{R}$ for the trivial line bundle $B \times \mathbb{R}$ over B. Thus $S(\mathbb{R} \oplus \xi)$ is the sphere bundle with fibre $S(\mathbb{R} \oplus \xi_b)$ at $b \in B$. We take $(1, 0)$ as basepoint in each fibre, and may then identify the pointed fibre bundle $S(\mathbb{R} \oplus \xi)$ in the usual way with the fibrewise one-point compactification ξ_B^+ of ξ.

Consider the fibrewise homotopy fibre $\Phi(\xi, \eta)$ of the inclusion $S(\mathbb{R} \oplus \xi) = \xi_B^+ \to S(\mathbb{R} \oplus \xi \oplus \eta) = (\xi \oplus \eta)_B^+$. The fibre over $b \in B$ is the path space

$$\Phi(\xi_b, \eta_b) = \{\omega : [0,1] \to S(\mathbb{R} \oplus \xi_b \oplus \eta_b) \mid \omega(0) = (1,0,0),\ \omega(1) \in S(\mathbb{R} \oplus \xi_b)\}.$$

THEOREM 3.1. *Let ξ and η be real vector bundles of dimension m and n, respectively, over a compact ENR B. Then the fibrewise homotopy fibre $\Phi(\xi,\eta)$ of the inclusion of sphere bundles $S(\mathbb{R}\oplus\xi)\to S(\mathbb{R}\oplus\xi\oplus\eta)$ has a natural fibrewise stable splitting as the wedge:*

$$\Phi(\xi,\eta)_{+B}\simeq (B\times S^0)\vee_B \bigvee_{B\,l\geq 1}(l\xi)_B^+\wedge_B S(\eta)_B^{(l-1)\zeta},$$

where ζ is the fibrewise tangent bundle of $S(\eta)\to B$.

PROOF. The result is trivial if η is the zero-vector bundle. Assume, therefore, that $\eta\neq 0$. The naturality of the constructions in Section 2 permits them to be carried out fibrewise to give fibrewise stable maps

$$\phi_0 : B\times S^0\to \Phi(\xi,\eta)_{+B} \quad\text{and}\quad \phi_l : (l\xi)_B^+\wedge S(\eta)_B^{(l-1)\zeta}\to \Phi(\xi,\eta)_{+B}$$

for $l\geq 1$. The fibrewise wedge

$$\phi_0\vee\bigvee_{l\geq 1}\phi_l : (B\times S^0)\vee_B \bigvee_{B\,l\geq 1}(l\xi)_B^+\wedge_B S(\eta)_B^{(l-1)\zeta}\to \Phi(\xi,\eta)_{+B}$$

restricts to a stable equivalence on fibres, by Proposition 2.1, and hence, by Dold's theorem, is a fibrewise stable homotopy equivalence. □

By collapsing the space B of fibrewise basepoints to a single point we obtain immediately a stable decomposition of the space $\Phi(\xi,\eta)_+$.

COROLLARY 3.2. *There is a stable decomposition*

$$\Phi(\xi,\eta)_+\simeq B_+\vee\bigvee_{l\geq 1}S(\eta)^{l\xi\oplus(l-1)\zeta}. \quad\square$$

Of course, the direct sum $\mathbb{R}\oplus\zeta$ of the fibrewise tangent bundle of $S(\eta)$ with a trivial line bundle is just the lift of η and we may identify the $(l-1)$-fold suspension of the Thom space of $l\xi\oplus(l-1)\zeta$ with the Thom space of $(2l-1)\xi$ over $S(\eta)$.

REMARK 3.3. The case in which $\xi=0$ and $n=1$ is anomalous: the connectivity of the summands does not increase with l. In fact, it is easy to see directly that $\Phi(\xi,\eta)$ is homotopy equivalent to a disjoint union of B and countably many copies of the double cover $S(\eta)$ indexed by $l\geq 1$.

4. Projective bundles and free loop spaces

We consider three cases in which ξ and η are vector bundles over a compact ENR B. In each case η is non-zero.

(a) *Real projective space*

We take $\xi=0$ to be the zero vector bundle and let η be a real vector bundle; Consider the fibrewise loop space $\Omega_B\mathbb{R}\mathrm{P}(\mathbb{R}\oplus\eta)$, with fibre at $b\in B$ the space $\Omega\mathbb{R}\mathrm{P}(\mathbb{R}\oplus\eta_b)$ of loops on the real projective space $\mathbb{R}\mathrm{P}(\mathbb{R}\oplus\eta_b)$ with basepoint $[1,0]$. We can identify $S(\mathbb{R}\oplus\xi)$ with the bundle of groups $B\times S(\mathbb{R})$ (with fibre the group $\{1,-1\}$ of real numbers with absolute value 1),

LEMMA 4.1. *There is an obvious identification:*

$$\Phi(\xi,\eta)\xrightarrow{\cong}\Omega_B\mathbb{R}\mathrm{P}(\mathbb{R}\oplus\eta). \quad\square$$

Proof of Theorem 1.1. Let V be an orthogonal real vector space. The free loop space $\mathcal{L}P(V) = \mathrm{map}(S^1, \mathbb{R}\mathrm{P}(V))$ fibres over $\mathbb{R}\mathrm{P}(V)$ by evaluation of a loop at the basepoint $1 \in S^1$. The fibre at $L \in \mathbb{R}\mathrm{P}(V)$ is the based loop space $\Omega(\mathbb{R}\mathrm{P}(V), L)$ of $\mathbb{R}\mathrm{P}(V)$ with L as basepoint. Now

$$\mathbb{R} \oplus \mathrm{Hom}(L, L^{\perp}) = \mathrm{Hom}(L, L \oplus L^{\perp}) = \mathrm{Hom}(L, V) = L^* \otimes V,$$

where $L^{\perp}$ is the orthogonal complement of L in V and $L^* = \mathrm{Hom}(L, \mathbb{R})$, and we have a bijection, given by tensoring a line in V with L^*, from $\mathbb{R}\mathrm{P}(V)$ to $\mathbb{R}\mathrm{P}(L^* \otimes V) = \mathbb{R}\mathrm{P}(\mathbb{R} \oplus \mathrm{Hom}(L, L^{\perp}))$, under which L corresponds to $\mathbb{R}$. We write H for the Hopf line bundle over $\mathbb{R}\mathrm{P}(V)$, with fibre L at $L \in \mathbb{R}\mathrm{P}(V)$. Then $\mathrm{Hom}(H, H^{\perp})$ is the tangent bundle of $\mathbb{R}\mathrm{P}(V)$. The discussion above identifies $\mathcal{L}\mathbb{R}\mathrm{P}(V) \to \mathbb{R}\mathrm{P}(V)$ with the fibrewise loop space $\Omega_B \mathbb{R}\mathrm{P}(\mathbb{R} \oplus \eta)$, where η is the tangent bundle of $B = \mathbb{R}\mathrm{P}(V)$.

Theorem 1.1 follows at once from Corollary 3.2, with $\xi = 0$, and the identification in Lemma 4.1. □

(b) *Complex projective space*

Take ξ to be an orthogonal real line bundle. Let $\mathbb{C}_\xi$ be the associated bundle of fields $\mathbb{R} \oplus \xi$ with multiplication $xy = -\langle x, y\rangle 1$ for $x, y \in \xi_b$. Thus, each fibre of $\mathbb{C}_\xi$ is isomorphic to $\mathbb{C}$, but not canonically so. Suppose that η is a $\mathbb{C}_\xi$-bundle. We may form the associated twisted complex projective bundle $\mathbb{C}\mathrm{P}_\xi(\mathbb{C}_\xi \oplus \eta)$ with basepoint $[1, 0]$ in each fibre.

The sphere bundle $S(\mathbb{C}_\xi)$ is a bundle of groups with fibres isomorphic to the group of complex numbers of modulus 1.

Lemma 4.2. *The natural map*

$$\Phi(\xi, \eta) \xrightarrow{\simeq} \Omega_B \mathbb{C}\mathrm{P}_\xi(\mathbb{C}_\xi \oplus \eta)$$

to the fibrewise loop space of the twisted complex projective bundle is a fibre homotopy equivalence. □

Indeed, the map is a fibrewise homotopy equivalence of fibrewise Hopf spaces with an appropriate multiplication on $\Phi(\xi, \eta)$. (See [**6**] (II.15.33).)

Proof of Theorem 1.2. When V is a complex vector space, we can identify $\mathcal{L}\mathbb{C}\mathrm{P}(V)$, fibred over $B = \mathbb{C}\mathrm{P}(V)$, with $\Omega_B \mathbb{C}\mathrm{P}(\mathbb{C} \oplus \eta)$, where η is the complex tangent bundle of $\mathbb{C}\mathrm{P}(V)$. Taking ξ to be the trivial bundle $B \times \mathbb{R}$, we deduce Theorem 1.2 from Corollary 3.2 and Lemma 4.2. □

Remark 4.3. When $\dim_{\mathbb{C}} V = 2$, the complex projective line $\mathbb{C}\mathrm{P}(V)$ is a 2-sphere $S(W)$ on a 3-dimensional real vector space W and we already have a decomposition, described in [**3**] (Theorem 4.1), of $(\mathcal{L}S(W))_+$ as $S(W)_+ \vee \bigvee_{k \geq 1} S(\eta)^{(2k-1)\zeta}$. This is compatible with the decomposition in Theorem 1.2, as ζ is a trivial (real) line bundle.

(c) *Quaternionic projective space*

We take ξ to be an oriented orthogonal 3-dimensional real vector bundle. Let $\mathbb{H}_\xi = \mathbb{R} \oplus \xi$ be the associated bundle of quaternion algebras, with multiplication given in terms of the scalar and vector products as $xy = -\langle x, y\rangle 1 + x \times y$ for $x, y \in \xi_b$, so that each fibre is isomorphic to $\mathbb{H}$. Suppose that η is a (left) $\mathbb{H}_\xi$-vector bundle. Again we may form a twisted quaternionic projective bundle $\mathbb{H}\mathrm{P}_\xi(\mathbb{H}_\xi \oplus \eta)$ with the basepoint $[1, 0]$ in each fibre.

The sphere bundle $S(\mathbb{H}_\xi)$ is a bundle of groups (with fibre isomorphic to the group of unit quaternions).

LEMMA 4.4. *There is a fibre homotopy equivalence:*

$$\Phi(\xi,\eta) \xrightarrow{\simeq} \Omega_B \mathbb{H}\mathrm{P}_\xi(\mathbb{H}_\xi \oplus \eta). \quad \square$$

PROOF OF THEOREM 1.3. Let V be a (left) $\mathbb{H}$-vector space with a positive-definite inner product. We shall take η to be the tangent bundle of $\mathbb{H}\mathrm{P}(V) = B$, and ξ will be the 3-dimensional oriented orthogonal real vector bundle over $\mathbb{H}\mathrm{P}(V)$ associated with the adjoint representation of the group of quaternions of modulus 1. The Hopf line bundle H over $\mathbb{H}\mathrm{P}(V)$, defined as the tautological subbundle of the trivial bundle $\mathbb{H}\mathrm{P}(V)\times V$, is a left $\mathbb{H}$-module. Its dual $H^* = \mathrm{Hom}_{\mathbb{H}}(H,\mathbb{H})$ is a right $\mathbb{H}$-module, and the bundle of (right) $\mathbb{H}$-endomorphisms of H^* can be written as $\mathbb{H}_\xi$. The tangent bundle $\eta = \mathrm{Hom}_{\mathbb{H}}(H, H^\perp) = H^* \otimes_{\mathbb{H}} H^\perp$ is, thus, a left $\mathbb{H}_\xi$-module. In this way $\mathcal{L}\mathbb{H}\mathrm{P}(V)$, fibred over $B = \mathbb{H}\mathrm{P}(V)$ by the evaluation map, can be identified with $\Omega_B\mathbb{H}\mathrm{P}_\xi(\mathbb{H}_\xi \oplus \eta)$. The proof is completed as in the other two cases. $\square$

REMARK 4.5. When $\dim_{\mathbb{H}} V = 2$, $\mathbb{H}\mathrm{P}(V)$ is a 4-sphere $S(W)$ on a 5-dimensional real vector space W, and we also have a decomposition of $(\mathcal{L}S(W))_+$ as in Remark 4.3. Here $\eta = \mathrm{Hom}_{\mathbb{H}}(H, H^\perp)$ and $\mathrm{Hom}_{\mathbb{H}}(H,H)$, being the opposite algebra of $\mathbb{H}_\xi$, is just $\mathbb{R}\oplus\xi$. Over $S(\eta)$ we have an isomorphism $H \to H^\perp$, which provides an isomorphism between ζ and the pull-back of ξ. Thus we may identify $(2l-1)\zeta$ with $l\xi \oplus (l-1)\zeta$, and this decomposition is consistent with Theorem 1.3.

The fibrewise splittings obtained by applying Theorem 3.1, rather than Corollary 3.2, to the three examples are also of interest and lead to computations of the additive structure of the loop homology of projective spaces. See [**5**] (Example 3.11).

References

1. S. Bauer and M. C. Crabb, *Polynomial loops on spheres*, Quart. J. Math. Oxford **55** (2004), 391–409.
2. S. Bauer, M. C. Crabb, and M. Spreafico, *The classifying space of the gauge group of an* SO(3)*-bundle over* S^2, Proc. Roy. Soc. Edinburgh **131A** (2001), 767–783.
3. ———, *The space of free loops on a real projective space*, Contemp. Math. ("Groups of homotopy self-equivalences and related topics", edited by K. Maruyama and J.W. Rutter) **274** (2001), 33–38.
4. M. B. Bökstedt and I. Ottosen, *The suspended free loop space of a symmetric space*, arXiv: math.AT/0511086, (2005).
5. M. C. Crabb, *Loop homology as fibrewise homology*, Proc. Edinburgh Math. Soc. **51** (2008), 27–44.
6. M. C. Crabb and I. M. James, *Fibrewise homotopy theory*, Springer, London, 1998.
7. K. Yamaguchi, *Spaces of free loops on real projective spaces*, Kyushu J. Math. **59** (2005), 145–153.

INSTITUTE OF MATHEMATICS, UNIVERSITY OF ABERDEEN, ABERDEEN, AB24 3UE, UK
E-mail address: `m.crabb@maths.abdn.ac.uk`

Contemporary Mathematics
Volume **519**, 2010

Rational homotopy of symmetric products and Spaces of finite subsets

Yves Félix and Daniel Tanré

ABSTRACT. We describe Sullivan models of the symmetric products of spaces and of some symmetric fat diagonals. From the determination of models of some spaces of finite subsets, we verify a Tuffley's conjecture in low ranges for rational spaces. Extending a theorem of Handel, we prove also the triviality of the inclusion of a space X in the space of $(n+2)$-th finite subsets of X, when X is of Lusternik-Schnirelmann category less than or equal to n.

The properties of configuration spaces of a space X have deserved many studies where the use of algebraic models was limited by the fact that the homotopy type of configuration spaces of points in X is not a homotopy invariant of X. The spaces of finite subsets, $\exp^n X$, of a space X are a substitute of configuration spaces where this anomaly does not exist any more, see [**14**]. They are related to configuration spaces by the fact that the cofiber of the canonical inclusion $\exp^{n-1} X \hookrightarrow \exp^n X$ is the one point compactification of the configuration space of unordered n-uples of distinct points of a compact space X, see [**14**, Proposition 2.3].

Recall that the n-th finite subsets space, $\exp^n X$, of a topological (non-empty) space X is the space of non-empty subsets of size at most n, topologised as the quotient of X^n by the surjective map $X^n \to \exp^n X$, $(x_1, \dots, x_n) \mapsto \{x_1\} \cup \dots \cup \{x_n\}$. The introduction of the spaces of finite subsets of a space is due to K. Borsuk and S. Ulam [**3**], [**2**]. Their study was continued by R. Bott [**4**] and more recently by D. Handel [**14**], R. Biro [**1**], C. Tuffley [**26**], [**27**], [**28**], J. Mostovoy [**19**], [**20**], S. Rose [**22**], S. Kallel and D. Sjerve [**16**], [**17**]. As $X \mapsto \exp^n X$ are homotopy functors, the elaboration of rational models for them is an open challenge and we develop here the first steps in this direction.

First, we quote a non rational result inspired by a theorem of Handel. If X is pointed, we denote by $\exp^n_* X$ the subspace of $\exp^n X$ formed from the subsets that contain the basepoint and by $\iota^n_X \colon X \to \exp^n_* X$ the map that adjoins the basepoint to each subset. This space $\exp^n_* X$ is often used as a first step in the study of $\exp^n X$. For any arc-connected pointed space $(X,*)$, Handel shows in ([**14**, Theorem 4.1 et Theorem 4.2]) that the two maps $\pi_j(\exp^n_* X) \to \pi_j(\exp^{2n-1}_* X)$ and $\pi_j(\exp^n X) \to \pi_j(\exp^{2n+1} X)$, induced by the inclusion, are the zero map. From these results, Handel deduces that the space $\exp^\infty X = \cup_{n\geq 1} \exp^n X$ is weakly

2000 *Mathematics Subject Classification.* Primary 55P62 ; 55S15; 55M30.

©2010 American Mathematical Society

contractible. In this paper, we show that the triviality is also at the level of some inclusions between finite subset spaces.

THEOREM 1. *If the Lusternik-Schnirelmann category of a pointed CW-complex* $(X,*)$ *is less than or equal to* n, *the following inclusions are homotopically trivial.*

(1) $\iota_X^{n+2}\colon X \to \exp_*^{n+2} X$,
(2) $X \to \exp^{n+2} X$,
(3) $\exp_*^{k+1} X \to \exp_*^{(n+1)k+1} X$.

The rest of the paper is concerned with Sullivan models for which we refer to [**9**], [**10**] or [**25**]. As $\exp^n X$ can be inductively built from the symmetric product $\operatorname{Sp}^n X$ and its fat diagonal $\mathbb{\Delta}_S^n X$ (see Definition 3.1 and the pushout 4.2), we begin by a study of models of those spaces. First, we recall some known basic facts for the models of spaces with an action of a finite group. We apply it to symmetric products and prove the next result.

THEOREM 2. *Let* (A,d) *be a connected model of a connected space* X. *Then the cdga* $\wedge^n(A,d)$ *is a model of* $\operatorname{Sp}^n X$, *with a multiplicative law on* $\wedge^n A$ *given by*

$$(a_1 \wedge \cdots \wedge a_n) * (b_1 \wedge \cdots \wedge b_n) = \sum_{\sigma \in \mathcal{S}_n} \pm (a_1 \bullet b_{\sigma(1)}) \wedge \cdots \wedge (a_n \bullet b_{\sigma(n)}),$$

where $\bullet$ *is the multiplicative law of* A, $\mathcal{S}_n$ *is the symmetric group and* $\pm$ *means the Koszul sign.*

From this model, we deduce the (additive) rational cohomology of the symmetric product described by D. Zagier in [**30**]. For instance, we get a short proof for the determination of the Poincaré polynomial of $\operatorname{Sp}^n X$ in terms of the Betti numbers of X, previously obtained by I.G. MacDonald in [**18**]. We give a description of the rational homotopy type of $\operatorname{Sp}^n \Sigma X$ and compute a model of $\operatorname{Sp}^2 \mathbb{C}P^2$, see Proposition 2.5 and Example 2.6. We get also an inductive construction of the fat diagonals $\mathbb{\Delta}_S^n X$ (see Proposition 3.2) which allows the determination of $\mathbb{\Delta}_S^n S^{2k+1}$ for any n. This works ends with the determination of the rational homotopy type of $\exp^3 \Sigma X$ and $\exp^4 \Sigma X$. An important conjecture of the theory, due to Tuffley ([**28**]), states as follows.

TUFFLEY'S CONJECTURE. *If* X *is an* r*-connected CW-complex, the space* $\exp^n X$ *is* $(n+r-2)$*-connected.*

In fact, Tuffley proves that $\exp^n X$ is $(n-2)$-connected if X is connected and $(n-1)$-connected if X is simply connected. He shows also that it is sufficient to prove the conjecture for a wedge of spheres ([**29**, Theorem 2]). From that observation and from our results, we deduce that the Tuffley's conjecture is true for the rationalisation of $\exp^3 X$ and $\exp^4 X$. Tuffley's conjecture has also been verified for $n=3$ in recent work of S. Kallel and D. Sjerve, [**17**].

Our program is carried out in Sections 1-6 below, whose headings are self-explanatory.

CONTENTS

1. Algebraic models of G-spaces with G a finite group

In this section, we recall some basic facts on Sullivan models of spaces on which acts a finite group G.

By definition, a G-dga is a differential graded algebra (A, d_A), with $H^0(A, d_A) = \mathbb{Q}$, on which G acts by dga maps. If the algebra A is commutative graded, we use the expression G-cdga. The invariant subspace $(A, d_A)^G$ of a G-dga defines a sub-dga of (A, d_A) and if $f\colon (A, d_A) \to (B, d_B)$ is a G-equivariant quasi-isomorphism, we have the following well-known properties (see [**11**, Section 1] for instance):

- $f^G\colon (A, d_A)^G \to (B, d_B)^G$ is also a quasi-isomorphism,
- $H((A, d_A)^G) = (H(A, d_A))^G$.

If V is a graded $\mathbb{Q}$-vector space, we denote by $T(V)$ the free associative graded algebra on V and by $\wedge V$ the free commutative graded algebra on V. Any G-cdga, (A, d_A), admits a minimal model

$$\varphi\colon (\wedge V, d) \to (A, d_A)$$

with an action of G on $(\wedge V, d)$ making the map φ G-equivariant, see [**13**]. This model is unique up to G-isomorphisms and we call it the minimal G-model of (A, d_A). More generally, a *G-model* of (A, d_A) is any G-cdga having the same minimal G-model as (A, d_A).

We apply these algebraic data to spaces with a G-action. Let X be a simplicial complex with a (simplicial) action of G. Recall from G. Bredon ([**5**, Page 115]) that the action is *regular* if, for any $g_0, \dots, g_n$ in G and simplices $(v_0, \dots, v_n)$, $(g_0v_0, \dots, g_nv_n)$ of X, there exists an element $g \in G$ such that $gv_i = g_iv_i$ for all i. By [**5**, Proposition 1.1, Page 116], the induced action on the second barycentric subdivision is always regular. Here, *by a G-space, we mean a connected simplicial complex on which G acts regularly.*

Denote by $C(X)$ the oriented chain complex of a G-space X and observe that $C(X)$ is a module over the group ring $\mathbb{Z}[G]$ of G. The canonical simplicial map $\rho\colon X \to X/G$ induces $\rho_*\colon C(X) \to C(X/G)$. Define now $\sigma\colon C(X) \to C(X)$, $c \mapsto \sum_{g\in G} gc$. One has $\mathrm{Ker}\,\sigma = \mathrm{Ker}\,\rho_*$. Therefore σ induces $\overline{\sigma}$

$$\begin{array}{ccc} C(X) & \xrightarrow{\ \sigma\ } & C(X) \\ {\scriptstyle \rho_*}\big\downarrow & \nearrow_{\overline{\sigma}} & \\ C(X/G) & & \end{array}$$

such that $\overline{\sigma} \circ \rho_* = \sigma$. Bredon proves the next result.

PROPOSITION 1.1 ([**5**, Page 120]). *Let $\Bbbk$ be a field of characteristic that does not divide the order of G and let X be a G-space. Then there are isomorphisms*

$$C_*(X/G; \Bbbk) \cong C_*(X; \Bbbk)/G \text{ and } C^*(X/G; \Bbbk) \cong C^*(X; \Bbbk)^G.$$

When X is a G-space, the (finite) group G acts on the Sullivan algebra of PL-forms on X, $A_{PL}(X)$, [**24**]. As in Proposition 1.1, the cdga's $A_{PL}(X/G)$ and $A_{PL}(X)^G$ are isomorphic and this isomorphism gives models of the quotient X/G from G-models of X, see [**11**, Proposition 7].

PROPOSITION 1.2. *Let (A, d_A) be a G-model of the G-space X, then the cdga $(A, d_A)^G$ is a model of X/G.*

2. Symmetric products and Proof of Theorem 2

In this section, we describe Sullivan models of symmetric products of a space.

DEFINITION 2.1. Let X be a space. The *symmetric product,* $\mathrm{Sp}^n X$, is the quotient of the product X^n by the action of the symmetric group $\mathcal{S}_n$,

$$\sigma(x_1, \dots, x_n) = (x_{\sigma(1)}, \dots, x_{\sigma(n)}),$$

with $\sigma \in \mathcal{S}_n$ and $(x_1, \dots, x_n) \in X^n$. We denote by $\langle x_1, \dots, x_n\rangle \in \mathrm{Sp}^n X$ the class associated to the element $(x_1, \dots, x_n) \in X^n$ and by $\rho_n \colon X^n \to \mathrm{Sp}^n X$ the canonical projection.

As first examples of symmetric products, recall that $\mathrm{Sp}^n S^1$ is homotopy equivalent to S^1, $\mathrm{Sp}^n S^2$ diffeomorphic to $\mathbb{C}P^n$ and $\mathrm{Sp}^n \mathbb{R}P^2$ diffeomorphic to $\mathbb{R}P^{2n}$, see [**21**], [**19**] and [**8**] for more details and historical comments.

If the space X is pointed by $*$, the adding of $*$ gives an inclusion of $\mathrm{Sp}^{n-1} X$ in $\mathrm{Sp}^n X$. By definition, the infinite symmetric product $\mathrm{Sp}^\infty X$ is the direct limit of the spaces $\mathrm{Sp}^n X$. If X is connected, this infinite product is a product of Eilenberg-McLane spaces, $\mathrm{Sp}^\infty(X,*) \simeq \prod_i K(\tilde{H}_i(X;\mathbb{Z}), i)$, see [**7**]. By convention, we set $\mathrm{Sp}^0 X = *$.

PROOF OF THEOREM 2. For any cdga (A, d), and any integer $n \geq 1$, the symmetric group $\mathcal{S}_n$ acts on the tensor product $\otimes^n(A, d)$ by permuting the factors,

$$\sigma(a_1 \otimes \dots \otimes a_n) = \pm(a_{\sigma(1)} \otimes \dots \otimes a_{\sigma(n)}),$$

where the sign comes from the ordinary rule of permutation of graded objects. If (A, d) is a model of X, as a direct consequence of Proposition 1.2, the cdga $(\otimes^n(A, d))^{\mathcal{S}_n}$ is a model of $\mathrm{Sp}^n X$.

To make the structure of the cdga $(\otimes^n(A, d))^{\mathcal{S}_n}$ more precise, recall the existence of an isomorphism $\wedge^n A \cong (\otimes^n A)^{\mathcal{S}_n}$ defined by

$$a_1 \wedge \dots \wedge a_n \mapsto \sum_{\sigma \in \mathcal{S}_n} \pm a_{\sigma(1)} \otimes \dots \otimes a_{\sigma(n)}.$$

Denote by $\bullet$ the multiplicative law of A. The isomorphism above transforms the product on $(\otimes^n(A, d))^{\mathcal{S}_n}$ to the following law of algebra on $\wedge^n A$,

$$(a_1 \wedge \dots \wedge a_n) * (b_1 \wedge \dots \wedge b_n) = \sum_{\sigma \in \mathcal{S}_n} \pm(a_1 \bullet b_{\sigma(1)}) \wedge \dots \wedge (a_n \bullet b_{\sigma(n)}),$$

with $\frac{1}{n!}(1 \wedge \dots \wedge 1)$ as identity. □

Theorem 2 implies that the cohomology vector space of $\mathrm{Sp}^n X$ depends only on the cohomology vector space of X and gives the Poincaré polynomial of a symmetric product, a formula due to I.G. Macdonald.

COROLLARY 2.2. [**18**] *Denote by* $(b_i)_{i\geq 0}$ *the Betti numbers of a space* X. *Then the Poincaré polynomial of the space* $\mathrm{Sp}^n X$ *is the coefficient of* t^n *in the power series associated to:*

$$\frac{(1+xt)^{b_1}(1+x^3t)^{b_3}\dots}{(1-t)^{b_0}(1-x^2t)^{b_2}(1-x^4t)^{b_4}\dots} = \frac{\prod_{i=0}(1+x^{2i+1}t)^{b_{2i+1}}}{\prod_{i=0}(1-x^{2i}t)^{b_{2i}}}.$$

PROOF. This is a direct consequence of the well-known Poincaré polynomial of the algebra $\wedge a$ where a is a generator of odd or even degree. □

Theorem 2 implies also immediately the following property.

COROLLARY 2.3. *If* X *is formal, the symmetric product* $\mathrm{Sp}^n X$ *is formal, for any* $n \geq 1$.

In the particular case of a suspension, the model of Theorem 2 can be expressed in a simple manner.

COROLLARY 2.4. *If* X *is a suspension having rational cohomology* H, *we have an isomorphism of algebras* $\psi\colon (\wedge H^+/\wedge^{>n} H^+,0) \to (\wedge^n H,0)$, *defined by* $\psi(1) = \dfrac{1}{n!}(1\wedge\dots\wedge 1)$ *and* $\psi(x) = \dfrac{x\wedge 1\wedge\dots\wedge 1}{(n-1)!}$ *if* $x\in H^+$. *In particular, the cdga* $(\wedge H^+/\wedge^{>n} H^+,0)$ *is a model of* $\mathrm{Sp}^n X$.

PROOF. As $H = \mathbb{Q}\oplus H^+$ with $H^+ = \oplus_{i\geq 1}H^i$, the algebra $\wedge^n H$ is isomorphic, as vector space, to $\wedge^{\leq n}H^+ \cong \wedge H^+/\wedge^{>n} H^+$. The product being null on H^+, the law of algebra of the quotient $\wedge H^+/\wedge^{>n} H^+$ corresponds, by ψ, to the product $*$ defined on $\wedge^n H$ in Theorem 2 and the result follows. □

In the case of a sphere S^p, Corollary 2.4 gives $(\wedge a/\wedge^{>n} a,0)$, with a of degree p, as model of $\mathrm{Sp}^n S^p$. More precisely,

- if $p=2k+1$, $\mathbb{Q}\oplus a\mathbb{Q}$ is a model of S^{2k+1} and we have $\mathrm{Sp}^n S^{2k+1} \simeq_{\mathbb{Q}} S^{2k+1}$, for any $n\geq 1$;
- if $p=2k$, we denote by P^nS^{2k} the rational space having $(\wedge a/(a^{n+1}),0)$ as model, with a of degree $2k$. (Observe that $P^nS^2 = \mathbb{C}P^n$.) For any $n\geq 1$, we have $\mathrm{Sp}^n S^{2k} \simeq_{\mathbb{Q}} P^nS^{2k}$ and $P^\infty S^{2k} = K(\mathbb{Q},2k)$.

The next result gives the precise rational homotopy type of the symmetric product of a general suspension. To state it, we first define a filtration on a product of filtered spaces, as follows. If the spaces X and Y are given with an increasing filtration $X^{(0)} = * \subset X^{(1)} \subset \dots \subset X^{(l-1)} \subset X = X^{(l)}$, $Y^{(0)} = * \subset Y^{(1)} \subset \dots \subset Y^{(k-1)} \subset Y = Y^{(k)}$, the product space is given with the filtration

$$(X\times Y)^{(i)} = \bigcup_{i_1+i_2=i} X^{(i_1)}\times Y^{(i_2)}.$$

We endow the odd sphere with the trivial filtration, $(S^{2i+1})^{(1)} = S^{2i+1}$, and the space P^nS^{2k} with the filtration defined by: $(P^nS^{2k})^{(l)}$ is the $2kl$-skeleton of P^nS^{2k}.

PROPOSITION 2.5. *Let* X *be a suspension of cohomology* $H = H^*(X;\mathbb{Q})$. *Let* $(\alpha_1,\dots,\alpha_l)$ *be a basis of* H^{odd} *and* $(\beta_1,\dots,\beta_k)$ *be a basis of* H^{even}. *Then, the commutative graded algebra* $(\wedge H^+/\wedge^{>n} H^+,0)$ *is a model of* $\mathrm{Sp}^n X$ *and we have*

$$\mathrm{Sp}^n(X) \simeq_{\mathbb{Q}} \left(\prod_{i=1}^{l} S^{|\alpha_i|}\times\prod_{j=1}^{k} P^nS^{|\beta_j|}\right)^{(n)}.$$

Moreover a model of the projection $\rho_{n,q}\colon X^n \times \mathrm{Sp}^q X \to \mathrm{Sp}^{n+q} X$ *is the cdga's map* $\varphi_{n,q}\colon (\wedge H^+/\wedge^{>n+q} H^+, 0) \to (H^{\otimes n} \otimes (\wedge H^+/\wedge^{>n} H^+), 0)$, *which sends* $x \in H^+$ *on* $(x \otimes 1 \otimes \ldots \otimes 1) \otimes 1 + \cdots + (1 \otimes \ldots \otimes 1 \otimes x) \otimes 1 + (1 \otimes \ldots \otimes 1) \otimes x$.

For instance, Proposition 2.5 implies $\mathrm{Sp}^n(S^3 \vee S^5) \simeq_{\mathbb{Q}} S^3 \times S^5$ for any $n \geq 2$, $\mathrm{Sp}^n(S^3 \vee S^5 \vee S^7) \simeq_{\mathbb{Q}} (S^3 \times S^5 \times S^7)$ for any $n \geq 3$ and $\mathrm{Sp}^2(S^3 \vee S^5 \vee S^7)$ has the rational homotopy type of the fat wedge of (S^3, S^5, S^7). Observe also that Proposition 2.5 is coherent with results of Snaith and Ucci, see [**23**, Proposition 2.2] and of B. W. Ong on $\mathrm{Sp}^n(\vee S^1)$, [**21**].

PROOF. The first part of the statement comes directly from Corollary 2.4. For the second part, we start with the canonical surjection $\rho_n = \rho_{n,0}\colon X^n \to \mathrm{Sp}^n X$, of model the canonical inclusion $(H^{\otimes n})^{\mathcal{S}_n} \hookrightarrow H^{\otimes n}$. Now we replace $(H^{\otimes n})^{\mathcal{S}_n}$ by the cga $\wedge H^+/\wedge^{>n} H^+$ as in Corollary 2.4. The sequence of isomorphisms

$$\wedge H^+/\wedge^{>n} H^+ \xrightarrow{\cong} \wedge^n H \xrightarrow{\cong} (H^{\otimes n})^{\mathcal{S}_n}$$

is defined by $x \rightarrowtail \dfrac{x \wedge 1 \wedge \ldots \wedge 1}{(n-1)!} \rightarrowtail x \otimes 1 \otimes \ldots \otimes 1 + \cdots + 1 \otimes \ldots \otimes 1 \otimes x.$

Composed with these isomorphisms, the inclusion $(H^{\otimes n})^{\mathcal{S}_n} \hookrightarrow H^{\otimes n}$ becomes $\varphi_{n,0}\colon \wedge H^+/\wedge^{>n} H^+ \to H^{\otimes n}$, $\varphi_{n,0}(x) = x \otimes 1 \otimes \ldots \otimes 1 + \cdots + 1 \otimes \ldots \otimes x$.

In the general case of the projection $\rho_{n,q}\colon X^n \times \mathrm{Sp}^q X \to \mathrm{Sp}^{n+q} X$, we have a commutative diagram

$$\begin{array}{ccc}
X^{n+q} & = & X^n \times X^q \\
\downarrow{\scriptstyle \rho_{n+q}} & & \downarrow{\scriptstyle 1\times\rho_q} \\
\mathrm{Sp}^{n+q} X & \xleftarrow{\rho_{n,q}} & X^n \times \mathrm{Sp}^q X
\end{array}$$

which induces in cohomology the diagram

$$\begin{array}{ccc}
H^{\otimes(n+q)} & = & H^{\otimes n} \otimes H^{\otimes q} \\
\uparrow{\scriptstyle \varphi_{n+q,0}} & & \uparrow{\scriptstyle 1\otimes\varphi_{q,0}} \\
\wedge H^+/\wedge^{>n+q} H^+ & \xrightarrow{\varphi_{n,q}} & H^{\otimes n} \otimes (\wedge H^+/\wedge^{>q} H^+).
\end{array}$$

Since $\varphi_{n+q,0}$ and $\varphi_{q,0}$ are injective, the morphism $\varphi_{n,q}$ is given on $x \in H^+$ by $\varphi_{n,q}(x) = (x \otimes 1 \otimes \ldots \otimes 1) \otimes 1 + \cdots + (1 \otimes \ldots \otimes 1 \otimes x) \otimes 1 + (1 \otimes \ldots \otimes 1) \otimes x$. □

EXAMPLE 2.6. We study now $\mathrm{Sp}^2(\mathbb{C}P^2)$. Its vector space of cohomology is $\wedge^2(1, \beta_1, \beta_2)$ with $|\beta_1| = 2$ and $|\beta_2| = 4$. We denote $a = \beta_1 \wedge 1$, $b = \beta_2 \wedge 1$, $c = \beta_1 \wedge \beta_1$, $e = \beta_1 \wedge \beta_2$, $f = \beta_2 \wedge \beta_2$. The law of algebra is given by Theorem 2 as follows, $a^2 = (\beta_1 \wedge 1) * (\beta_1 \wedge 1) = (\beta_1 \bullet \beta_1) \wedge 1 + \beta_1 \wedge \beta_1 = \beta_2 \wedge 1 + \beta_1 \wedge \beta_1 = b + c$.

We compute similarly the other products and get $e = ab$, $ac = 2e$, $ae = b^2 = f$, $c^2 = 2f$ and zero for the other ones. The algebra structure can be described by $H^*(\mathrm{Sp}^2(\mathbb{C}P^2); \mathbb{Q}) = \{a, b, a^2, a^3, a^4\}\mathbb{Q}$, with $a^3 = 3ab$ and $b^2 = a^2 b = a^4/3$.

The projection $\{a, b, a^2, a^3, a^4\}\mathbb{Q} \to b\mathbb{Q}$ gives a map $S^4 \to \mathrm{Sp}^2 \mathbb{C}P^2$ whose rational homotopy cofiber is $\mathbb{C}P^4$; the associated long exact sequence of this cofibration splits in short ones.

REMARK 2.7. If (A, d_A) is a model of X, the canonical inclusion $\mathrm{Sp}^{n-1} X \to \mathrm{Sp}^n X$, obtained by adding a fixed point x_0, has for model the cdga's map, $\wedge^n H \to$

$\wedge^{n-1}H$, defined by $a_1 \wedge \cdots \wedge a_n \mapsto \sum_{i=1}^n \varepsilon(a_i)\, a_1 \wedge \ldots \wedge a_{i-1} \wedge a_{i+1} \wedge \ldots \wedge a_n$, where $\varepsilon\colon A \to \mathbb{Q}$ is the augmentation corresponding to $\{x_0\} \hookrightarrow X$. In the case of an odd sphere, the adding of x_0 is clearly a homotopy equivalence, $\mathrm{Sp}^m S^{2k+1} \cong \mathrm{Sp}^{m+1} S^{2k+1}$.

3. Fat diagonal of the symmetric product

This section contains an inductive description of the fat diagonal of a symmetric product and the determination of its rational homotopy type in the case of an odd sphere.

DEFINITION 3.1. The *fat diagonal*, $\mathbb{\Delta}^n_{\mathcal{S}} X$, of the symmetric space $\mathrm{Sp}^n X$ is defined by

$$\mathbb{\Delta}^n_{\mathcal{S}} X = \{\langle x_1, \ldots, x_n\rangle \in \mathrm{Sp}^n(X) \mid \text{there exist } i \in \{1,\ldots,n\} \text{ and } j \in \{1,\ldots,n\} \text{ such that } i \neq j \text{ and } x_i = x_j\}.$$

The fat diagonal can be easily determined in low ranges, the maps $X \to \mathbb{\Delta}^2_{\mathcal{S}} X$, $x \mapsto \langle x, x\rangle$, and $X \times X \to \mathbb{\Delta}^3_{\mathcal{S}} X$, $(x,y) \mapsto \langle x,x,y\rangle$, being homeomorphisms if X is a finite CW-complex. For the study of the general case, we introduce the subspace $\mathbb{\Delta}^{n,k}_{\mathcal{S}} X$ of $\mathrm{Sp}^n X$, defined as the image of the product $X^k \times X^{n-2k}$ by the map

$$(x_1,\ldots,x_k,y_1,y_2,\ldots,y_{n-2k}) \mapsto (x_1,x_1,x_2,x_2,\ldots,x_k,x_k,y_1,y_2,\ldots,y_{n-2k}).$$

We clearly have $\mathbb{\Delta}^n_{\mathcal{S}} X = \mathbb{\Delta}^{n,1}_{\mathcal{S}} X$, $\mathbb{\Delta}^{2k,k}_{\mathcal{S}} X \simeq \mathrm{Sp}^k X$ and $\mathbb{\Delta}^{2k+1,k}_{\mathcal{S}} X \simeq X \times \mathrm{Sp}^k X$. These diagonal spaces can be constructed inductively as follows.

PROPOSITION 3.2. *We have a homotopy pushout*

$$\begin{array}{ccc} \mathrm{Sp}^k X \times \mathbb{\Delta}^{n-2k}_{\mathcal{S}} X & \xhookrightarrow{\;g\;} & \mathrm{Sp}^k X \times \mathrm{Sp}^{n-2k} X \\ \varphi\big\downarrow & & \big\downarrow\psi \\ \mathbb{\Delta}^{n,k+1}_{\mathcal{S}} X & \xhookrightarrow{\;f\;} & \mathbb{\Delta}^{n,k}_{\mathcal{S}} X \end{array}$$

where g and f are canonical inclusions, φ and ψ consist to double the first k coordinates.

PROOF. This square is clearly a pushout. As $\mathbb{\Delta}^{n-2k}_{\mathcal{S}} X \hookrightarrow \mathrm{Sp}^{n-2k} X$ is a cofibration, it is also a homotopy pushout. □

This inductive construction allows the determination of the fat diagonal in the case of an odd sphere.

PROPOSITION 3.3. *The symmetric fat diagonals of an* odd-dimensional *sphere, S^n, satisfy the next properties.*

(1) *The canonical inclusions*

$$\mathbb{\Delta}^{2m,m}_{\mathcal{S}} S^n \subset \mathbb{\Delta}^{2m,m-1}_{\mathcal{S}} S^n \subset \cdots \subset \mathbb{\Delta}^{2m,1}_{\mathcal{S}} S^n \subset \mathrm{Sp}^{2m} S^n$$

are rational homotopy equivalences. This implies $\mathbb{\Delta}^{2m,k}_{\mathcal{S}} S^n \simeq_{\mathbb{Q}} S^n$ for all (m,k) with $m \geq k \geq 1$.

(2) *For any $m \geq 1$ and $1 \leq k \leq m$, one has*

$$\mathbb{\Delta}^{2m+1,k}_{\mathcal{S}} S^n \simeq_{\mathbb{Q}} S^n \times (*^{m-k+1} S^n) \simeq S^n \times S^{(n+1)(m+1-k)-1}.$$

Moreover, the inclusion of $\mathbb{\Delta}_S^{2m+1,k+1} S^n$ *in* $\mathbb{\Delta}_S^{2m+1,k} S^n$ *restricts to the identity on* S^n.

(3) *Let* x_0 *be a point in* S^n. *The map* $\delta_{x_0}\colon \mathbb{\Delta}_S^{2m-1,k} S^n \to \mathbb{\Delta}_S^{2m+1,k+1} S^n$, *obtained by adding* (x_0, x_0), *is a rational homotopy equivalence, for any* k, $1 \le k \le m-1$.

PROOF. (1) We work by induction on m. For $m = 1$, the inclusion $\mathbb{\Delta}_S^2 S^n \simeq S^n \to \mathrm{Sp}^2 S^n$ is a rational homotopy equivalence. Suppose that the result is true for q, $q < m$, i.e. $\mathbb{\Delta}_S^{2m-2i} S^n \to \mathrm{Sp}^{2m-2i} S^n$ is a rational homotopy equivalence for any i, $1 \le i \le m$. In the pushout,

$$\begin{array}{ccc}
\mathrm{Sp}^i S^n \times \mathbb{\Delta}_S^{2m-2i} S^n & \hookrightarrow & \mathrm{Sp}^i S^n \times \mathrm{Sp}^{2m-2i} S^n \\
\downarrow & & \downarrow \\
\mathbb{\Delta}_S^{2m,i+1} S^n & \longrightarrow & \mathbb{\Delta}_S^{2m,i} S^n
\end{array}$$

the top map is a rational homotopy equivalence. Thus, the bottom map is a rational homotopy equivalence also. The composite of the canonical maps

$$\mathrm{Sp}^m S^n \to \mathbb{\Delta}_S^{2m,m} S^n \to \mathbb{\Delta}_S^{2m,m-1} S^n \to \cdots \to \mathbb{\Delta}_S^{2m} S^n \to \mathrm{Sp}^{2m} S^n$$

is the square map $\mathrm{Sp}^m S^n \to \mathrm{Sp}^{2m} S^n$. The commutativity of the diagram

$$\begin{array}{ccc}
(S^n)^m & \xrightarrow{\text{square}} & (S^n)^{2m} \\
{\scriptstyle \rho_m}\downarrow & & \downarrow{\scriptstyle \rho_{2m}} \\
\mathrm{Sp}^m S^n & \xrightarrow{\text{square}} & \mathrm{Sp}^{2m} S^n
\end{array}$$

shows that the map induced in cohomology by the square map,

$$H^*(\text{square})\colon H^*(\mathrm{Sp}^{2m} S^n) = \wedge x \to H^*(\mathrm{Sp}^m S^n) = \wedge x,$$

is the multiplication by 2 and the square map is a rational homotopy equivalence. Therefore, the inclusion $\mathbb{\Delta}_S^{2m} S^n \to \mathrm{Sp}^m S^n$ is also a rational homotopy equivalence.

(2) As we observed before, the result is true for $k = m$ and we can begin with $k = m-1$. We consider the homotopy pushout

$$\begin{array}{ccc}
\mathrm{Sp}^{m-1} S^n \times \mathbb{\Delta}_S^3 S^n & \xrightarrow{g_1} & \mathrm{Sp}^{m-1} S^n \times \mathrm{Sp}^3 S^n \\
{\scriptstyle g_2}\downarrow & & \downarrow{\scriptstyle g_4} \\
\mathbb{\Delta}_S^{2m+1,m} S^n & \hookrightarrow & \mathbb{\Delta}_S^{2m+1,m-1} S^n,
\end{array}$$

where g_2 is the square map on the factor $\mathrm{Sp}^{m-1} S^n$. With the notation of Proposition 2.5, the map g_2 can be expressed as a product of $\rho_{m-1,1}$ with the identity map:

$$\mathrm{Sp}^{m-1} S^n \times \mathbb{\Delta}_S^3 S^n \simeq (\mathrm{Sp}^{m-1} S^n) \times S^n \times S^n \to \mathbb{\Delta}_S^{2m+1,m} S^n \simeq (\mathrm{Sp}^m S^n) \times S^n.$$

Therefore, a model of g_2 is given by $\varphi_2\colon (\wedge(u,v),0) \to (\wedge(x,y,z),0)$, $\varphi_2(u) = x+y$, $\varphi_2(v) = z$, where x, y, z, u and v are of degree n. The previous question gives a model of g_1 as $\varphi_1\colon (\wedge(a,b),0) \to (\wedge(x,y,z),0)$, $\varphi_1(a) = x$, $\varphi_1(b) = y+z$.

By making the change of generators, $u' = u+v$, $v' = v$, $a' = a$, $b' = a+b$, $x' = x$, $y' = x+y+z$, $z' = z$, we get $\varphi_2(u') = y'$, $\varphi_2(v') = z'$, $\varphi_1(a') = x'$ and

$\varphi_1(b') = y'$. This implies that $g_1,\, g_2 \colon S^n \times S^n \times S^n \to S^n \times S^n$ are respectively the projection on the two first factors and on the two last factors.

Therefore the homotopy pushout $\boldsymbol{\Delta}_S^{2m+1,m-1} S^n$ has the same rational homotopy type than $S^n \times (S^n * S^n)$.

For a general k, $1 \leq k \leq m-1$, we use a descending induction. The result is already proved for $k = m-1$. For the induction, we consider the homotopy pushout:

$$\begin{array}{ccc}
\mathrm{Sp}^k S^n \times \boldsymbol{\Delta}_S^{2m+1-2k} S^n & \xrightarrow{g_1} & \mathrm{Sp}^k S^n \times \mathrm{Sp}^{2m+1-2k} S^n \\
{\scriptstyle g_2}\downarrow & & \downarrow{\scriptstyle g_4} \\
\boldsymbol{\Delta}_S^{2m+1,k+1} S^n & \longrightarrow & \boldsymbol{\Delta}_S^{2m+1,k} S^n.
\end{array}$$

As in the previous case, using the induction hypothesis, we show that this square is rationally homotopic to a homotopy pushout

$$\begin{array}{ccc}
S^n \times (S^n \times (*^{m-k-1} S^n)) & \xrightarrow{g_1} & S^n \times S^n \\
{\scriptstyle g_2}\downarrow & & \downarrow \\
S^n \times (*^{m-k-1} S^n) & \longrightarrow & \boldsymbol{\Delta}_S^{2m+1,k} S^n,
\end{array}$$

where g_1 and g_2 are the product of the identity map with a projection. Thus the homotopy pushout $\boldsymbol{\Delta}_S^{2m+1,k} S^n$ has the same rational homotopy type than $S^n \times (*^{m-k+1} S^n) \simeq S^n \times S^{(n+1)(m+1-k)-1}$.

(3) Statement (2) implies that the two spaces, $\boldsymbol{\Delta}_S^{2m-1,k} S^n$ and $\boldsymbol{\Delta}_S^{2m+1,k+1} S^n$, have the same rational homotopy type. We show that a rational homotopy equivalence between them can be obtained by adding (x_0, x_0) where x_0 is a point of S^n. We prove this result by a descending induction for k, starting from $m-1$ and ending to 1. For $k = m-1$, we consider the commutative diagram

$$\begin{array}{ccc}
\boldsymbol{\Delta}_S^{2m-1,m-1} S^n & \xrightarrow{g_1} & \boldsymbol{\Delta}_S^{2m+1,m} S^n \\
\uparrow{\scriptstyle \cong} & & {\scriptstyle \cong}\uparrow \\
\mathrm{Sp}^{m-1} S^n \times S^n & \xrightarrow{g_2 \times \mathrm{id}} & \mathrm{Sp}^m S^n \times S^n,
\end{array}$$

where g_1 is obtained by adding (x_0, x_0) and g_2 by adding x_0. The map g_2 is a rational homotopy equivalence (see Remark 2.7) so is also g_1. For the inductive step, we consider the next cube.

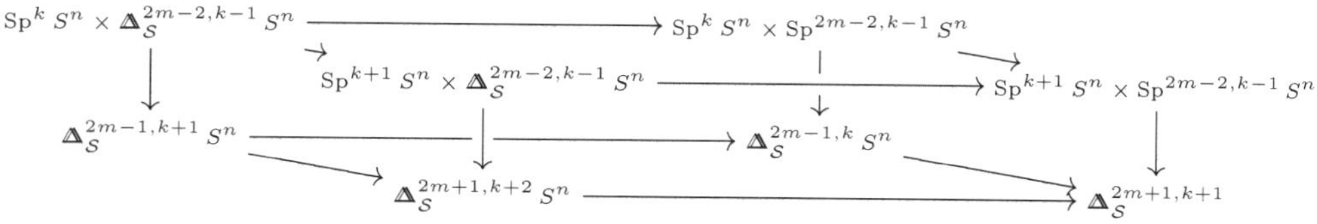

By Proposition 3.2, the front and the back faces are homotopy pushouts. The morphisms between these two squares are represented by oblique arrows in the previous diagram. The oblique arrows on the top consist to add the point x_0 and are rational homotopy equivalences. The oblique arrows on the bottom are the adding of (x_0, x_0). The result follows by induction. □

4. Finite subsets spaces and Proof of Theorem 1

In this section, we present a generalization of a theorem of Handel concerning the inclusion of a space X in its n-th finite subsets space, $\exp^n X$.

DEFINITION 4.1. Let X be a non-empty space. *The n-th finite subsets space of X* is the space $\exp^n X$ of non-empty subsets of size at most n, topologised as the quotient of X^n by the surjective map $X^n \to \exp^n X$, $(x_1, \ldots, x_n) \mapsto \{x_1\} \cup \cdots \cup \{x_n\}$.

If X is pointed, we denote by $\exp^n_* X$ the subspace of $\exp^n X$ formed by the subsets that contain the basepoint. The space $\exp^n_* X$ is pointed by $\{*\}$.

The correspondence $X \mapsto \exp^n X$ is a homotopy functor. C. Tuffley and R.A. Biro have determined its value on some spheres. For $n \geq 2$, we have:

- $\exp^{2k}(S^1) = \exp^{2k-1}(S^1) = S^{2k-1}$, [**26**, Theorem 4],
- $\exp^n(S^2) \simeq_{\mathbb{Q}} S^{2n} \vee S^{2n-2}$, $\exp^n_*(S^2) \simeq_{\mathbb{Q}} S^{2n-2}$,[**29**, Theorem 1],
- $\exp^{2i+1}(S^{2k+1}) \simeq_{\mathbb{Q}} S^{(2k+1)(i+1)+i}$, $\exp^{2i+2}(S^{2k+1}) \simeq_{\mathbb{Q}} S^{(2k+1)(i+1)+i}$, $\exp^n(S^{2k}) \simeq_{\mathbb{Q}} S^{2kn} \vee S^{2k(n-1)}$, [**1**, Lemma 6.3.2. and Lemma 6.3.8.].

Let X be a CW-complex. The space $\exp^n X$ can be described as the following pushout

4.2.
$$\begin{array}{ccc} \Delta^n_S X & \longrightarrow & \mathrm{Sp}^n X \\ \downarrow & & \downarrow \\ \exp^{n-1} X & \longrightarrow & \exp^n X. \end{array}$$

As the map $\Delta^n_S X \to \mathrm{Sp}^n X$ is the inclusion of a sub-CW-complex in a CW-complex, this pushout is also a homotopy pushout. Observe that this pushout is the key in the study of the finite subsets spaces of surfaces done by C. Tuffley in [**29**, Lemma 4].

A *homotopy cofibration* is a sequence $A \xrightarrow{g} X \xrightarrow{j} Y$ where $j\colon X \to Y$ is the homotopy cofiber of $g\colon A \to X$. In the sequel, we do not make any distinction between a map and its homotopy class. The proof of Theorem 1 uses the two next lemmas.

LEMMA 4.3. *For any homotopy cofibration of pointed spaces, $A \xrightarrow{g} X \xrightarrow{j} Y$, if the inclusion $\iota^n_X\colon X \to \exp^n_* X$ is homotopically trivial, then the inclusion $\iota^{n+1}_Y\colon Y \to \exp^{n+1}_* Y$ is also homotopically trivial.*

PROOF. Any homotopy cofibration $A \xrightarrow{g} X \xrightarrow{j} Y$ gives a coaction of ΣA on Y, denoted by $\nabla\colon Y \to Y \vee \Sigma A$ and defined by pinching the cone CA in $Y \simeq X \cup CA$. For any space T, and any couple of maps $f\colon Y \to T$, $\mu\colon \Sigma A \to T$, we denote by (f, μ) the composition of $f \vee \mu\colon Y \vee \Sigma A \to T \vee T$ with the folding map $T \vee T \to T$.

From two maps f and μ, the coaction gives a map $(f, \mu) \circ \nabla\colon Y \to T$ denoted by f^μ. This correspondence satisfies (cf. [**15**, Chapter 15]) :

$$(f^\mu)^\nu = f^{\mu+\nu} \text{ et } f^* = f,$$

where $*\colon \Sigma A \to T$ is the constant map on the basepoint and ν is a map from ΣA to T. Moreover, one can construct a long exact sequence

$$A \xrightarrow{g} X \xrightarrow{j} Y \xrightarrow{\partial} \Sigma A \xrightarrow{\Sigma g} \Sigma X \longrightarrow \cdots$$

where each pair of consecutive maps is a homotopy cofibration. Therefore, for any space T, we get an exact sequence

$$\cdots \longrightarrow [\Sigma X, T] \xrightarrow{(\Sigma g)_*} [\Sigma A, T] \xrightarrow{\partial_*} [Y, T] \xrightarrow{j_*} [X, T] \xrightarrow{g_*} [A, T].$$

When the maps are homomorphisms of groups, the word *exact* means the usual definition. Between pointed sets, it means that the image of the first application coincides with the preimage of the basepoint by the second one. P. Hilton studies the properties of this exact sequence in [**15**] and proves in particular:

(1) if f and h are elements of $[Y, T]$ such that $j_*(f) = j_*(h)$, then there exists $\mu\colon \Sigma A \to T$ such that $h = f^\mu$, cf. [**15**, 15.5],
(2) the map ∂_* satisfies $\partial_*(\mu) = *^\mu$, thus $(\partial_*(\mu))^\nu = \partial_*(\mu + \nu)$, cf. [**15**, Proof of 15.3],
(3) $\partial_*(\mu) = \partial_*(\nu)$ if, and only if, $\partial_*(\mu - \nu) = *$, cf. [**15**, Proof of 15.6].

We begin now with the proof of the lemma. The commutativity of the next diagram in full lines

$$\begin{array}{ccccc}
X & \xrightarrow{\iota_X^n} & \exp_*^n X & & \\
{\scriptstyle j}\downarrow & & \downarrow{\scriptstyle \exp_*^n j} & & \\
Y & \xrightarrow[\iota_Y^n]{} & \exp_*^n Y & \overset{\iota}{\rightarrowtail} & \exp_*^{n+1} Y \\
{\scriptstyle \partial}\downarrow & \nearrow_{k} & & & \\
\Sigma A & & & &
\end{array}$$

implies the existence of a map $k\colon \Sigma A \to \exp_*^n Y$ such that $\iota_Y^n = k \circ \partial$. If we compose ι_Y^n with the inclusion $\iota\colon \exp_*^n Y \hookrightarrow \exp_*^{n+1} Y$, we have $\iota_Y^{n+1} = \iota \circ \iota_Y^n = \iota \circ k \circ \partial = \partial_*(\iota \circ k)$. Consider now the commutative diagram:

$$\begin{array}{ccccccc}
 & \overset{\nabla}{\nearrow} & Y \vee \Sigma A & \xrightarrow{\mathrm{id}_Y \vee k} & Y \vee \exp_*^n Y & \overset{\psi'}{\searrow} & \\
 & & \downarrow & & \downarrow & & \\
Y & \xrightarrow[\mathrm{id}_Y \times \partial]{} & Y \times \Sigma A & \xrightarrow[\mathrm{id}_Y \times k]{} & Y \times \exp_*^n Y & \xrightarrow[\psi]{} & \exp_*^{n+1} Y,
\end{array}$$

where $\psi(y, \Gamma) = \{y\} \cup \Gamma$ and ψ' is the restriction of ψ. Using the previous determination of $\iota \circ \iota_Y^n$, we see that the composition of the maps located on the bottom line is $\iota \circ \iota_Y^n = \partial_*(\iota \circ k)$. We have also

$$\psi' \circ (\mathrm{id}_Y \vee k) \circ \nabla = (\iota \circ \iota_Y^n, \iota \circ k) \circ \nabla = (\iota \circ \iota_Y^n)^{\iota \circ k} = (\partial_*(\iota \circ k))^{\iota \circ k} = \partial_*(2(\iota \circ k)).$$

Therefore, we get $\partial_*(\iota \circ k) = \partial_*(2(\iota \circ k))$ and $\partial_*(\iota \circ k) = *$. Thus $\iota_Y^{n+1} = \iota \circ \iota_Y^n = \partial_*(\iota \circ k) = *$ as required. $\square$

LEMMA 4.4. *For any space X, the inclusion $\Sigma X \to \exp_*^3 \Sigma X$ is homotopically trivial.*

PROOF. Let $*$ be the basepoint of S^1. Theorem 4.1 of Handel ([**14**]) implies that the inclusion $\exp_*^2 S^1 \cong S^1 \to \exp_*^3 S^1$ induces the zero map between the homotopy groups. Therefore there exists a map

$$G\colon S^1 \times [0,1] \to \exp_*^3 S^1,\ G(v,s) = \{G_1(v,s), G_2(v,s), G_3(v,s)\}$$

such that

- $G(*, s) = \{*\}$, pour tout $s \in [0, 1]$,
- $G(v, 0) = \{v\}$, pour tout $v \in S^1$,
- $G(v, 1) = \{*\}$, pour tout $v \in S^1$.

We can now define a homotopy H between the inclusion of $\Sigma X = S^1 \wedge X$ in $\exp_*^3 \Sigma X$ and the constant map by

$$H((v, x), s) = \{G_1(v, s) \wedge x, G_2(v, s) \wedge x, G_3(v, s) \wedge x\}, \text{ for } v \in S^1,\ x \in X,\ s \in [0, 1].$$

□

Recall now that the LS-category of a space X is the least integer n such that X can be covered by $(n + 1)$ open sets contractible in X. For CW-complexes, an equivalent definition is given using Ganea fibrations $q_n \colon G_n(X) \to X$, of fiber $i_n \colon F_n(X) \to G_n(X)$, defined as follows: q_0 is the path space fibration and q_{n+1} is the fibration associated to the extension $G_n(X) \cup_{i_n} CF_n(X) \to X$ of q_n sending the cone $CF_n(X)$ on the basepoint. In [**12**], Ganea proved that $\operatorname{cat}(X) \leq n$ if, and only if, q_n admits a section. For other equivalent definitions and basic properties of the Lusternik-Schnirelmann category (LS-category), we send the reader to [**6**].

PROOF OF THEOREM 1.

(1) First, we check that the inclusion $G_n(X) \to \exp_*^{n+2} G_n(X)$ is homotopically trivial. This is true for $n = 1$ grants to Lemma 4.4 because $G_1(X) = \Sigma\Omega X$. Suppose now that the result is true for $G_n(X)$. As $G_{n+1}(X)$ is the cofibre of a map with values in $G_n(X)$, the result comes directly from Lemma 4.3.

Suppose now $\operatorname{cat}(X) = n$ and let $r \colon X \to G_n(X)$ be a section of q_n. The commutativity of the next diagram implies the homotopy triviality of the inclusion of X in $\exp_*^{n+2} X$:

$$\begin{array}{ccc} X & \xrightarrow{\iota_X} & \exp_*^{n+2} X \\ {\scriptstyle r}\downarrow & & \uparrow{\scriptstyle \exp_*^{n+2} q_n} \\ G_n(X) & \longrightarrow & \exp_*^{n+2} G_n(X) \end{array}$$

(2) For a space of LS-category n, the diagonal map $\Delta \colon X \to X^{n+1}$ factorizes through the fat wedge. Thus the inclusion of X in $\exp^{n+2} X$ factorizes through $\exp_*^{n+2} X$ and the result follows from (1).

(3) Let $\iota_X^{n+2} \colon X \to \exp_*^{n+2} X$ be the inclusion. The inclusion $\exp_*^{k+1} X \to \exp_*^{(n+1)k+1} X$ factorizes as

$$\exp_*^{k+1} X \xrightarrow{\exp_*^{k+1}(\iota_X^{n+2})} \exp_*^{k+1}(\exp_*^{n+2} X) \xrightarrow{\psi} \exp_*^{k(n+1)+1} X,$$

where the map ψ is the union. The triviality of ι_X^{n+2} implies the result.

□

5. Rational homotopy of n-th finite subsets spaces for $n = 3$

The space $\exp^2 X$ being homeomorphic to $\operatorname{Sp}^2 X$, the first interesting case is $\exp^3 X$.

PROPOSITION 5.1. *Let X be a finite CW-complex. The finite subsets space having at most 3 elements,* $\exp^3 X$*, is the homotopy pushout*

$$\begin{array}{ccc} X\times X & \xrightarrow{g} & \mathrm{Sp}^3 X \\ {\scriptstyle \rho_2}\downarrow & & \downarrow \\ \mathrm{Sp}^2 X & \longrightarrow & \exp^3 X, \end{array}$$

where $g(x,y)=\langle x,x,y\rangle$.

PROOF. We already know that $\exp^3(X)$ is the homotopy pushout

$$\begin{array}{ccc} \Delta^3_{\mathcal{S}}(X) & \longrightarrow & \mathrm{Sp}^3 X \\ \downarrow & & \downarrow \\ \mathrm{Sp}^2 X & \longrightarrow & \exp^3 X. \end{array}$$

The result comes now from the homeomorphism $X\times X\xrightarrow{\cong}\Delta^3_{\mathcal{S}} X$, induced by $X\times X\to X\times X\times X$, $(x,y)\mapsto(x,x,y)$. ☐

If V is a graded vector space, we denote by sV the suspension of V defined by $(sV)^n=V^{n-1}$.

PROPOSITION 5.2. *If ΣX is an r-connected suspension of cohomology $H = H^*(\Sigma X;\mathbb{Q})$, the space* $\exp^3\Sigma X$ *is a $(2r+1)$-connected suspension. Its rational cohomology is given by*

$$H^+(\exp^3\Sigma X;\mathbb{Q})\cong\wedge^3H^+\oplus\wedge^2H^+\oplus s\mathrm{Sym}^2(H^+),$$

where $\mathrm{Sym}^2(H^+)$ *is the symmetric power of* H^+.

PROOF. From Proposition 5.1 and classical constructions in rational homotopy theory, a model of $\exp^3 X$ is given by the kernel of a surjective map $\varphi=\overline{\mu}+\overline{\rho}$, where $\overline{\mu}$ and $\overline{\rho}$ are respective models of g and ρ_2. Since the map g is the composition

$$\Sigma X\times\Sigma X\xrightarrow{\Delta\times\mathrm{id}}\Sigma X\times\Sigma X\times\Sigma X\xrightarrow{\rho_3}\mathrm{Sp}^3\Sigma X,$$

a model of g is given by $\overline{\mu}\colon(\wedge H^+/\wedge^{>3}H^+,0)\to(H\otimes H,0)$, $\overline{\mu}(a)=2a\otimes1+1\otimes a$ for $a\in H^+$. A model of $\rho_2\colon\Sigma X\times\Sigma X\to\mathrm{Sp}^2(\Sigma X)$ has already be made explicit; we modify it in a surjective map $\overline{\mu}$ as follows:

$$\overline{\mu}\colon(A,D)=(\mathrm{Sym}^2(H^+)\oplus s\mathrm{Sym}^2(H^+)\oplus\wedge^2H,D)\to(H\otimes H,0),$$

$\overline{\mu}$ is the inclusion on $\mathrm{Sym}^2(H^+)\oplus\wedge^2H$ and zero on $s\mathrm{Sym}^2(H^+)$. The differential D is given by $D(w)=sw$ for $w\in\mathrm{Sym}^2(H^+)$ and zero on $s\mathrm{Sym}^2(H^+)\oplus\wedge^2H$. The map $\varphi=\overline{\mu}+\overline{\rho}$ is clearly surjective. If we write an element of $A\oplus(\wedge H^+/\wedge^{>3}H^+)$ as a couple (a,b), the kernel of φ is the vector space

$$(s\mathrm{Sym}^2(H^+),0)\oplus(0,\wedge^3H^+)\oplus\left\{(x,0)-2(0,x)\mid x\in\wedge^2H^+\right\}.$$

The product and the differential being null, the space $\exp^3\Sigma X$ is a rational suspension. All the elements in this kernel having a degree greater than or equal to $2r+2$, the space $\exp^3\Sigma X$ is rationally $(2r+1)$-connected. ☐

6. Rational homotopy of n-th finite subsets spaces for $n = 4$

PROPOSITION 6.1. *Let X be a finite CW-complex. The finite subsets space having at most 4 elements is the homotopy pushout*

$$\begin{array}{ccc} X \times \mathrm{Sp}^2 X & \xrightarrow{g_2'} & \mathrm{Sp}^4 X \\ {\scriptstyle g_1'}\downarrow & & \downarrow \\ \mathrm{Sp}^3 X & \longrightarrow & \exp^4 X, \end{array}$$

where $g_1'(x, \langle x_1, x_2\rangle) = \langle x, x_1, x_2\rangle$ and $g_2'(x, \langle x_1, x_2\rangle) = \langle x, x, x_1, x_2\rangle$.

PROOF. Consider the next diagram, where the map $\mathbb{\Delta}^4_{\mathcal{S}}(X) \to \exp^3 X$ is induced by $X \times \mathrm{Sp}^2 X \to \mathrm{Sp}^3 X$, $(x, \langle x_1, x_2\rangle) \mapsto \langle x, x_1, x_2\rangle$.

$$\begin{array}{ccccc} X \times X & \xrightarrow{g_1} & X \times \mathrm{Sp}^2 X & \longrightarrow & \mathrm{Sp}^3 X \\ {\scriptstyle g_2}\downarrow & & {\scriptstyle g_4}\downarrow & & \downarrow \\ \mathrm{Sp}^2(X) & \xrightarrow[g_3]{} & \mathbb{\Delta}^4_{\mathcal{S}}(X) & \longrightarrow & \exp^3 X \\ & & \downarrow & & \downarrow \\ & & \mathrm{Sp}^4 X & \longrightarrow & \exp^4 X \end{array}$$

The bottom square is a homotopy pushout by definition of $\exp^4 X$. The top rectangle is a homotopy pushout by Proposition 5.1 and the top left square also by Proposition 3.2. Then the top right square and the square of the statement are homotopy pushouts from classical manipulations with homotopy pushouts. □

PROPOSITION 6.2. *If ΣX is an r-connected suspension of cohomology $H = H^*(\Sigma X; \mathbb{Q})$, the space $\exp^4 \Sigma X$ is a $(2r+1)$-connected suspension.*

PROOF. A model for the map $g_2' \colon X \times \mathrm{Sp}^2(\Sigma X) \to \mathrm{Sp}^4(\Sigma X)$ is the morphism $\overline{\mu} \colon (\wedge H^+ / \wedge^{>4} H^+, 0) \to (H \otimes \wedge^2 H, 0)$, $\overline{\mu}(a) = 2a \otimes 1 + 1 \otimes a$.

We choose an ordered basis of H^+ and denote by W the subvector space of $H \otimes \wedge^2 H$ generated by the elements $a \otimes b$, with $a < b$, the elements $a \otimes a$, with a of odd degree, the elements $a \otimes bc$ with $a < b$ or $a < c$ and the elements $a \otimes ac$ with a of odd degree. We then form a surjective cdga model $\overline{\rho}$ of g_1' as follows:

$$\overline{\rho} \colon (A, D) = (\wedge^3 H \oplus W \oplus dW, D) \to (H \otimes \wedge^2 H, 0),$$

with all the products on A null except $(a \otimes b) \cdot c = (a \otimes bc)$, and $d(a \otimes b) \cdot c = d(a \otimes bc)$ for $a \otimes b \in W$, $c \in H^+$ and $a \otimes bc \in W$. The differential D is defined by $D(w) = dw$ if $w \in W$ and zero otherwise. The map $\overline{\rho}$ is the injection on W and is zero on dW. Moreover $\overline{\rho}(a) = a \otimes 1 + 1 \otimes a$ for $a \in H^+$. As the sum

$$\varphi = (\overline{\rho} + \overline{\mu}) : (A, D) \oplus (\wedge^4 H, 0) \to (H \otimes \wedge^2 H, 0)$$

is a surjective map, its kernel is a model for $\exp^4 \Sigma X$. This kernel is the sum $(dW, 0) \oplus (0, \wedge^4 H^+) \oplus \{2(x, 0) - (0, x)\,, x \in \wedge^3 H^+\}$. The product and the differential being trivial on this kernel, the space $\exp^4 \Sigma X$ is a suspension. All the elements in this kernel having a degree greater than or equal to $2r + 2$, the space $\exp^4 \Sigma X$ is rationally $(2r+1)$-connected. □

References

1. Ross A. Biro, *The structure of cyclic configuration spaces and the homotopy type of Kontsevich's orientation space*, Ph.D. thesis, Stanford, 1994.
2. Karol Borsuk, *On the third symmetric potency of the circumference*, Fund. Math. **36** (1949), 236–244. MR MR0035987 (12,42a)
3. Karol Borsuk and Stanislaw Ulam, *On symmetric products of topological spaces*, Bull. Amer. Math. Soc. **37** (1931), no. 12, 875–882. MR MR1562283
4. Raoul Bott, *On the third symmetric potency of S^1*, Fund. Math. **39** (1952), 264–268 (1953). MR MR0054954 (14,1003e)
5. Glen E. Bredon, *Introduction to compact transformation groups*, Academic Press, New York, 1972, Pure and Applied Mathematics, Vol. 46. MR MR0413144 (54 #1265)
6. Octav Cornea, Gregory Lupton, John Oprea, and Daniel Tanré, *Lusternik-Schnirelmann category*, Mathematical Surveys and Monographs, vol. 103, American Mathematical Society, Providence, RI, 2003. MR **MR1990857 (2004e:**55001)
7. Albrecht Dold and René Thom, *Quasifaserungen und unendliche symmetrische Produkte*, Ann. of Math. (2) **67** (1958), 239–281. MR MR0097062 (20 #3542)
8. J. L. Dupont and G. Lusztig, *On manifolds satisfying* $w_1\mathrm{Sp}2 = 0$, Topology **10** (1971), 81–92. MR MR0273631 (42 #8508)
9. Yves Félix, Stephen Halperin, and Jean-Claude Thomas, *Rational homotopy theory*, Graduate Texts in Mathematics, vol. 205, Springer-Verlag, New York, 2001. MR **MR1802847 (2002d:**55014)
10. Yves Félix, John Oprea, and Daniel Tanré, *Algebraic models in geometry*, Oxford Graduate Texts in Mathematics, vol. 17, Oxford University Press, Oxford, 2008.
11. Yves Félix and Daniel Tanré, *The cohomology algebra of unordered configuration spaces*, J. London Math. Soc. (2) **72** (2005), no. 2, 525–544. MR **MR2156668 (2006d:**55017)
12. Tudor Ganea, *Lusternik-Schnirelmann category and strong category*, Illinois J. Math. **11** (1967), 417–427. MR MR0229240 (37 #4814)
13. Karsten Grove, Stephen Halperin, and Micheline Vigué-Poirrier, *The rational homotopy theory of certain path spaces with applications to geodesics*, Acta Math. **140** (1978), no. 3-4, 277–303. MR **MR496895 (80g:**58024)
14. David Handel, *Some homotopy properties of spaces of finite subsets of topological spaces*, Houston J. Math. **26** (2000), no. 4, 747–764. MR **MR1823966 (2002d:**55016)
15. Peter Hilton, *Homotopy theory and duality*, Gordon and Breach Science Publishers, New York, 1965. MR MR0198466 (33 #6624)
16. Sadok Kallel and Denis Sjerve, *Finite subset spaces and a spectral sequence of Biro*, 2008.
17. ———, *Remarks on finite subset spaces*, 2009.
18. I. G. Macdonald, *The Poincaré polynomial of a symmetric product*, Proc. Cambridge Philos. Soc. **58** (1962), 563–568. MR MR0143204 (26 #764)
19. Jacob Mostovoy, *Geometry of truncated symmetric products and real roots of real polynomials*, Bull. London Math. Soc. **30** (1998), no. 2, 159–165. MR **MR1489327 (99a:**57021)
20. ———, *Lattices in $\mathbb{C}$ and finite subsets of a circle*, Amer. Math. Monthly **111** (2004), no. 4, 357–360. MR MR2057192
21. Boon W. Ong, *The homotopy type of the symmetric products of bouquets of circles*, Internat. J. Math. **14** (2003), no. 5, 489–497. MR **MR1993792 (2004g:**55010)
22. S. C. F. Rose, *A hyperbolic approach to* $\exp^3 S^1$, 2007.
23. V. P. Snaith and J. J. Ucci, *Three remarks on symmetric products and symmetric maps*, Pacific J. Math. **45** (1973), 369–377. MR MR0367988 (51 #4230)
24. Dennis Sullivan, *Infinitesimal computations in topology*, Inst. Hautes Études Sci. Publ. Math. (1977), no. 47, 269–331 (1978). MR MR0646078 (58 #31119)
25. Daniel Tanré, *Homotopie rationnelle: modèles de Chen, Quillen, Sullivan*, Lecture Notes in Mathematics, vol. 1025, Springer-Verlag, Berlin, 1983. MR **MR764769 (86b:**55010)
26. Christopher Tuffley, *Finite subset spaces of S^1*, Algebr. Geom. Topol. **2** (2002), 1119–1145 (electronic). MR **MR1998017 (2004f:**54008)
27. ———, *Finite subset spaces of graphs and punctured surfaces*, Algebr. Geom. Topol. **3** (2003), 873–904 (electronic). MR **MR2012957 (2004i:**55023)
28. ———, *Connectivity of finite subset spaces of cell complexes*, Pacific J. Math. **217** (2004), no. 1, 175–179. MR **MR2105772 (2005g:**55019)

29. ______, *Finite subset spaces of closed surfaces*, 2007.
30. Don Bernard Zagier, *Equivariant Pontrjagin classes and applications to orbit spaces. Applications of the G-signature theorem to transformation groups, symmetric products and number theory*, Lecture Notes in Mathematics, Vol. 290, Springer-Verlag, Berlin, 1972. MR MR0339202 (49 #3965)

Département de Mathématiques, Université Catholique de Louvain, 2, Chemin du Cyclotron, 1348 Louvain-La-Neuve, Belgium
E-mail address: felix@math.ucl.ac.be

Département de Mathématiques, UMR 8524, Université des Sciences et Technologies de Lille, 59655 Villeneuve d'Ascq Cedex, France
E-mail address: Daniel.Tanre@univ-lille1.fr

Contemporary Mathematics
Volume **519**, 2010

Derivations, Hochschild cohomology and the Gottlieb group

J.-B Gatsinzi

ABSTRACT. We consider a finite simply connected CW-complex X of which the rational homotopy group $\pi_*(X)\otimes\mathbb{Q}$ is finite dimensional. We show that the Hochschild cohomology of a Sullivan model of X contains a polynomial algebra over the Gottlieb group of X.

1. Introduction

Let $\mathbf{k}$ be a commutative ring and (A, d) a graded cochain algebra over $\mathbf{k}$, that is, $A = \{A^i\}_{i\geq 0}$ and $d : A^i \to A^{i+1}$. Let $\mathcal{L} = \mathrm{Der}(A)$ denote the $\mathbb{Z}$-graded differential Lie algebra of derivations whose differential δ is defined by $\delta\theta = [d, \theta]$, for $\theta \in \mathrm{Der}(A)$. Here $\mathrm{Der}_i(A) \subset \prod_{j\geq 0} \mathrm{Hom}(A^j, A^{j-i})$. We denote by $C^*(A; A)$ (resp. $HH^*(A; A)$) the Hochschild complex (resp. cohomology) of the cochain algebra A with coefficients in A [**13**]. Moreover, Gerstenhaber defined a bracket on $C^*(A; A)$ inducing a graded Gerstenhaber structure on $HH^*(A; A)$ [**13**]. Our aim is to prove the following results.

THEOREM A. *There is a canonical injective homomorphism of differential graded Lie algebras*

$$\imath : \mathcal{L} \to sC^*(A; A)$$

when $C^*(A; A)$ is equipped with the Gerstenhaber bracket.

Let (M, d) a graded differential A-module. The A-tensor algebra $T_A(M)$ is defined by $T_A(M) = A \oplus (\oplus_{k\geq 1} T^k_A(M))$, where

$$T^k_A(M) = M \otimes_A M \otimes \cdots \otimes_A M \quad (k \text{ factors}).$$

The exterior algebra $\wedge_A M$ is the graded differential commutative algebra obtained as the quotient of $T_A(M)$ by the ideal generated by elements of the form $x \otimes y - (-1)^{|x||y|} y \otimes x$, $x, y \in T_A(M)$. The exterior product induces

2000 *Mathematics Subject Classification.* Primary 55P62; Secondary 55M35.
Key words and phrases. Hochschild cohomology, Gottlieb group, elliptic space.
Partial support by the Abdus Salam ICTP and A. von Humboldt Foundation.

©2010 American Mathematical Society

a graded commutative algebra structure on $\wedge_A M$. The differential defined by

$$d(m_1 \wedge \cdots \wedge m_k) = \sum_{i=1}^{k} (-1)^{\epsilon(i)} m_1 \wedge \cdots \wedge d(m_i) \cdots \wedge m_k, \quad \epsilon(i) = \sum_{j<i} |m_j|,$$

turns $\wedge_A^k M$ into a differential graded A-algebra.

There is a natural structure of $\mathcal{L}$ as a graded differential A-module, namely $(a\theta)(x) = a\theta(x)$, for $a, x \in A$ and $\theta \in \mathcal{L}$. The differential δ' on $L = s^{-1}\mathcal{L}$ uniquely extends into an algebra differential d_0 on $\Lambda_A L$.

We define a bracket of degree 1 on L by $\{x, y\} = s^{-1}[sx, sy]$. Moreover, there is also an extension of the Lie bracket to $\Lambda_A L$, the Schouten-Nijenhuis bracket defined by

$$\{x_1 \wedge \cdots \wedge x_m, y_1 \wedge \cdots \wedge y_n\} = \sum_{i,j} (-1)^{\epsilon_{ij}} x_1 \wedge \cdots \hat{x}_i \wedge x_m \wedge \{x_i, y_j\} \wedge \cdots \wedge \hat{y}_j \wedge \cdots y_n,$$

where ˆ means omitted and $\epsilon_{ij} = \sum_{k>i} |x_k|(|x_i|+1) + \sum_{k<j}(|y_k|+1)|y_j|$. This turns $\Lambda_A L$ into a differential graded Gerstenhaber algebra, hence $H_*(\Lambda_A L, d_0)$ inherits a graded Gerstenhaber structure as well.

THEOREM B. *If* $\mathbf{k} \supset \mathbb{Q}$ *then* $\imath$ *induces a canonical homomorphism of graded Gerstenhaber algebras*

$$H(\phi) : H(\wedge_A L, d_0) \to HH^*(A; A).$$

A Sullivan algebra is a commutative cochain algebra of the form $(\wedge V, d)$, where $V = \oplus_{i \geq 1} V^i$. Moreover $V = \bigcup_{k \geq 0} V(k)$, where $V(0) \subset V(1) \ldots$ such that $d(V(0)) = 0$ and $d(V(k)) \subset \wedge V(k-1)$. It is called minimal if $dV \subset \wedge^{\geq 2} V$. For any commutative differential graded algebra, (A, d) of which the cohomology is connected and finite in each degree, there is a unique minimal Sullivan algebra $(\wedge V, d)$ equipped with a quasi-isomorphism $(\wedge V, d) \to (A, d)$.

THEOREM C. *If* $\mathbf{k} = \mathbb{Q}$ *and* $A = (\wedge V, d)$ *is a minimal Sullivan algebra then there is an isomorphism of graded algebras*

$$H(\wedge_A L, d_0) \to HH(A; A).$$

Moreover, if $\dim H(\wedge V, d) < \infty$ *and* $\dim V < \infty$ *then* $H(\wedge_A L, d_0)$ *contains a non trivial polynomial sub algebra.*

Let X be a closed oriented m-manifold and $LX = \mathrm{map}(S^1, X)$ the space of free loops on X. The loop homology of X is the homology of LX with a shift of degrees, $\mathbb{H}_*(LX) = H_{*+m}(LX)$ and an associative and graded commutative product

$$\mu : \mathbb{H}_p(LX) \otimes \mathbb{H}_q(LX) \to \mathbb{H}_{p+q}(LX)$$

called *loop product* [**3**]. When coefficients are taken in a field there is an isomorphism of graded vector spaces [**17**]

$$HH_*(C^*X; C^*X) \cong H^*(LX)$$

which dualizes in

$$HH^*(C^*X; C_*X) \cong H_*(LX).$$

Using the cap product, Félix et al. [**11**] construct a linear isomorphism

$$\Phi : \mathbb{H}_*(LX) \to HH^*(C^*X; C^*X).$$

If $\mathbf{k}$ is of characteristic 0 and X is simply connected, the same authors show that Φ is a homomorphism of graded algebras [**10**].

Moreover given a Sullivan minimal model $(\wedge V, d)$ of X, one has an isomorphism

$$HH^*(\wedge V; \wedge V) \cong HH^*(C^*X; C^*X).$$

Therefore we have the following result.

THEOREM D. *If X is a simply connected closed oriented m-manifold such that $\pi_*(X) \otimes \mathbb{Q} < \infty$ then $\mathbb{H}_*(LX, \mathbb{Q})$ contains a graded polynomial algebra.*

2. Hochschild cohomology

Let (A, d) be an augmented differential graded algebra over $\mathbf{k}$ and $\bar{A} = \ker(A \stackrel{\epsilon}{\to} \mathbf{k})$. If (M, d) and (N, d) are differential graded A-modules, respectively right and left A-module, the two-sided bar construction $\mathbb{B}(M; A; N)$ is defined by

$$\mathbb{B}_k(M; A; N) = M \otimes T^k(s\bar{A}) \otimes N.$$

An element $m[a_1|a_2|\cdots a_k]n$ is of degree $|m|+|n|+\sum_{i=1}^k |sa_i|$. The differential $d = d_0 + d_1$ is defined as follows (see for instance [**8**]).

$$d_0 : \mathbb{B}_k(M; A; N) \to \mathbb{B}_k(M; A; N), \quad d_1 : \mathbb{B}_k(M; A; N) \to \mathbb{B}_{k-1}(M; A; N),$$

$$\begin{aligned} d_0(m[a_1|a_2|\cdots|a_k]n) &= (dm)[a_1|a_2|\cdots|a_k]n - \textstyle\sum_{i=1}^k (-1)^{\epsilon(i)} m[a_1|\cdots|da_i|\cdots|a_k]n \\ &\quad +(-1)^{\epsilon(k+1)} m[a_1|a_2|\cdots|a_k](dn), \end{aligned}$$

$$\begin{aligned} d_1(m[a_1|a_2|\cdots|a_k]n) = &\ (ma_1)[a_2|\cdots|a_k]n - \textstyle\sum_{i=2}^k (-1)^{\epsilon(i)} m[a_1|\cdots|a_{i-1}a_i|\cdots|a_k]n \\ &-(-1)^{\epsilon(k+1)} m[a_1|a_2|\cdots|a_{k-1}](a_k n). \end{aligned}$$

Here $\epsilon(i) = |m| + \sum_{j<i} |sa_j|$.

If A is a free $\mathbf{k}$-module then the canonical projection

$$\mathbb{B}(A; A; A) \to A$$

is a semi-free resolution of A as a graded differential $A \otimes A^{op}$-module. Therefore the normalized Hochschild cochain complex is

$$(C^*(A; A), D) = \mathrm{Hom}_{A\otimes A^{op}}(\mathbb{B}(A; A; A), A) \cong (\mathrm{Hom}(T(s\bar{A}), A), D_0 + D_1),$$

with the differential defined as follows [**8**].

$$\begin{aligned} (D_0 f)([a_1|a_2|\dots|a_k]) = &\ d(f([a_1|a_2|\dots|a_k])) \\ &- \textstyle\sum_{i=1}^k (-1)^{\epsilon(i)} f([a_1|\dots|da_i|\dots|a_k]) \end{aligned}$$

and

$$\begin{aligned}(D_1f)([a_1|a_2|\dots|a_k]) = \quad & -(-1)^{|sa_1||f|}a_1f([a_2|\dots|a_k]) \\ & +(-1)^{\epsilon(k)}f([a_1|\dots|a_{k-1}])a_k \\ & -\textstyle\sum_{i=2}^{k}(-1)^{\epsilon(i)}f([a_1|\dots|a_{i-1}a_i|\dots|a_k]),\end{aligned}$$

where $\epsilon(i) = |f| + |sa_1| + \dots + |sa_{i-1}|$.

As $T(s\bar{A})$ is a graded coalgebra, the complex $C^*(A;A)$ is endowed with a product, turning it into a differential graded algebra. For $f \in C^p(A;A)$ and $g \in C^q(A;A)$,

$$(fg)(a_1|\cdots|a_{p+q}) = (-1)^{\epsilon(p)}f(a_1|\cdots|a_p)g(a_{p+1}|\cdots|a_{p+q}), \tag{1}$$

where $\epsilon(p) = |g|(|sa_1| + \dots + |sa_p|)$. This product induces a graded commutative algebra structure on $HH^*(A;A)$[**13**].

Moreover, Gerstenhaber defined a Lie bracket on $C^*(A;A)$, inducing a graded Gerstenhaber algebra structure on $HH^*(A,A)$ [**13**]. The Lie bracket is defined by the formula

$$\{f,g\} = f\bar{\circ}g - (-1)^{|f||g|}g\bar{\circ}f, \tag{2}$$

where

$$(f\bar{\circ}g)[a_1|a_2|\dots|a_k]) = \sum_{0\le i\le j\le k}(-1)^{\epsilon(i)}f([a_1|\dots|a_i|g([a_{i+1}|\dots|a_j])|a_{j+1}\dots|a_k]),$$

and $\epsilon(i) = |g|(|sa_1| + \cdots |sa_i|)$. If $f \in C^p(A;A)$ and $g \in C^q(A;A)$, then $\{f,g\} \in C^{p+q-1}(A;A)$. As $C^1(A;A)$ is closed under this bracket, $HH^1(A;A)$ is a sub Lie algebra of $HH^*(A;A)$.

Let (A,d) be a differential graded algebra over **k** with $d : A^n \to A^{n+1}$, and $\mathcal{L}$ the differential graded Lie algebra defined in the Introduction. Define

$$F_1C^*(A;A) = \{f \in C^*(A;A)|f(T^{>1}(sA)) = 0\}.$$

Consider the composition mapping

$$\imath_A : s^{-1}\mathcal{L} \hookrightarrow F_1C^*(A;A) \subset C^*(A;A).$$

Explicitly if $\theta \in \mathcal{L}$, then $(\imath_A(s^{-1}\theta))([a]) = (-1)^{|\theta|}\theta(a)$. Denote by δ' the differential on $s^{-1}\mathcal{L}$ induced by δ. It is defined by the relation $\delta'(s^{-1}\theta) = -s^{-1}(\delta\theta) = -s^{-1}[d,\theta]$.

THEOREM 1. *If A is a differential graded commutative* **k***-algebra, then the inclusion $\imath_A : (s^{-1}\mathcal{L}, \delta') \to (C^*(A;A), D_0 + D_1)$ induces an injective map in homology.*

PROOF. First we show that $\imath_A$ commutes with differentials. As $\delta'(s^{-1}\theta) = -s^{-1}\delta\theta$ for $\theta \in \mathcal{L}$, then

$$\imath_A(\delta' s^{-1}\theta)([a]) = \imath_A(-s^{-1}\delta\theta)[a] = -(-1)^{|\theta|+1}(\delta\theta)(a) = (-1)^{|\theta|}(\delta\theta)(a). \tag{3}$$

If $f = \imath_A(s^{-1}\theta)$ then $Df = 0$ on $T^{>2}(s\bar{A})$. Moreover,

$$(D_0 + D_1)f([a]) = D_0f([a]) = df([a]) + (-1)^{|f|}f([da]) = (-1)^{|\theta|}(\delta\theta)(a), \tag{4}$$

(5)
$$\begin{aligned}(D_0+D_1)f([a|b]) &= (D_1f)([a|b])\\ &= -(-1)^{|f||sa|}af([b])+(-1)^{|f|+|sa|}f([a])b-(-1)^{|f|+|sa|}f([ab])\\ &= (-1)(-1)^{|f||sa|}((-1)^{|\theta||a|}af([b])+f([a])b-f([ab])).\end{aligned}$$

As θ is a derivation this shows that $D_1f=0$. Hence $\imath_A$ commutes with differentials. Moreover, Equation (5) means that $D_1f=0$ for $f\in F_1C^*(A;A)$ if and only if f is in the image of $\imath_A$.

Furthermore, Formula (2) shows that $f\bar{\circ}g=f\circ g$ for $f\in F_1C^*(A;A)$, hence $\imath_A$ preserves the brackets.

Assume that a homology class $[s^{-1}\theta]\in H_*(s^{-1}\mathcal{L})$ is in the kernel of $H_*(\imath_A)$. If $f=i_A(s^{-1}\theta)$ then there is $g=g_0+g_1+\ldots\in C^*(A;A)$ such that $g_i\in C^i(A;A)$ and $(D_0+D_1)g=f$. Therefore $D_0g_0=0$, $D_1g_0+D_0g_1=f$ and $Dg_k=0$ for $k\geq 2$. A simple computation shows that $(D_1g_0)([a])=(-1)^{|m|}(ma-(-1)^{|a||m|}am)$, where $m=g_0(1)$. Hence $D_1g_0=0$ as A is graded commutative. Moreover $D_0g_1=f$ if and only $\theta\in\mathcal{L}$ is a boundary, consequently $[s^{-1}\theta]=0$. □

REMARK 2. Consider the non unital bar construction $\tilde{B}(A)$. It is endowed with a graded coalgebra structure. Let Coder $\tilde{B}(A)$ be the differential graded Lie algebra of coderivations over $\tilde{B}(A)$. There is an isomorphism of complexes of degree one $\beta_A : C^*(A;A)\to \operatorname{Coder}\tilde{B}(A)$ [**19**]. The composition with $\imath_A:\mathcal{L}\to sC^*(A;A)$ gives rise to an injective map of differential graded Lie algebras $\alpha_A:\mathcal{L}\to\operatorname{Coder}\tilde{B}(A)$. If A is graded commutative, α_A induces an injective map in homology, by Theorem 1.

Recall that a Sullivan model of a simply connected topological space X is a Sullivan algebra $(\wedge V,d)$ which algebraically models the rational homotopy type of X (see [**20, 7**] for details). It is called minimal if $(\wedge V,d)$ is minimal.

Let $A=(\Lambda V,d)$ be a Sullivan model of a simply connected space X. Denote by $\tilde{\mathcal{L}}=\{\tilde{\mathcal{L}}_i\}_{i\geq 1}$ the differential graded sub Lie algebra of $\mathcal{L}=\operatorname{Der}(A)$ such that $\tilde{\mathcal{L}}_i=\mathcal{L}_i$ for $i\geq 2$, $\tilde{\mathcal{L}}_1=\mathcal{L}_1\cap\ker\delta$ and $\tilde{\mathcal{L}}_i=0$ for $i\leq 0$. Let $autX$ denote the monoid of self homotopy equivalences of X and aut_1X the path-connected component of the identity. If either $\pi_*(X)\otimes\mathbb{Q}$ or $H^*(X,\mathbb{Q})$ is finite dimensional, then the graded differential Lie algebra $(\tilde{\mathcal{L}},\delta)$ is a Lie model for $Baut_1(X)$ as the universal cover of $BautX$ [**20**]. In particular $\pi_*(\Omega Baut_1(X))\otimes\mathbb{Q}\cong H_*(\tilde{\mathcal{L}},\delta))$.

COROLLARY 3. *If X is a simply connected space, there is an injective map $\gamma:\pi_*(\Omega aut_1(X))\otimes\mathbb{Q}\to HH^*(A;A)$.*

REMARK 4. If $\mathbf{k}$ is a field of characteristic 0, Félix et al. construct an injective map [**9**],

(6)
$$\pi_*(\Omega aut_1(X))\otimes\mathbf{k}\to H_{*+d}(LX,\mathbf{k})$$

into the Chas-Sullivan loop space homology of the free loop space [**3**]. If X is a simply connected closed oriented manifold, there is an isomorphism of graded algebras [**10**],

$$\mathbb{H}_*(LX,\mathbb{Q}) = H_{*+d}(LX,\mathbb{Q}) \overset{\cong}{\to} HH^*(C^*(X);C^*(X)).$$

The inclusion (6) holds also when $\mathbf{k}$ is a principal ideal domain [**18**]. L. Menichi actually proves a stronger result, namely that if the integral homology $H_*(\Omega aut_1 X)$ is torsion free, then there is a map of BV-algebras

$$H_*(\Omega aut_1 X)\otimes \mathbb{H}_*(X) \to \mathbb{H}_*(LX).$$

Now we consider the composition map

$$\pi_*(\Omega aut_1 X) \overset{hur}{\to} H_*(\Omega aut_1 X) \hookrightarrow H_*(\Omega aut_1 X)\otimes \mathbb{H}_*(X) \to \mathbb{H}_*(LX).$$

For $\mathbf{k}=\mathbb{Q}$, this yields the inclusion of Corollary 3 (see Remark 9 below).

We consider the commutative graded algebra $\Lambda_A L$ endowed with the Schouten-Nijenhuis bracket defined in the Introduction. The differential d_0 on $\wedge_A L$ is the unique extension of δ' on L to $\Lambda_A L$. We first show the following result.

LEMMA 5. *$(\Lambda_A L, d_0)$ is a differential graded Gerstenhaber algebra.*

PROOF. As d is an element of degree -1 in $\mathcal{L}=\mathrm{Der}\wedge V$ then $\bar{d}=s^{-1}d$ has degree -2 in L. Moreover $\delta'\theta = -\{\bar{d},\theta\}$ for $\theta\in L$. In the same way $d_0\alpha = -\{\bar{d},\alpha\}$ for $\alpha\in\wedge_A L$. Now $\wedge_A L$ is a Gerstenhaber algebra, hence $\{\bar{d},-\}$ is a derivation of degree -1 in $\wedge_A L$. We deduce that d_0 is compatible with the multiplication in $\wedge_A L$. From the Jacobi rule in $\wedge_A L$, we conclude that d_0 respects the bracket in $\wedge_A L$ as well. □

Therefore $H_*(\Lambda_A L, d_0)$ is a graded Gerstenhaber algebra.

If $\mathbf{k}\supset\mathbb{Q}$, the injection $\imath : L = s^{-1}\mathcal{L}\hookrightarrow C^*(A;A)$ can be canonically extended into a map $\phi:\Lambda_A L\to C^*(A;A)$ by the formula

$$\phi(\alpha_1\wedge\cdots\wedge\alpha_n) = \frac{1}{n!}\sum_{\sigma\in S_n}\epsilon(\sigma)\imath(\alpha_{\sigma(1)})\cdots\imath(\alpha_{\sigma(n)}), \tag{7}$$

where $\epsilon(\sigma)$ is the Koszul sign of the permutation $(\alpha_1,\ldots,\alpha_n)\to(\alpha_{\sigma(1)},\ldots,\alpha_{\sigma(n)})$.

LEMMA 6. [**2**, Lemma 6.1] *The map ϕ is a homomorphism of differential graded modules.*

PROOF. It is easily seen that

$$\phi(\{\alpha,\beta\}) = \{\phi(\alpha),\phi(\beta)\}, \quad \text{for } \alpha\in L \text{ and } \beta\in\wedge_A L. \tag{8}$$

Therefore $\phi(d_0\alpha) = -\phi(\{\bar{d},\alpha\}) = -\{\phi(\bar{d}),\phi(\alpha)\} = D_0\phi(\alpha)$. In the course of the proof of Theorem 1, it was shown that $f\in F_1C^*(A;A)$ is in the image of $\imath_A$ if and only $D_1 f = 0$. Therefore D_1 zero on the image of ϕ. We deduce that ϕ commutes with differentials. □

Moreover we have the following result.

PROPOSITION 7. *The above homomorphism $\phi : (\Lambda_A L, d_0) \to (C^*(A;A), D_0 + D_1)$ induces a morphism of graded Gerstenhaber algebras*

$$H(\phi) : H(\Lambda_A L, d_0) \to H^*(A;A).$$

PROOF. Although ϕ is not a morphism of graded algebras, $H(\phi)$ respects the product structures because $HH^*(A;A)$ is graded commutative. For instance, if α_1 and α_2 are cycles in L, then the cohomology class $[\phi(\alpha_1)\phi(\alpha_2)] = (-1)^{|\alpha_1||\alpha_2|}[\phi(\alpha_2)\phi(\alpha_1)]$. Hence

$$H(\phi)([\alpha_1]\wedge[\alpha_2]) = \frac{1}{2}([\phi(\alpha_1)\phi(\alpha_2)] + (-1)^{|\alpha_1||\alpha_2|}[\phi(\alpha_2)\phi(\alpha_1)] = [\phi(\alpha_1)][\phi(\alpha_2)].$$

This generalizes to cocycles β_1 and β_2 in $\wedge_A^p L$ and $\wedge_A^q L$ respectively.

To prove that $H(\phi)$ is compatible with brackets, we use Equation 8, that is, $\phi(\{\alpha, \beta\}) = \{\phi(\alpha), \phi(\beta)\}$, for $\alpha \in \wedge_A L$ and $\beta \in L$. Assuming that the differential d on A is zero, ϕ induces an algebra homomorphism $\phi_2 : \wedge_A L \to H(C^*(A;A), D_1)$. As $\wedge_A L$ is a Gerstenhaber algebra, then $\{\alpha, -\}$ is an algebra derivation, that is,

$$\{\alpha, \beta_1 \wedge \beta_2\} = \{\alpha, \beta_1\} \wedge \beta_2 + (-1)^{(|\alpha|+1)|\beta_1|}\beta_1 \wedge \{\alpha, \beta_2\},$$

for β_1, β_2 in L. Applying ϕ_2 to both sides of the above equation, one deduces easily that

$$\phi_2(\{\alpha, \beta_1 \wedge \beta_2\}) = \{\phi_2(\alpha), \phi_2(\beta_1 \wedge \beta_2)\}.$$

An induction argument yields that

$$\phi_2 : \wedge_A L \to H(C^*(A;A), D_1)$$

is a morphism of graded Gerstenhaber algebras. See also [**14**, Theorem 5] for the non graded case.

For arbitrary cycles $\alpha, \beta \in (\wedge_A L, d_0)$, $\phi(\{\alpha, \beta\}) = \{\phi(\alpha), \phi(\beta)\}$ up to homotopy [**2**]. Hence $H(\phi)$ is compatible with the brackets.

□

3. Hochschild cohomology of Sullivan algebras

Let $A = (\wedge V, d)$ be a Sullivan minimal algebra. As A is graded commutative, then $A \otimes A^{op} \cong A \otimes A$. Consider the following Sullivan relative model of the multiplication m [**7**].

$$\begin{array}{ccc}
(\wedge V, d) \otimes (\wedge V, d) & \xrightarrow{\quad m \quad} & (\wedge V, d) \\
& \searrow & \simeq \uparrow \varphi \\
& & (\wedge V \otimes \wedge V' \otimes \wedge \bar{V}, D),
\end{array}$$

where $V' \cong V$, $\bar{V}^n = (sV)^n = V^{n+1}$, $D\bar{v} = v - v' + \alpha$ and $\alpha \in \wedge^+(V \oplus V') \otimes \bar{V}$. Hence

$$(\wedge V \otimes \wedge V' \otimes \wedge \bar{V}, D) \xrightarrow[\simeq]{\varphi} (\wedge V, d)$$

is a free resolution of $A = (\wedge V, d)$ as a graded differential $A \otimes A$-module (see also [**1**]).

A model of the free loop space is obtained as a push out in the following diagram.

$$\begin{array}{ccc} (\wedge V, d) \otimes (\wedge V, d) & \xrightarrow{m} & (\wedge V, d) \\ \downarrow & & \downarrow \\ (\wedge V \otimes \wedge V' \otimes \wedge \bar{V}, D) & \longrightarrow & (\wedge V \otimes \wedge \bar{V}, D). \end{array}$$

We deduce isomorphisms of vector spaces

$$\begin{aligned} HH^*(A; A) \cong \mathrm{Ext}_{A \otimes A}(A, A) &= H_*(\mathrm{Hom}_{\wedge V \otimes \wedge V}(\wedge V \otimes \wedge V \otimes \wedge \bar{V}, \wedge V), D) \\ &\cong (\mathrm{Hom}_{\wedge V}(\wedge V \otimes \wedge sV, \wedge V), \tilde{D}) \\ &\cong H_*(\mathrm{Hom}_{\mathbb{Q}}(\wedge \bar{V}, \wedge V), \bar{D}). \end{aligned}$$

Moreover the differential D verifies the condition $D(\wedge V \otimes \wedge^n sV) \subset \wedge V \otimes \wedge^n sV$ [**1**]. Hence each subspace $\mathrm{Hom}_{\wedge V}(\wedge V \otimes \wedge^n sV, \wedge V)$ is a sub complex of $\mathrm{Hom}_{\wedge V}(\wedge V \otimes \wedge sV, \wedge V)$.

Let A be a commutative differential graded **k**-algebra and M a differential graded A-module. A derivation of A with values in M is a **k**-linear map $\theta : A \to M$ such that $\theta(ab) = \theta(a)b + (-1)^{|\theta|} a\theta(b)$. Let $s : \wedge V \to \wedge V \otimes sV$ be the derivation defined $s(v) = sv$, for $v \in V$. Define

$$\mu : (\mathrm{Hom}_{\wedge V}(\wedge V \otimes sV, \wedge V), \tilde{D}) \to (Der(\wedge V), \delta)$$

by $\mu(f)(x) = (-1)^{|f|} f(sx)$, for $x \in \wedge V$. It is easily seen that μ is a bijective linear map of degree 1.

LEMMA 8. *The map μ commutes with differentials.*

PROOF. Let $f \in \mathrm{Hom}_{\wedge V}(\wedge V \otimes sV, \wedge V)$.

$$\begin{aligned} (\tilde{D}f)(sv) &= d(f(sv)) - (-1)^{|f|} f(D(sv)) \\ &= d(f(sv)) + (-1)^{|f|} f(sdv)), \end{aligned}$$

hence $(\mu(\tilde{D}f))(v) = -(-1)^{|f|}(d(f(sv)) + (-1)^{|f|} f(sdv)))$. But

$$\begin{aligned} (\delta\mu(f))(v) &= d(\mu(f)(v)) - (-1)^{|\mu(f)|} \mu(f)(dv) \\ &= (-1)^{|f|} d(f(sv)) + f(sdv) \\ &= (-1)^{|f|}(d(f(sv)) + (-1)^{|f|} f(sdv)). \end{aligned}$$

Therefore $\delta\mu(f) = -\mu(\tilde{D}f)$. □

We deduce that

$$\psi_1 = \mu^{-1} \circ s : (s^{-1}\mathcal{L}, \delta') \to (\mathrm{Hom}_{\wedge V}(\wedge V \otimes \bar{V}, \wedge V), \tilde{D})$$

is an isomorphism of chain complexes.

REMARK 9. The inclusion of Corollary 3 is given by the following composition mapping [**9**]:

$$\pi_*(\Omega aut_1 X)\otimes\mathbb{Q}\xrightarrow[\cong]{H(\psi_1)} H(\mathrm{Hom}_{\wedge V}(\wedge V\otimes sV,\wedge V))\hookrightarrow H(\mathrm{Hom}_{\wedge V}(\wedge V\otimes \wedge sV,\wedge V))\ .$$

If $\theta\in\mathcal{L}$, then $s^{-1}\theta\in L$ will be denoted by $\bar{\theta}$. We define a chain map

$$\psi_n:(\wedge_A^n L,d_0)\to(\mathrm{Hom}_{\wedge V}(\wedge V\otimes\wedge^n\bar{V},\wedge V),\tilde{D})$$

by

$$\psi_n(\bar{\theta}_1\wedge\cdots\wedge\bar{\theta}_n)(\bar{v}_1\wedge\cdots\wedge\bar{v}_n)=\sum_{\sigma\in S_n}\epsilon(\sigma)(\psi_1(\bar{\theta}_1))(\bar{v}_{\sigma(1)})\cdots(\psi_1(\bar{\theta}_n))(v_{\sigma(n)}).$$

As a $\wedge V$-module, $\mathcal{L}$ is generated by derivations θ such that $\theta(v)=1$ and zero on other elements of a basis of V. We denote by $(\bar{v}_{i_1}\wedge\ldots\wedge\bar{v}_{i_n},x)$ the element $\alpha\in\mathrm{Hom}_{\wedge V}(\wedge V\otimes\wedge^n\bar{V},\wedge V)$ such that $\alpha(\bar{v}_{i_1}\wedge\ldots\wedge\bar{v}_{i_n})=x$ and zero on other elements of a basis of $\wedge^n\bar{V}$. Now consider the derivations $\theta_{i_k}=(v_{i_k},1)$ and corresponding $\bar{\theta}_{i_k}=s^{-1}\theta_{i_k}$ in L. We define a map

$$\zeta_n:\mathrm{Hom}_{\wedge V}(\wedge V\otimes\wedge^n\bar{V},\wedge V)\to\wedge_A^n L$$

by $\zeta_n(\alpha)=x.(\bar{\theta}_{i_1}\wedge\ldots\wedge\bar{\theta}_{i_n})$. As ζ_n is the inverse of ψ_n, we deduce that each ψ_n is a chain isomorphism, for each $n\geq 1$.

We then define $\psi:\wedge_A L\to\mathrm{Hom}_{\wedge V}(\wedge V\otimes\wedge\bar{V},\wedge V)$ by $\psi|_{\wedge_A^n L}=\psi_n$. We have proved the following result.

THEOREM 10. *Let $A=(\wedge V,d)$ be a minimal Sullivan algebra, $\mathcal{L}$ the $\wedge V$-module of derivations of $\wedge V$ and $L=s^{-1}\mathcal{L}$. Then*

$$\psi:(\wedge_A L,d_0)\to(\mathrm{Hom}_{\wedge V}(\wedge V\otimes\wedge\bar{V},\wedge V),\tilde{D})$$

is an isomorphism of chain complexes.

In order to prove Theorem C of the Introduction it remains to prove that ψ is a homomorphism of graded algebras.

If $A=\wedge V$, we define the $A\otimes A$-module map

$$\jmath:\wedge V\otimes\wedge V\otimes\wedge\bar{V}\to\bar{B}(A;A;A)\quad\text{by}\quad\jmath(\bar{v}_1\wedge\ldots\wedge\bar{v}_n)=\frac{1}{n!}\sum_{\sigma\in S_n}\epsilon(\sigma)[v_{\sigma(1)}|\ldots|v_{\sigma(n)}],$$

where $v_i\in V$.

We consider the commutative diagram

$$\begin{array}{ccc}\wedge V\otimes\wedge V\otimes\wedge\bar{V} & \xrightarrow{\ \jmath\ } & \bar{B}(A;A;A)\\ & \searrow^{\simeq}\quad\swarrow^{\simeq} & \\ & \wedge V & \end{array}$$

Therefore $\jmath$ is also a quasi-isomorphism. Moreover $\jmath$ induces a mapping $\mathrm{Hom}(\wedge\bar{V},\wedge V)\to\mathrm{Hom}(T(\bar{A});\wedge V)$. We recall that $\wedge\bar{V}$ is endowed with a coalgebra structure defined by

$$\bar{\Delta}(\bar{v}_1\wedge\cdots\wedge\bar{v}_n)=\sum_{p=1}^{n-1}\sum_{\sigma}\epsilon(\sigma)(\bar{v}_{\sigma(1)}\wedge\cdots\wedge\bar{v}_{\sigma(p)})\otimes(\bar{v}_{\sigma(p+1)}\wedge\cdots\wedge\bar{v}_{\sigma(n)}),$$

where σ extends to all $(p,n-p)$-shuffles.

As the restriction of $\jmath$ to $\wedge\bar{V}$ induces a coalgebra morphism to $T(\bar{A})$, we deduce that

$$\mathrm{Hom}(\wedge\bar{V},\wedge V)\to\mathrm{Hom}(T(\bar{A}),\wedge V)$$

is a morphism of algebras. Moreover, the following diagram is commutative.

$$\begin{array}{ccc}
(\wedge_A L,d_0) & \xrightarrow{\quad\phi\quad} & (\mathrm{Hom}(T(\bar{A}),\wedge V),D)\\
& {\scriptstyle\psi}\searrow \quad \swarrow{\scriptstyle\mathrm{Hom}(\jmath)} & \\
& (\mathrm{Hom}(\wedge\bar{V},\wedge V),\tilde{D}) &
\end{array}$$

Therefore $H(\psi):H(\wedge_A L,d_0)\to H(\mathrm{Hom}(\wedge\bar{V},\wedge V),\tilde{D})\cong HH(\wedge V;\wedge V)$ is an isomorphism of algebras.

4. The Gottlieb group

Recall that $\alpha\in\pi_n(X)$ is a Gottlieb element of X if $(\alpha,id):S^n\vee X\to X$ extends to $\tilde{\alpha}:S^n\times X\to X$ [**15**]. Gottlieb elements form a subgroup $G_*(X)$ of $\pi_*(X)$. For a simply connected space X, let $\mathcal{L}_X=\pi_*(\Omega X)\otimes\mathbb{Q}$ denote its homotopy Lie algebra. If $\alpha\in\pi_n(\Omega X)\otimes\mathbb{Q}\cong\pi_{n+1}(X_{\mathbb{Q}})$ represents a Gottlieb element, then it is in the center of $\mathcal{L}_X$. Here $X_{\mathbb{Q}}$ denotes the rationalization of X. Moreover for a finite CW-complex X, $G_*(X_{\mathbb{Q}})$ is concentrated in odd degrees. If $(\wedge V,d)$ is the minimal Sullivan model of X, then an element $v\in V^n\simeq\mathrm{Hom}_{\mathbb{Z}}(\pi_n(X),\mathbb{Q})$ represents a Gottlieb element of $\pi_n(X_{\mathbb{Q}})$ if and only if there is a derivation θ of $\wedge V$ verifying $\theta(v)=1$ and such that $[d,\theta]=0$ [**6**]. Such a derivation represents a non zero homology class in $(\mathrm{Der}(\wedge V),\delta)$.

Let A be a graded connected $\mathbf{k}$-algebra and M a graded A-module. We say that $m\in M$ is A-decomposable if $m\in\bar{A}.M$, where $\bar{A}=\oplus_{i>0}A^i$. Otherwise m is called A-indecomposable. For $A=\wedge V$, Gottlieb elements are A-indecomposable in $\mathrm{Der}\wedge V$.

Recall that a simply connected space X is called elliptic if both $\pi_*(X)\otimes\mathbb{Q}$ and $H^*(X,\mathbb{Q})$ are finite dimensional. We have the following result.

THEOREM 11. *If X is an elliptic space and $A=(\wedge V,d)$ is its minimal Sullivan model, then the graded commutative algebra $HH^*(A;A)$ contains a polynomial algebra as a graded sub algebra.*

PROOF. Let $A=(\wedge V,d)$ be the minimal model of an elliptic space X. As $\pi_*(X)\otimes\mathbb{Q}$ is finite dimensional, then V has a basis $\{v_1,\dots,v_n\}$ where

$|v_1| \leq |v_2| \leq \cdots \leq |v_n|$. Moreover v_n represents a non zero Gottlieb element of $X_{\mathbb{Q}}$, hence $G_*(X_{\mathbb{Q}}) \neq 0$.

Let θ be a derivation representing a Gottlieb element $v \in V$. In this case θ is odd and $\bar{\theta} = s^{-1}\theta$ represents a non zero homology class in $H_*(s^{-1}\mathcal{L})$. As θ is an A-indecomposable element of $\mathcal{L}$ as a $\wedge V$-module, so is $\bar{\theta}^k \in \wedge_A L$ for $k \geq 1$. Moreover the image of d_0 is A-decomposable in $\Lambda_A L$. Therefore $\bar{\theta}^k$ cannot be a boundary in $(\Lambda_A L, d_0)$. In fact the sub algebra of $\wedge_A L$ generated by Gottlieb elements is A-indecomposable. We deduce that $H_*(\Lambda_A L, d_0)$ contains a sub algebra isomorphic to $\wedge W$, where $W = s^{-1}G_*(X_{\mathbb{Q}})$. □

As $HH^*(A; A)$ and $\mathbb{H}_*(LX, \mathbb{Q})$ are isomorphic, we deduce the following result.

COROLLARY 12. *If X is an elliptic space, then $\mathbb{H}_*(LX, \mathbb{Q})$ contains a polynomial algebra.*

The BV-homomorphism $H_*(\Omega aut_1 X) \otimes \mathbb{H}_*(X) \to \mathbb{H}_*(LX)$ restricts to a morphism of graded algebras $\Gamma : H_*(\Omega aut_1 X, \mathbb{Q}) \to \mathbb{H}_*(LX, \mathbb{Q})$. This map is not injective in general [**9**], but the above Corollary asserts that Γ becomes injective if restricted to the sub algebra generated by $s^{-1}G_*(X_{\mathbb{Q}})$. In the case $G_*(X_{\mathbb{Q}}) = 0$, one expects cup lengths of $\mathbb{H}_*(LX, \mathbb{Q})$ and of $H^*(X, \mathbb{Q})$ to be related as in the next Example.

EXAMPLE 13. Consider $X = S^3 \vee S^3$ of which the Sullivan minimal model is given by $(A, d) = (\wedge(x, y, z, t, u \dots), d)$ where $dx = dy = 0$, $dz = xy$, $dt = xz$, $du = yz$, $\dots$ The Lie algebra $H_*(\mathcal{L}, \delta)$ is infinite dimensional [**12**]. Moreover $G_*(X_{\mathbb{Q}}) = 0$, hence each element of $H_*(\mathcal{L}, \delta)$ is represented by a derivation with value in $\wedge^+(x, y, \dots)$. As X is formal, there is a quasi-isomorphism $\mathrm{Der}(A, d) \to \mathrm{Der}(A, H^*(A, d))$ induced by the formality quasi-isomorphism $(A, d) \to H^*(A, d) = H$. As $H^+.H^+ = 0$, we deduce that $\alpha\beta = 0$, for $\alpha \in H_*(L, \delta')$ and $\beta \in H_*(\Lambda_A^+ L, d_0)$. However we do not know whether $H_*(\Lambda_A^{\geq 2} L, d_0) = 0$.

References

[1] D. Burghelea and M. Vigué-Poirrier, *Cyclic homology of commutative algebras I*, Lecture Notes in Mathematics 1318 (1988), 51–72.

[2] A. Cattaneo and G. Felder, *Relative formality theorem and quantisation of coisotropic submanifolds*, Adv. Math. 208, No. 2 (2007) , 521-548.

[3] M. Chas and D. Sullivan, *String topology*, preprint math GT/9911159.

[4] R.L. Cohen and J.D.S Jones, *A homotopy theoretic realization of string topology*, Math. Ann. 324(4), (2002), 773-798.

[5] Y. Félix, *La dichotomie elliptic hyperbolic en homotopie rationnelle*, Astérisque vol. 176 (1989), Société Mathématique de France.

[6] Y. Félix and S. Halperin, *Rational LS category and its applications*, Trans. Amer. Math. Soc. 273 (1982), 575–588.

[7] Y. Félix, S. Halperin and J.-C. Thomas, *Rational Homotopy Theory*, Graduate Texts in Mathematics, vol. 205, Springer, New York, 2001.

[8] Y. Félix, L. Menichi and J.-C. Thomas, *Gerstenhaber duality in Hochschild cohomology*, J. Pure Appl. Algebra 199 (2005), 43–59.

[9] Y. Félix and J.-C. Thomas, *Monoid of self-equivalences of free loop spaces*, Proc. Amer. Math. Soc. 132 (2004), 305–312.

[10] Y. Félix, J.-C. Thomas and M. Vigué-Poirrier, *Rational String Topology*, J. Eur. Math. Soc. (JEMS) 9 (2007), 123–156.

[11] Y. Félix, J.-C. Thomas and M. Vigué-Poirrier, *The Hochschild cohomology of a closed manifold*, Publ. Math. IHES. 99 (2004), 235-252.

[12] J.-B. Gatsinzi, *LS-category of classifying spaces*, Bull. Belg. Math. Soc. 2 (1995), 121–126.

[13] M. Gerstenhaber, *The cohomology structure of an associative ring*, Annals of Math. 78 (1963), 267–288.

[14] M. Gerstenhaber and S. Schack, *Algebras, bialgebras, quantum groups, and algebraic deformations*, Deformation theory and quantum groups with applications to mathematical physics, Contemp. Math. 134 (1992), 51-92.

[15] D. Gottlieb, *Evaluation subgroups of homotopy groups*, Amer. J. of Math. 91 (1969), 729–756.

[16] G. Hochschild, B. Kostant and A. Rosenberg, *Differential forms on regular affine algebras*, Trans. Amer. Math. Soc. 102 (1962), 383–408.

[17] J. D. S. Jones, *Cyclic homology and equivariant homology*, Inv. Math. 87 (1987), 403–423.

[18] L. Menichi, *A Batalin-Vilkovisky algebra morphism from double loop spaces in free loops*, preprint math AT/0908.1883, August 2009.

[19] J.D. Stasheff, *The intrinsic bracket on the deformation complex of an associative algebra*, J. Pure Appl. Algebra 89 (1993), 231–235.

[20] D. Sullivan, *Infinitesimal computations in topology,* Publ. I.H.E.S. 47 (1977), 269–331.

University of Botswana, Private Bag 0022 Gaborone, Botswana
E-mail address: gatsinzj@mopipi.ub.bw

Contemporary Mathematics
Volume **519**, 2010

Rational Homotopy Groups of Function Spaces.

J.-B. Gatsinzi and R. Kwashira

ABSTRACT. Given a map $f : X \to Y$ between simply connected finite CW-complexes and $f_* : (\mathbb{L}(V), \delta_V) \to (\mathbb{L}(W), \delta_W)$ its Quillen model, we show that $\pi_*(\mathrm{map}(X, Y; f)) \otimes \mathbb{Q} \cong \mathrm{Ext}_{TV}(\mathbb{Q}, \mathbb{L}(W))$. Moreover, if $\pi_*(\Omega Y) \otimes \mathbb{Q}$ is finite dimensional then $\pi_*(\mathrm{map}(X, Y; f)) \otimes \mathbb{Q}$ and $\pi_*(\mathrm{map}_*(X, Y; f)) \otimes \mathbb{Q}$ are both finite dimensional.

1. Introduction

Through out this paper, spaces are assumed to be 1-connected finite CW-complexes. A map between spaces $f : X \to Y$ induces a map of Quillen models which (abusing notation) we denote $f : \mathbb{L}(V) \to \mathbb{L}(W)$. Lupton and Smith [**6**] showed that from the Lie bracket of $\mathbb{L}(V)$ and $\mathbb{L}(W)$, one can extend the notion of derivation of a differential graded Lie algebra to a derivation with respect to a map of differential graded Lie algebras. Also, they proved the following vector space isomorphisms,

$$\pi_n(\mathrm{map}_*(X, Y; f)) \otimes \mathbb{Q} \xrightarrow{\cong} H_n(\mathrm{Der}\,(\mathbb{L}(V), \mathbb{L}(W); f)),$$

$$\pi_n(\mathrm{map}(X, Y; f)) \otimes \mathbb{Q} \xrightarrow{\cong} H_n(s\mathbb{L}(W) \oplus \mathrm{Der}\,(\mathbb{L}(V), \mathbb{L}(W); f)),$$

for a map $X \xrightarrow{f} Y$ and $\mathbb{L}(V) \xrightarrow{f} \mathbb{L}(W)$ its Quillen minimal model.

2000 *Mathematics Subject Classification.* Primary 55P62; Secondary 16E40, 55P35.

Key words and phrases. function space, generalized derivation, Gottlieb group.

The first author was partially supported by the Abdus-Salam ICTP and MPI für Mathematik.

The second author was partially supported by OEA of the Abdus-Salam ICTP.

©2010 American Mathematical Society

Gatsinzi [**3**] proved that $\mathrm{Hom}_{TV}(TV\otimes(\mathbb{Q}\oplus sV),\mathbb{L}(V))$ and $\mathbb{L}(V)\oplus \mathrm{Der}\,\mathbb{L}(V)$ are isomorphic as graded vector spaces. This can be extended for a map $f:\mathbb{L}(V)\to\mathbb{L}(W)$.

Our main Theorem states.

THEOREM. *There is an isomorphism of degree* 1

$$F:\mathrm{Hom}_{TV}(TV\otimes(\mathbb{Q}\oplus sV),\mathbb{L}(W))\to s\mathbb{L}(W)\oplus \mathrm{Der}\,(\mathbb{L}(V),\mathbb{L}(W);f).$$

2. Generalized Lie derivations

For a simply connected space X, there exists a differential graded Lie algebra $(\mathbb{L}(V),\delta_V)$, where $\mathbb{L}(V)$ denotes the free graded Lie algebra on the rational vector space V, called a *Quillen model* and whose homology is isomorphic to $\pi_*(\Omega X)\otimes\mathbb{Q}$, the *(rational) homotopy Lie algebra* of X.

A map of spaces $f:X\to Y$ induces a map of Quillen models which we denote abusively by $f:(\mathbb{L}(V),\delta_V)\to(\mathbb{L}(W),\delta_W)$.

A *f-derivation of degree n* is a linear map $\theta:\mathbb{L}(V)\to\mathbb{L}(W)$ that increases degree by n and satisfies $\theta([x,y])=[\theta(x),f(y)]+(-1)^{n|x|}[f(x),\theta(y)]$ for $x,y\in\mathbb{L}(V)$. Denote by $\mathrm{Der}_n\,(\mathbb{L}(V),\mathbb{L}(W);f)$ the space of all f-derivations of degree n from $\mathbb{L}(V)$ to $\mathbb{L}(W)$. Define

$$\tilde{D}:\mathrm{Der}_n\,(\mathbb{L}(V),\mathbb{L}(W);f)\to\mathrm{Der}_{n-1}\,(\mathbb{L}(V),\mathbb{L}(W);f)$$

by $\tilde{D}(\theta)=\delta_W\theta-(-1)^{|\theta|}\theta\delta_V$. Then $(\mathrm{Der}_*(\mathbb{L}(V),\mathbb{L}(W);f),\tilde{D})$ is a differential graded vector space.

The adjoint map associated to f is

$$\mathrm{ad}_f:\mathbb{L}(W)\to\mathrm{Der}(\mathbb{L}(V),\mathbb{L}(W);f),$$

where $\mathrm{ad}_f(w)(x)=[w,f(x)]$, $w\in\mathbb{L}(W)$.

It is known that for a connected differential graded algebra $(T(V),d)$, there is an acyclic differential module of the form $(T(V)\otimes(\mathbb{Q}\oplus sV),\bar{d})$ [**1**],[**5**]. The differential $\bar{d}$ is defined by

$$\bar{d}v=dv\otimes 1,\ \ \bar{d}(sv)=v\otimes 1-S(dv\otimes 1),$$

where S is the $\mathbb{Q}$-graded vector space map (of degree 1) defined by

$$S(v\otimes 1)=1\otimes sv,\ \ S(1\otimes(\mathbb{Q}\oplus sV))=0,$$

$$S(ax\otimes 1)=(-1)^{|x|}aS(x),\ \forall a\in TV,\ |x|>0.$$

Recall that if A is a differential graded algebra and M and N are A-differential graded modules, then $\mathrm{Hom}_A(M,N)$ is a differential vector space. For $f \in \mathrm{Hom}_A(M,N)$, the differential is given by, $Df = d_M \circ f - (-1)^{|f|} f \circ d_N$.

From [**3**], consider the map

$$\Phi : \mathrm{Hom}_{TV}(TV \otimes (\mathbb{Q} \oplus sV), \mathbb{L}(V)) \xrightarrow{\cong} s\mathbb{L}(V) \oplus \mathrm{Der}\, \mathbb{L}(V)$$

defined by $\Phi(f) = \theta + (-1)^{|f|} sf(1)$, where θ is the derivation of $\mathbb{L}(V)$ defined by $\theta(v) = f(sv)$. Moreover $\theta(x) = f(S(x \otimes 1))$, for $x \in \mathbb{L}(V) \subset T(V)$. Clearly Φ is an isomorphism of complexes. This can be extended for a map $f : \mathbb{L}(V) \to \mathbb{L}(W)$. Denote by $Uf : TV \to TW$ the enveloping algebra of f. The adjoint action of TW on $\mathbb{L}(W)$ combined with Uf induces a $T(V)$-module structure on $\mathbb{L}(W)$. Then $(s\mathbb{L}(W) \oplus \mathrm{Der}\,(\mathbb{L}(V), \mathbb{L}(W); f), \tilde{D})$ is a differential graded vector space with the differential $\tilde{D}$ extended to $s\mathbb{L}(W)$ by $\tilde{D}(sw) = -s\delta w + \mathrm{ad}_f(w)$.

Define a map

(1)
$$F : \mathrm{Hom}_{TV}(TV \otimes (\mathbb{Q} \oplus sV), \mathbb{L}(W)) \to s\mathbb{L}(W) \oplus \mathrm{Der}(\mathbb{L}(V), \mathbb{L}(W); f)$$

as follows. For $k \in \mathrm{Hom}_{TV}(TV \otimes (\mathbb{Q} \oplus sV), \mathbb{L}(W))$, $F(k) = (-1)^{|k|} sk(1) + \theta$ where $\theta \in \mathrm{Der}\,(\mathbb{L}(V), \mathbb{L}(W); f)$ is defined by

$$\theta(v) = k(sv) = k(S(v \otimes 1)).$$

Here θ has degree $|k|+1$ and verifies the relation $\theta(x) = k(S(x \otimes 1))$, where $x \in \mathbb{L}(V)$.

THEOREM 1. *The map F is an isomorphism of differential graded vector spaces.*

PROOF. It is straightforward that F is a one-one morphism of complexes. We show that F commutes with differentials. Let $k \in \mathrm{Hom}_{TV}(TV \otimes (\mathbb{Q} \oplus sV), \mathbb{L}(W))$. On one hand,

$$\begin{aligned} \tilde{D}(Fk) &= \tilde{D}((-1)^{|k|} sk(1) + \theta) \\ &= (-1)^{|k|} \tilde{D}(sx) + \tilde{D}\theta, \text{ where } \tilde{D}\theta = [\delta, \theta] = \delta_W \theta - (-1)^{|\theta|} \theta \delta_V,\ x = k(1) \\ &= -(-1)^{|k|} s\delta x + (-1)^{|k|}\, \mathrm{ad}_f\ x + [\delta, \theta]. \end{aligned}$$

On the other hand

$$\begin{aligned} F(Dk) &= -(-1)^{|k|} s(Dk)(1) + \theta' \\ &= -(-1)^{|k|} s\delta(k(1)) + \theta', \\ &\quad \text{where } \theta' \in \mathrm{Der}\,(\mathbb{L}(V), \mathbb{L}(W)) \text{ is defined by } \theta'(v) = (Dk)(sv). \end{aligned}$$

In order to verify that $\tilde{D}(Fk) = F(Dk)$ we first note that $\tilde{D}k(1) = \delta k(1) = \delta x$.
Next,

$$\begin{aligned}
\theta'(v) &= (Dk)(sv)\\
&= \delta_W k(sv) - (-1)^{|k|}k(\bar{d}(sv))\\
&= \delta_W k(sv) - (-1)^{|k|}k(v\otimes 1 - S(dv\otimes 1))\\
&= \delta_W \theta(v) - (-1)^{|k|}k(v\otimes 1) + (-1)^{|k|}k(S(dv\otimes 1))\\
&= \delta_W \theta(v) - (-1)^{|k|}(-1)^{|k||v|}[f(v),k(1)] + (-1)^{|k|}\theta(dv)\\
&= \delta_W \theta(v) - (-1)^{|k|+|k||v|}(-1)(-1)^{|k||v|}[k(1),f(v)] + (-1)^{|k|}\theta(dv)\\
&= \delta_W \theta(v) + (-1)^{|k|}[k(1),f(v)] + (-1)^{|\theta|+1}\theta(dv)\\
&= [\delta,\theta](v) + (-1)^{|k|}(\mathrm{ad}_f\ x)(v).
\end{aligned}$$

Then $\theta' = (-1)^{|k|}\,\mathrm{ad}_f\ x + [\delta,\theta]$, resulting in $\tilde{D}(Fk) = F(Dk)$. Therefore F is an isomorphism of differential graded vector spaces. □

COROLLARY 2. *The graded vector space $\pi_*(map(X,Y;f))\otimes\mathbb{Q}$ is isomorphic to $Ext_{TV}(\mathbb{Q},\mathbb{L}(W))$, where $\mathbb{L}(W)$ is considered as a $T(V)$-module via the map Uf and the adjoint action of TW on $\mathbb{L}(W)$.*

PROOF. Recall from [**6**] that

$$\pi_n(\mathrm{map}(X,Y;f))\otimes\mathbb{Q} \xrightarrow{\cong} H_n(s\mathbb{L}(W)\oplus \mathrm{Der}\,(\mathbb{L}(V),\mathbb{L}(W);f),\tilde{D}).$$

Since

$$\begin{aligned}
\mathrm{Ext}_{TV}(\mathbb{Q},\mathbb{L}(W)) &\cong H_*(s\mathbb{L}(W)\oplus \mathrm{Der}\,(\mathbb{L}(V),\mathbb{L}(W);f),\tilde{D})\\
&\cong \pi_*(\mathrm{map}(X,Y;f))\otimes\mathbb{Q},
\end{aligned}$$

we derive the result. □

A mapping $g:\mathbb{L}(W)\to L$ induces a homomorphism

$$g_*:\mathrm{Der}_*\,(\mathbb{L}(V),\mathbb{L}(W);f)\to \mathrm{Der}_*\,(\mathbb{L}(V),L;g\circ f).$$

PROPOSITION 3. *If g is a quasi-isomorphism then,*

(1) $g_*:\mathrm{Der}\,(\mathbb{L}(V),\mathbb{L}(W)) \xrightarrow{\simeq} \mathrm{Der}\,(\mathbb{L}(V),L)$
and
(2) $\tilde{g}_*: s\mathbb{L}(W)\oplus\mathrm{Der}\,(\mathbb{L}(V),\mathbb{L}(W)) \xrightarrow{\simeq} sL\oplus \mathrm{Der}\,(\mathbb{L}(V),L)$

are quasi-isomorphisms.

PROOF. For the second assertion consider the following commutative diagram:

$$\begin{array}{ccc}
\mathrm{Hom}_{TV}(P,\mathbb{L}(W)) & \xrightarrow[\tilde{g}]{\simeq} & \mathrm{Hom}_{TV}(P,L)\\
{\scriptstyle F}\downarrow{\scriptstyle\simeq} & & {\scriptstyle F'}\downarrow{\scriptstyle\simeq}\\
s\mathbb{L}(W)\oplus\mathrm{Der}\,(\mathbb{L}(V),\mathbb{L}(W);f) & \xrightarrow{\tilde{g}_*} & sL\oplus\mathrm{Der}\,(\mathbb{L}(V),L;g\circ f)
\end{array}$$

where $P = TV \otimes (\mathbb{Q} \oplus sV)$. By Theorem 1, F and F' are isomorphisms. As $\tilde{g}$ is a quasi-isomorphism, we deduce that $\tilde{g}_*$ is also a quasi-isomorphism.

For the first assertion we have the following commutative diagram

$$\begin{array}{ccccc}
\mathbb{L}(W) & \xrightarrow{\mathrm{ad}_f} & \mathrm{Der}\,(\mathbb{L}(V), \mathbb{L}(W)) & \longrightarrow & s\mathbb{L}(W) \oplus \mathrm{Der}\,(\mathbb{L}(V), \mathbb{L}(W)) \\
\downarrow g & & \downarrow g_* & & \downarrow \tilde{g}_* \\
L & \xrightarrow{\mathrm{ad}_{g\circ f}} & \mathrm{Der}\,(\mathbb{L}(V), L) & \longrightarrow & sL \oplus \mathrm{Der}\,(\mathbb{L}(V), L)
\end{array}$$

and the corresponding long exact sequence in homology. As g and $\tilde{g}_*$ are quasi-isomorphisms, so is g_*, by applying the "five lemma". □

COROLLARY 4. *If $\pi_*(\Omega Y)\otimes\mathbb{Q}$ is finite dimensional, then $\pi_*(\mathrm{map}(X, Y; f))\otimes \mathbb{Q}$ and $\pi_*(\mathrm{map}_*(X, Y; f)) \otimes \mathbb{Q}$ are both finite dimensional.*

PROOF. Since $\dim \pi_*(\Omega Y) \otimes \mathbb{Q} < \infty$, $H_k(\mathbb{L}(W), \delta) = 0$ for $k \geq N$, for some N. Denote by I the acyclic differential ideal $I = \mathbb{L}(W)_{\geq N+1} \oplus (\mathrm{Ker}\,\delta)_N$. Therefore $(\mathbb{L}(W), \delta)$ is quasi-isomorphic to the quotient differential algebra $(L, \delta) = (\mathbb{L}(W)/I, \delta)$. Hence we have the following isomorphisms.

$$\begin{aligned}
\mathrm{Ext}_{TV}(\mathbb{Q}, L) &\cong H_*((\mathrm{Hom}_{TV}(TV \otimes (\mathbb{Q} \oplus sV), L), D)) \\
&\cong H_*((\mathrm{Hom}_{\mathbb{Q}}(\mathbb{Q} \oplus sV, L), D)).
\end{aligned}$$

As V and L are both finite dimensional, we deduce that $\mathrm{Ext}_{TV}(\mathbb{Q}, L)$ is finite dimensional. Now $\mathrm{Ext}_{TV}(\mathbb{Q}, L) \cong H_*(sL \oplus \mathrm{Der}\,(\mathbb{L}(V), L))$ and $\pi_*(\mathrm{map}_*(X, Y; f)) \cong \mathrm{Der}\,(\mathbb{L}(V), L)$, consequently $\pi_*(\mathrm{map}(X, Y; f)) \otimes \mathbb{Q}$ and $\pi_*(\mathrm{map}_*(X, Y; f)) \otimes \mathbb{Q}$ are both finite dimensional. □

EXAMPLE 5. Consider the formal space X for which $L = \pi_*(\Omega X)\otimes \mathbb{Q} =< z_1, z_3, z_4, z_6 >$, with $z_6 = [z_3, z_3], z_4 = [z_1, z_3]$ and other brackets are zero. Let $(\mathbb{L}(x_1, x_3), \delta)$, $\delta x_1 = 0$, $\delta x_3 = [x_1, x_1]$ be the Quillen minimal model of $\mathbb{C}P(2)$. Consider the mapping $f : (\mathbb{L}(x_1, x_3), \delta) \to (L, 0)$ defined by $f(x_1) = z_1$ and $f(x_3) = 0$. Our aim is to compute $\mathrm{Ext}_{TV}(\mathbb{Q}, L)$. As $\mathrm{Hom}_{TV}(TV \otimes (\mathbb{Q} \oplus sV), L) \cong \mathrm{Hom}(\mathbb{Q} \oplus sV, L)$, we will denote by $< x, y >$ the mapping θ such that $\theta(x) = y$ and zero on other elements of the basis. We have the following elements.

$$\begin{array}{lll}
\alpha_1 =< 1, z_1 >, & \alpha_3 =< 1, z_3 >, & \alpha_4 =< 1, z_4 >, \\
\alpha_6 =< 1, z_6 >, & \beta_1 =< sx_1, z_3 >, & \beta_2 =< sx_1, z_4 >, \\
\beta_4 =< sx_1, z_6 >, & \gamma_0 =< sx_3, z_4 >, & \gamma_2 =< sx_3, z_6 > .
\end{array}$$

The computation shows that $\mathrm{Ext}_{TV}(\mathbb{Q}, L)$ is spanned as a vector space by α_1, α_4, α_6, β_4 and γ_2.

3. The Gottlieb groups

The Gottlieb group $G_*(X)$ of X is the image of the map induced in homotopy by the evaluation $\operatorname{aut} X \to X$ [**4**], or equivalently, the image of the connecting map of the long exact sequence of homotopy groups of the universal fibration $X \xrightarrow{i} B\operatorname{aut}^\bullet X \to B\operatorname{aut} X$, that is,

$$\cdots \to \pi_{n+1}(B\operatorname{aut} X) \xrightarrow{\delta} \pi_n(X) \xrightarrow{i_\sharp} \pi_n(B\operatorname{aut}^\bullet X) \to \pi_n(B\operatorname{aut} X) \to \cdots$$

As $G_n(X) = \operatorname{Image}\{\pi_{n+1}(B\operatorname{aut} X) \xrightarrow{\delta} \pi_n(X)\}$, and $\pi_*(B\operatorname{aut} X) \cong H_*(s\mathbb{L}(V) \oplus \operatorname{Der}\mathbb{L}(V))$ [**3**], therefore $G_n(X) = \operatorname{Image}\{H_n(s\mathbb{L}(V) \oplus \operatorname{Der}(\mathbb{L}(V)) \xrightarrow{\delta} H_{n-1}(\mathbb{L}(V))\}$ [**9**]. Moreover, given $\mathbb{L}(V) \xrightarrow{\mathrm{ad}} \operatorname{Der}\mathbb{L}(V) \to s\mathbb{L}(V) \oplus \operatorname{Der}\mathbb{L}(V)$, $G_n(X) = \ker(H_*(\mathrm{ad}))$. Let $f : X \to Y$ be a based map of connected spaces, the evaluation at the base point of X determines a fibration $\omega : \operatorname{map}(X, Y; f) \to Y$. The nth evaluation subgroup of f is the subgroup

$$G_n(Y, X; f) = \operatorname{Image}\{\omega_\sharp : \pi_n(\operatorname{map}(X, Y; f)) \to \pi_n(Y)\}$$

of $\pi_n(Y)$. In particular the nth Gottlieb group $G_n(X)$ of a space X is the group $G_n(X, X; id)$. Lupton and Smith extend the description of the rationalized Gottlieb group to a description of the rationalized evaluation subgroup of a map [**6**], they proved that $G_n(Y, X; f) \otimes \mathbb{Q} = \ker H_{n-1}(\mathrm{ad}_f)$.

Following [**3**], we define an evaluation map

$$\mathrm{ev} : \operatorname{Hom}_{TV}(TV \otimes (\mathbb{Q} \oplus sV), \mathbb{L}(W)) \to \mathbb{L}(W)$$

by $\mathrm{ev}(k) = k(1)$. Consider the induced map in homology

$$H_*(\mathrm{ev}) : \operatorname{Ext}_{TV}(\mathbb{Q}, \mathbb{L}(W)) \to H_*(\mathbb{L}(W), \delta).$$

PROPOSITION 6.

$$G_n(Y, X; f) \otimes \mathbb{Q} \cong \operatorname{Im}(H_*(\mathrm{ev})).$$

PROOF. It is an immediate consequence of the following commutative diagram,

$$\begin{array}{ccc}
s\mathbb{L}(W) \oplus \operatorname{Der}(\mathbb{L}(V), \mathbb{L}(W); f) & \xrightarrow{\quad p \quad} & \mathbb{L}(W), \\
\downarrow \simeq & & \| \\
\operatorname{Hom}_{TV}(TV \otimes (\mathbb{Q} \oplus sV), \mathbb{L}(W)) & \xrightarrow{\quad \mathrm{ev} \quad} & \mathbb{L}(W)
\end{array}$$

where p is the canonical projection. □

EXAMPLE 7. Consider the mapping $\mathbb{C}P(2) \to X$ as defined in Example 5. Clearly $\operatorname{Im} H_*(\mathrm{ev})$ is spanned by $\{z_1, z_4, z_6\}$. As X is coformal,

$G_*(X)\otimes\mathbb{Q}$ is the centre of L, hence it is spanned by $\{z_4, z_6\}$. Here $G_*(X,\mathbb{C}P(2))\otimes\mathbb{Q}$ strictly contains $G_*(X)\otimes\mathbb{Q}$.

4. The Lie bracket

Recall that TV is endowed with a Hopf algebra structure. The diagonal map $\Delta : TV\to TV\otimes TV$ is defined by $\Delta(v)=v\otimes 1+1\otimes v$, for $v\in V$. If M and N are A-modules, then $M\otimes N$ is clearly an $A\otimes A$-module by the action $(a_1\otimes a_2)(m\otimes n)=(-1)^{|a_2||m|}(a_1m)\otimes(a_2n)$ for $a_i\in A$, $m\in M$ and $n\in N$. Therefore, given differential TV-modules M and N, $M\otimes N$ is also a differential TV-module by the action induced by the diagonal map. Let (P,d) be a semi-free resolution of $\mathbb{Q}$ as a TV-module. Clearly $P\otimes P\to\mathbb{Q}$ is a quasi-isomorphism, hence $P\otimes P$ a semi-free resolution of $\mathbb{Q}$ as a TV-module. Consequently there is a quasi-isomorphism of TV- modules $\rho : P\to P\otimes P$. Let $\phi,\ \psi\in\mathrm{Hom}_{TV}(P,\mathbb{L}(W))$. The composition mapping

$$P\xrightarrow{\rho}P\otimes P\xrightarrow{\phi\otimes\psi}\mathbb{L}(W)\otimes\mathbb{L}(W)\xrightarrow{[-,-]}\mathbb{L}(W)\tag{2}$$

defines a bilinear map

$$\mathrm{Hom}_{TV}(P,\mathbb{L}(W))\otimes\mathrm{Hom}_{TV}(P,\mathbb{L}(W))\to\mathrm{Hom}_{TV}(P,\mathbb{L}(W)).$$

This yields a Lie bracket on $\mathrm{Hom}_{TV}(P,\mathbb{L}(W))$.

Assume that $(\mathbb{L}(V),\delta)=(\mathbb{L}(V),\delta_1+\delta_2)$, where $\delta_1V\subset V$ and $\delta_2V\subset\mathbb{L}^2(V)$. Following [**2**], we define a Lie bracket on $\mathrm{Der}(\mathbb{L}(V),\mathbb{L}(W);f)$ as follows. Take $\phi,\psi\in\mathrm{Der}(\mathbb{L}(V),\mathbb{L}(W);f)$. For $v\in V$, write $\delta_2(v)=\sum_{ij}\alpha_{ij}[v_i,v_j]$, where $\{v_i\}$ forms a basis of V. Define

$$[\phi,\psi](v)=\sum_{ij}(-1)^{|v_i||\psi|}\alpha_{ij}[\phi(v_i),\psi(v_j)]-(-1)^{|v_i||\phi|+|\phi||\psi|}\alpha_{ij}[\psi(v_i),\phi(v_j)].\tag{3}$$

Consider $(P=TV\otimes(\mathbb{Q}\oplus sV),d)$. For $v\in V\subset TV$, $d_1v=\sum\beta_jv_j$, $d_2v=\sum_{ij}\alpha_{ij}v_iv_j$, where $\{v_j\}$ is a basis of V. Hence $d(sv)=v\otimes 1+\sum\beta_j(sv_j)+\sum_{ij}\alpha_{ij}(-1)^{|v_i|}v_i\otimes(sv_j)$. From time to time sv and $v\otimes sw$ will be denoted by $\bar v$ and $v\bar w$ respectively. Moreover $a\otimes b\in P\otimes P$ will be written $a|b$ whenever necessary to avoid confusion between respective elements of P and $P\otimes P$.

We give an explicit definition of $\rho : P\to P\otimes P$ referred to in Equation 2. Define $\rho(1)=1\otimes 1$ and $\rho(\bar v)=\bar v|1+1|\bar v-\sum(-1)^{|v_i|}a_{ij}(\bar v_i|\bar v_j)$. We extend ρ as a morphism of TV-modules. We define $S\otimes S : V\otimes V\to sV\otimes sV\subset P\otimes P$ by $(S\otimes S)(\sum v_i\otimes w_i)=\sum(-1)^{|v_i|}(\bar v_i|\bar w_i)$. Hence $\rho(\bar v)=\bar v|1+1|\bar v-(S\otimes S)(d_2v)$.

LEMMA 8. *The map ρ is a morphism of chain complexes.*

PROOF. We need to verify that $\rho d = d\rho$.

$$\begin{aligned}
\rho(d(sv)) &= \rho(v - sd_1v - s(d_2v)) \\
&= v.\rho(1) - (sd_1v|1) - (1|sd_1v) + (S\otimes S)(d_2(d_1v)) - \rho(sd_2v) \\
&= (v\otimes 1 + 1\otimes v)(1|1) - (sd_1v|1) - (1|sd_1v) \\
&\quad -(S\otimes S)(d_2(d_1v)) - \rho(sd_2v) \\
&= v|1 + 1|v - (sd_1v|1) - (1|sd_1v) + (S\otimes S)(d_2(d_1v)) - \rho(sd_2v).
\end{aligned}$$

But

$$\begin{aligned}
(S\otimes S)(d_2d_1v) &= -(S\otimes S)(d_1d_2v) \\
&= -(S\otimes S)(d_1 \textstyle\sum_{ij} a_{ij}v_iv_j) \\
&= -(S\otimes S)(\textstyle\sum_{ij} a_{ij}((d_1v_i)v_j + (-1)^{|v_i|}v_i(d_1v_j)) \\
&= \textstyle\sum_{ij}(-1)^{|v_i|}a_{ij}(sd_1v_i)|\bar{v}_j - \sum_{ij} a_{ij}\bar{v}_i|(sd_1v_j).
\end{aligned}$$

Moreover,

$$\begin{aligned}
\rho(sd_2v) &= \rho(\textstyle\sum_{ij}(-1)^{|v_i|}a_{ij}v_i\bar{v}_j) = \sum_{ij}(-1)^{|v_i|}a_{ij}v_i.\rho(\bar{v}_j) \\
&= \textstyle\sum_{ij}(-1)^{|v_i|}a_{ij}(v_i\otimes 1 + 1\otimes v_i)(\bar{v}_j|1 + 1|\bar{v}_j - (S\otimes S)(d_2v_j)) \\
&= \textstyle\sum_{ij}(-1)^{|v_i|}a_{ij}(v_i\bar{v}_j|1 + 1|v_i\bar{v}_j + (-1)^{|v_i|(|v_j|+1)}\bar{v}_j|v_i + v_i|\bar{v}_j) \\
&\quad - \textstyle\sum_{ij}(-1)^{|v_i|}a_{ij}(v_i\otimes 1 + 1\otimes v_i)(S\otimes S)(d_2v_j) \\
&= sd_2v|1 + 1|sd_2v + \textstyle\sum_{ij}(-1)^{|v_i|}a_{ij}v_i|\bar{v}_j + \sum_{ij} a_{ij}(-1)^{|v_i||v_j|}\bar{v}_j|v_i \\
&\quad - \textstyle\sum_{ij}(-1)^{|v_i|}a_{ij}v_i(S\otimes S)(d_2v_j).
\end{aligned}$$

Therefore

$$\begin{aligned}
\rho(d(sv)) &= v|1 + 1|v - sd_1v|1 - 1|sd_1v - sd_2v|1 - 1|sd_2v \\
&\quad + \textstyle\sum_{ij}(-1)^{|v_i|}a_{ij}(sd_1v_i)|\bar{v}_j - \sum_{ij} a_{ij}\bar{v}_i|(sd_1v_j) \\
&\quad - \textstyle\sum_{ij} a_{ij}(-1)^{|v_i|}v_i|\bar{v}_j - \sum_{ij}(-1)^{|v_i||v_j|}a_{ij}\bar{v}_j|v_i \\
&\quad + \textstyle\sum_{ij}(-1)^{|v_i|}a_{ij}v_i(S\otimes S)(d_2v_j).
\end{aligned}$$

On the other hand

$$\begin{aligned}
d(\rho(\bar{v})) &= d(\bar{v}|1 + 1|\bar{v} - (S\otimes S)(d_2v)) \\
&= (v - sd_1v - sd_2v)|1 + 1|(v - sd_1v - sd_2v) - d(S\otimes S)(d_2v) \\
&= v|1 - sd_1v|1 - sd_2v|1 + 1|v - 1|sd_1v - 1|sd_2v - d(S\otimes S)(d_2v).
\end{aligned}$$

But

$$\begin{aligned}
d(S\otimes S)(d_2v) &= d(S\otimes S)(\textstyle\sum_{ij} a_{ij}v_iv_j) \\
&= d(\textstyle\sum_{ij}(-1)^{|v_i|}a_{ij}\bar{v}_i|\bar{v}_j) \\
&= \textstyle\sum_{ij}(-1)^{|v_i|}a_{ij}(d\bar{v}_i|\bar{v}_j - (-1)^{|v_i|}\bar{v}_i|d\bar{v}_j) \\
&= \textstyle\sum_{ij}(-1)^{|v_i|}a_{ij}(v_i - sd_1v_i - sd_2(v_i))|\bar{v}_j \\
&\quad - \textstyle\sum_{ij} a_{ij}\bar{v}_i|(v_j - sd_1v_j - sd_2(v_j)) \\
&= \textstyle\sum_{ij}(-1)^{|v_i|}a_{ij}(v_i|\bar{v}_j - s(d_1v_i)|\bar{v}_j - sd_2(v_i)|\bar{v}_j) \\
&\quad - \textstyle\sum_{ij} a_{ij}(\bar{v}_i|v_j - \bar{v}_i|(sd_1v_j) - \bar{v}_i|(sd_2v_j)).
\end{aligned}$$

Hence

$$\begin{aligned} d(\rho(\bar{v})) &= v|1 + 1|v - sd_1v|1 - 1|sd_1v - sd_2v|1 - 1|sd_2v \\ &\quad - \sum_{ij}(-1)^{|v_i|}a_{ij}(v_i|\bar{v}_j - s(d_1v_i)|\bar{v}_j - s(d_2v_i)|\bar{v}_j) \\ &\quad + \sum_{ij} a_{ij}(\bar{v}_i|v_j - \bar{v}_i|(sd_1v_j) - \bar{v}_i|(sd_2v_j)). \end{aligned}$$

Using the relation $d_2^2v = 0$, one deduces that

$$\sum_{ij} a_{ij}s(d_2v_i)|\bar{v}_j) = \sum_{ij}(-1)^{|v_i|}a_{ij}(v_i \otimes 1)(S \otimes S)(d_2v_j).$$

In the same way

$$-\sum_{ij} a_{ij}\bar{v}_i|(sd_2v_j) = \sum_{ij}(-1)^{|v_i|}a_{ij}(1 \otimes v_i)(S \otimes S)(d_2v_j).$$

Hence ρ commutes with differentials. □

As ρ is a quasi-isomorphism, it provides an explicit formula to compute the Lie bracket on $\mathrm{Hom}_{TV}(TV \otimes (\mathbb{Q} \oplus sV), \mathbb{L}(W))$.

Now we consider the inclusion

$$G : \mathrm{Der}(\mathbb{L}(V), \mathbb{L}(W); f) \to \mathrm{Hom}_{TV}(TV \otimes (\mathbb{Q} \oplus sV), \mathbb{L}(W)).$$

It is the restriction of F^{-1} to $\mathrm{Der}(\mathbb{L}(V), \mathbb{L}(W); f)$, as defined by Equation (1). Hence it is a morphism of chain complexes of degree -1. Moreover if $\theta, \theta' \in \mathrm{Der}(\mathbb{L}(V), \mathbb{L}(W); f)$, then $G([\theta, \theta']) = (-1)^{|\theta|}[G(\theta), G(\theta')]$. We deduce the following result.

THEOREM 9. *The inclusion*

$$\mathrm{Der}(\mathbb{L}(V), \mathbb{L}(W); f) \to \mathrm{Hom}_{TV}(TV \otimes (\mathbb{Q} \oplus sV), \mathbb{L}(W))$$

is a morphism of differential Lie algebras.

References

[1] J.F. Adams and P.J. Hilton, *On the chain algebra of a loop space*, Comment. Math. Helv., 30 (1956), 305–330.

[2] U. Buijs and A. Murillo, *The rational homotopy Lie algebra of function spaces*, Comment. Math. Helv. 83 (2008), 723–739.

[3] J.-B. Gatsinzi, *The homotopy Lie algebra of classifying spaces*, J. Pure Appl. Algebra, 120 (1997), 281-289.

[4] D.H. Gottlieb, *Evaluation subgroups of homotopy groups*, Amer. J. Math., 91 (1969), 729–756.

[5] J.-M. Lemaire, *Algèbres connexes et homologie des espaces de lacets*, Lecture Notes in Mathematics 422, Springer, Berlin, 1974.

[6] G. Lupton and S.B. Smith, *Rationalized evaluation subgroups of a map II: Quillen models and adjoint maps*, J. Pure Appl. Algebra, 209, (2007), 173–188.

[7] D. Quillen, *Rational homotopy theory*, Ann. Math. (2), 90 (1969), 205–295.

[8] M. Schlessinger and J. Stasheff, *Deformations theory and rational homotopy type*, preprint, 1982.

[9] D. Tanré, *Homotopie rationnelle; modèles de Chen, Quillen, Sullivan*, Lecture Notes in Mathematics 1025, Springer, Berlin, 1983.

Department of Mathematics, University of Botswana, Private Bag 0022 Gaborone, Botswana.
E-mail address: gatsinzj@mopipi.ub.bw

Department of Mathematics, University of Botswana, Private Bag 0022 Gaborone, Botswana.
E-mail address: rkwashira@gmail.com

Contemporary Mathematics
Volume **519**, 2010

Formality of the framed little 2-discs operad and semidirect products

Jeffrey Giansiracusa and Paolo Salvatore

ABSTRACT. We prove that the operad of framed little 2-discs is formal. Tamarkin and Kontsevich each proved that the unframed 2-discs operad is formal. The unframed 2-discs is an operad in the category of S^1-spaces, and the framed 2-discs operad can be constructed from the unframed 2-discs by forming the operadic semidirect product with the circle group. The idea of our proof is to show that Kontsevich's chain of quasi-isomorphisms is compatible with the circle actions and so one can essentially take the operadic semidirect product with the homology of S^1 everywhere to obtain a chain of quasi-isomorphisms between the homology and the chains of the framed 2-discs.

1. Introduction

We begin by recalling two closely related operads. First, let D_2 denote the *little 2-discs operad* of Boardman and Vogt. In arity n it is the space of embeddings of the union of n discs into a standard disc, where each disc is embedded by a map which is a translation composed with a dilation. At the level of spaces, group complete algebras over this operad are 2-fold loop spaces, and at the level of homology an algebra over $H_*(D_2)$ is precisely a Gerstenhaber algebra.

A variant of the D_2 operad is the *framed little 2-discs* operad, denoted fD_2, introduced by Getzler [**2**]. Here the little discs are allowed to be embedded by a composition of a dilation, rotation, and translation. The (unframed) little 2-discs operad D_2 is an operad in the category of S^1-spaces, where the circle acts by conjugation. Markl and Salvatore-Wahl [**9**] presented the framed little 2-discs operad as the semidirect product of the circle group S^1 with D_2. In particular, $fD_2(n) = D_2(n) \times (S^1)^n$. Getzler observed that algebras over the homology operad $H_*(fD_2)$ are precisely Batalin-Vilkovisky algebras, and at the space level Salvatore-Wahl proved that a group complete algebra over fD_2 is a 2-fold loop space on a based space with a circle action.

The operad D_2 is homotopy equivalent to the Fulton-MacPherson operad $FM = FM_2$ [**8**] (we drop the subscript since we will only be discussing 2-discs in this note);

2000 *Mathematics Subject Classification.* Primary: 18D50; Secondary: 55P48, 81Q30, 81T45.

Key words and phrases. semidirect product operad, framed little discs, operad, formality, graph complex.

©2010 American Mathematical Society

the space $FM(n)$ is a compactification of the configuration space of n ordered distinct points in the plane modulo translations and positive dilations. As with D_2, the circle acts on FM by rotations. The semidirect product construction for this action gives the *framed Fulton-MacPherson operad* $fFM = fFM_2$, which is homotopy equivalent to fD_2, and such that $fFM(n) = FM(n) \times (S^1)^n$. Both FM and fFM are operads of semi-algebraic sets.

Tamarkin [**11**] and Kontsevich [**6, 5**] proved the following formality theorem.

THEOREM 1.1. *The operad $C_*(FM)$ of chains on FM with real coefficients is quasi-isomorphic to its homology operad $H_*(FM)$, the Gerstenhaber operad.*

(One can also use singular or semi-algebraic chains in the statement; we will return to this point later.) Kontsevich's proof seems more geometric and has the advantage of extending to a proof of formality for the little k-discs for all $k \geq 2$; this proof has been explained in greater detail by Lambrechts and Volic [**7**].

In general, formality of an operad is a powerful property with many theoretical and computational applications. The above operad formality theorem plays an important role in Tamarkin's proof [**12**] of Kontsevich's deformation quantization theorem. Our purpose in writing this note is to show that Kontsevich's proof of formality of the operad FM can be adapted to show the formality of the operad fFM. Our main result is:

THEOREM 1.2. *The operad $C_*(fFM)$ of chains on fFM with real coefficents is quasi-isomorphic to its homology operad $H_*(fFM) = BV$, the Batalin-Vilkovisky operad.*

An independent proof of this formality, built from Tamarkin's method rather than Kontsevich's, is due to Severa [**10**]. One interesting application of this formality result is given in [**1**], where it is used to construct homotopy BV algebra structures on objects such as the chains on double loop spaces.

Unline fD_2, the operad fFM in fact has the structure of a *cyclic operad*, and so it is natural to ask if the formality can be made compatible with the cyclic structure. After the present work was completed we found a proof [**3**] of the stronger result that fFM is formal as a cyclic operad, although that proof is significantly more involved and required the introduction of a new type of graph complex.

Kontsevich's proof showed formality of the little k-discs operad for *all* k, and so it is reasonable to ask if the framed k-discs operads are all formal as well (as operads, or better yet as cyclic operads). We plan to address this question in future work. The proofs given in this paper, [**10**], and [**3**] address only the case $k = 2$. These arguments do not work for $k > 2$ for various reasons. In the Tamarkin formality argument it is essential that the operad spaces are $K(\pi, 1)$s, and this is no longer true for $k > 2$. The argument in this paper does not immediately extend to higher k becasue one would have to replace the group $S^1 = SO(2)$ with $SO(k)$ and find a quasi-isomorphism $H^*SO(k) \to \Omega^*SO(k)$ that is compatible with Kontsevich's integration map; we do not know if this is possible. There are similar obstacles to adapting the argument in [**3**] to higher k.

1.1. Outline of the proof. First recall the outline of Kontsevich's proof of Theorem 1.1. It goes by constructing a certain DG-algebra $G(n)$ of graphs together with a quasi-isomorphism $I : G(n) \to \Omega^*(FM(n))$ to the DG-algebra of semi-algebraic forms, and a projection $G(n) \to H^*(G(n)) = H^*(FM(n))$ that is also a quasi-isomorphism. Both of these quasi-isomorphisms are essentially morphisms of DGA cooperads (this is not quite true — the subtleties here are discussed nicely in [7]). By dualizing one can obtain from this a chain of quasi-isomorphisms giving formality of FM.

What we show in this note is that Kontsevich's formality proof is in compatible in a precise sense with the circle action. The circle action on FM makes $H^*(FM)$ into a cooperad in $H := H^*(S^1)$-comodules. Kontsevich's DGA cooperad of admissible graphs G has a differential given by contracting edges; we define a degree -1 derivation Δ on $G(n)$ given by deleting edges. This derivation defines a H-comodule structure on $G(n)$; we check that that this comodule structure is compatible with the cooperad structure and that the projection $G(n) \to H^*(FM(n))$ is a morphism of H-comodules. Using the quasi-isomorphism $H^*(S^1) \to \Omega^*(S^1)$, this morally allows us to form a diagram of quasi-isomorphisms of semidirect product cooperads

$$\Omega^*(FM \rtimes S^1) \leftarrow G \rtimes H \to H^*(FM) \rtimes H \cong H^*(FM \rtimes S^1).$$

However, the proof is not quite so simple because the functor of semi-algebraic forms is contravariant monoidal and so $\Omega^*(FM \rtimes S^1)$ is not a cooperad on the nose. Nevertheless, this issue can be overcome easily, exactly as discussed in [7] in the unframed case.

2. A degree -1 derivation on admissible graphs

Consider the circle action $\rho_n : S^1 \times FM(n) \to FM(n)$, and let H denote the coalgebra $H^*(S^1) = \mathbb{R}[d\theta]$. The circle action induces an H-comodule structure on $H^*(FM(n))$; the coaction is given by the formula

$$\rho_n^*(x) = [d\theta] \otimes \Delta(x) + 1 \otimes x$$

where

$$\Delta : H^*(FM(n)) \to H^{*-1}(FM(n))$$

is a degree -1 derivation. Clearly the H-comodule structure and the derivation Δ determine each other.

We shall now lift the H-comodule structure to Kontsevich's admissible graph complex $G(n)$ by lifting the derivation Δ. Recall that $G(n)$ is the complex of real vector spaces spanned by admissible graphs on n external vertices [5]. The grading is defined by

$$(\#\text{ edges}) - 2(\#\text{ internal vertices}).$$

Graphs are *admissible* if they are at least trivalent at each internal vertex and satisfy a few additional conditions. Each graph is equipped with a total ordering of its edges, and a permutation of the edges acts on the corresponding generator of $G(n)$ by its sign. Given a graph g with edges $e_1, \ldots, e_k$, the differential is defined by

$$dg = \sum_i (-1)^i g/e_i$$

where g/e_i is the graph obtained from g by collapsing the edge e_i. Any non-admissible terms occuring in the sum are set to zero. Recall that the complex $G(n)$ has a graded commutative algebra structure given by disjoint union of internal vertices and the union of edges. The Kontsevich integral defines a morphism of differential graded algebras $I : G(n) \to \Omega^*(FM(n))$ (the target is the algebra of semi-algebraic forms defined in [4]) and this is a quasi-isomorphism.

DEFINITION 2.1. *A linear operator $\Delta : G(n) \to G(n)$ of degree -1 is defined as follows. Given a graph $g \in G(n)$ with ordered set of edges $e_1, \ldots, e_k$,*

$$\Delta(g) = \sum_i (-1)^{i+1}(g - e_i)$$

where $g - e_i$ is the graph obtained by deleting the edge e_i from g (without identifying the endpoints together). If a summand is a non-admissible graph then we set it to zero.

PROPOSITION 2.2. *The operator Δ satisfies:*

(1) $\Delta^2 = 0$;
(2) *it is a derivation of the algebra $G(n)$;*
(3) *it graded commutes with the differential d of $G(n)$, i.e. $d\Delta = -\Delta d$.*

Hence the rule

$$g \mapsto 1 \otimes g + [d\theta] \otimes \Delta(g)$$

gives $G(n)$ the structure of a DG-comodule over the coalgebra $H := H^*(S^1)$.

Let $\theta_{ij} : FM(n) \to S^1$ be the map measuring the angle of the line from the i-th to the j-th point with the first coordinate axis. The algebra $G(n)$ is freely generated by *indecomposable* graphs, those that do not get disconnected by removing a small neighbourhood of the set of external vertices. If g is indecomposable with no internal vertices then it has only an edge between some vertices i and j, and we denote it $g = \alpha_{ij}$. Kontsevich considers the algebra map

$$q_n : G(n) \to H^*(G(n)) = H^*(FM(n))$$

sending all graphs with internal vertices to 0, and such that

$$q_n(\alpha_{ij}) = \theta_{ij}^*(d\theta) := d\theta_{ij}.$$

PROPOSITION 2.3 (Kontsevich, Lambrechts-Volic). *The collection of maps $\{q_n\}$ assemble to a quasi-isomorphism of DG-cooperads $q : G \to H^*(FM)$.*

PROPOSITION 2.4. *The projection $q_n : G(n) \to H^*(FM(n))$ is a map of H-comodules, i.e. $q \circ \Delta = \Delta \circ q$.*

PROOF. For any graph g, the summands of $\Delta(g)$ have the same number of internal vertices as g. Therefore if g is indecomposable with some internal vertices then $q(\Delta(g)) = \Delta(q(g)) = 0$. If g is indecomposable with no internal vertices, then $g = \alpha_{ij}$ for some i, j, and $\Delta(g) = 1$ (the unit of the algebra $G(n)$ is the graph on n external vertices with no edges). Then $q(g) = [d\theta_{ij}]$. Since the map θ_{ij} is S^1-equivariant, we have that

$$\Delta([d\theta_{ij}]) = \theta_{ij}^*(\Delta([d\theta])) = \theta_{ij}^*(1) = 1$$

and so $\Delta(q(g)) = 1 = q(\Delta(g))$. □

3. Compatibility of Δ with the Kontsevich integral and the cooperad structures

We show in the next lemma that the operator Δ is compatible with the integration map $I : G(n) \to \Omega^*(FM(n))$.

LEMMA 3.1. *For $g \in G(n)$,*

$$\rho_n^*(I(g)) = d\theta \times I(\Delta(g)) + 1 \times I(g) \in \Omega^*(S^1 \times FM(n)).$$

PROOF. If $g = \alpha_{ij}$, then $I(g) = d\theta_{ij}$, $\Delta(g) = 1$, and $\rho_n^*(d\theta_{ij}) = d\theta \times 1 + 1 \times d\theta_{ij}$. If g is a graph with no internal vertices, then it is a product of forms $d\theta_{ij}$ and the result follows by multiplying the corresponding expressions, since ρ_n^*, I are algebra maps and Δ is a derivation. If g has k internal vertices then $I(g) = p_*(I(h))$, for some $h \in G(n+k)$, where p_* denotes the push-forward along the semi-algebraic bundle projection $p : F(n+k) \to F(n)$. It follows from the definition of I that $p_*(I(\Delta(h))) = I(\Delta(g))$. The diagram

$$\begin{array}{ccc} S^1 \times F(n+k) & \xrightarrow{\rho_{n+k}} & F(n+k) \\ \downarrow{\scriptstyle S^1 \times p} & & \downarrow{\scriptstyle p} \\ S^1 \times F(n) & \xrightarrow{\rho_n} & F(n) \end{array}$$

is a pullback of semi-algebraic sets. By Proposition 8.13 in [**4**]

$$(\rho_n)^* \circ p_* = (S^1 \times p)_* \circ \rho_{n+k}^*.$$

Since h has no internal vertices

$$\rho_{n+k}^*(I(h)) = d\theta \times I(\Delta(h)) + 1 \times I(h)$$

and so

$$\begin{aligned} \rho_n^*(I(g)) &= \rho_n^*(p_*(I(h))) = (S^1 \times p)_*(\rho_{n+k}^*(I(h))) \\ &= d\theta \times p_*(I(\Delta(h))) + 1 \times p_*(I(h)) \\ &= d\theta \times I(\Delta(g)) + 1 \times I(g). \end{aligned}$$ □

We show next that the operator Δ is compatible with the cooperad structure of G constructed by Kontsevich. The tensor product of H-comodules is a H-comodule, such that Δ on the tensor product is defined by the Leibniz rule.

PROPOSITION 3.2. *The cooperad structure map*

$$\circ_i : G(m+n-1) \to G(m) \otimes G(n)$$

(with $1 \leq i \leq m$) commutes with Δ; i.e. it is a map of H-comodules.

PROOF. Given a graph $g \in G(m+n-1)$,

$$\circ_i(g) = \sum_j (-1)^{s(j)} g'_j \otimes g''_j,$$

where j ranges over partitions of the set V of internal vertices into two sets V'_j and V''_j. Then $g''_j \in G(n)$ is the full subgraph of g containing the external vertices $\{i, \dots, i+n-1\}$ (relabelled) and the internal vertices in V''_j . The graph g'_j is obtained from g by collapsing g''_j to a single external vertex, and relabelling external vertices. The sign $s(j)$ is the sign of the permutation moving the edges from the ordering of g to the ordering of g'_j followed by the ordering of g''_j. If such graphs have repeated edges or are not admissible then they are identified to zero. From the definition one sees that

$$\circ_i(\Delta(g)) = \sum_j (-1)^{s(j)}(\Delta(g'_j) \otimes g''_j + (-1)^{|g'_j|} g'_j \otimes \Delta(g''_j)) = \Delta(\circ_i(g)).$$ □

4. From H-comodules to semidirect product cooperads

By Proposition 3.2 above one can form the semidirect product cooperad $G \rtimes H$ with $(G \rtimes H)(n) = G(n) \otimes H^{\otimes n}$, by extending to the differential graded setting the construction in section 4 of [**9**], and dualizing it. Explicitly the cooperad structure maps

$$\circ_i : (G \rtimes H)(m+n-1) \to (G \rtimes H)(m) \otimes (G \rtimes H)(n)$$

are the algebra maps defined by sending

$$\begin{aligned} d\theta_k &\mapsto d\theta'_i + d\theta''_{k-i+1} \text{ for } i \leq k \leq n+i-1, \\ d\theta_k &\mapsto d\theta'_k \text{ for } k < i, \\ d\theta_k &\mapsto d\theta'_{k-i+1} \text{ for } k \geq n+i, \end{aligned}$$

and for $g \in G(m+n-1)$,

$$g \mapsto \sum_j (-1)^{s(j)} (g'_j \otimes d\theta'_i \otimes \Delta(g''_j) + g'_j \otimes g''_j).$$

By Proposition 2.4 and Proposition 2.3 the collection q induces a quasi-isomorphism of cooperads

$$G \rtimes H \to H^*(FM) \rtimes H = H^*(fFM).$$

The semi-algebraic differential forms on FM do not exactly constitute a cooperad because semi-algebraic forms is a contravariant monoidal functor. The cross product of forms $\Omega^*(FM(m)) \otimes \Omega^*(FM(n)) \to \Omega^*(FM(m) \times FM(n))$ (which is a quasi-isomorphism) and the operad composition $\circ_i : FM(m) \times FM(n) \to FM(m+n-1)$ induce a zigzag

$$\Omega^*(FM(m+n-1)) \to \Omega^*(FM(m) \times FM(n)) \leftarrow \Omega^*(FM(m)) \otimes \Omega^*(FM(n)).$$

Nevertheless, there is a compatibility rule between operadic composition in G and FM.

LEMMA 4.1 (Lemma 8.19 of [**7**]). *The pullback along the operad composition map $\circ_i^{FM} : FM(m) \times FM(n) \to FM(m+n-1)$, for $g \in G(m+n-1)$, gives*

$$(\circ_i^{FM})^*(I(g)) = \sum_j (-1)^{s(j)} I(g'_j) \times I(g''_j)$$

where $\circ_i^G(g) = \sum_j (-1)^{s(j)} g'_j \otimes g''_j$.

We state next an analogous compatibility condition for the *framed* case. There are quasi-isomorphisms

$$\begin{aligned} \Omega^*(FM(n)) \otimes H^{\otimes n} \to \Omega^*(FM(n)) \otimes \Omega^*(S^1)^{\otimes n} \to \Omega^*(FM(n) \times (S^1)^n) \\ = \Omega^*(fFM(n)). \end{aligned}$$

The first map sends fundamental classes of circles to volume forms, and the second map is the cross product of forms. The composition with the Kontsevich integral gives a quasi-isomorphism of algebras

$$I' : G(n) \otimes H^{\otimes n} = (G \rtimes H)(n) \to \Omega^*(fFM(n)).$$

LEMMA 4.2. *The diagram*

$$\begin{array}{ccccc} (G \rtimes H)(m+n-1) & & \xrightarrow{\circ_i^{G\rtimes H}} & & (G \rtimes H)(m) \otimes (G \rtimes H)(n) \\ \downarrow I' & & & & \downarrow I'\otimes I' \\ \Omega^*(fFM(m+n-1)) & \xrightarrow{(\circ_i^{fFM})^*} & \Omega^*(fFM(m) \times fFM(n)) & \longleftarrow & \Omega^*(fFM(m)) \otimes \Omega^*(fFM(n)) \end{array}$$

commutes.

PROOF. By definition of semidirect product the composition in fFM is

$$(x, z_1, \ldots, z_m) \circ_i^{fFM} (y, w_1, \ldots, w_n)$$
$$= (x \circ_i^{FM} \rho_m(z_i, y), z_1, \ldots, z_{i-1}, z_i w_1, \ldots, z_i w_n, z_{i+1}, \ldots, z_m).$$

The lemma follows from this, Lemma 3.1 and Lemma 4.1. $\square$

We proceed similarly as in section 10 of [**7**] observing that integration on semi-algebraic chains of forms associated to graphs defines pairings

$$C_*(fFM(n)) \otimes (G \rtimes H)(n) \to \mathbb{R}$$

sending $c \otimes g \mapsto \int_c I(g)$, and their adjoints give a quasi-isomorphism of operads

$$C_*(fFM) \to (G \rtimes H)^*.$$

This together with the fact that $q^* : H_*(fFM) \to (G \rtimes H)^*$ is a quasi-isomorphism of operads establishes theorem 1.2.

References

[1] I. Gálvez-Carrillo, A. Tonks, and B. Vallette, Homotopy Batalin-Vilkovisky algebras. arXiv:0907.2246.

[2] E. Getzler, Batalin-Vilkovisky algebras and two-dimensional topological field theories. Comm. Math. Phys. 159 (1994), no. 2, 265–285.

[3] J. Giansiracusa and P. Salvatore, cyclic formality of the framed 2-discs operad and genus zero stable curves, arXiv:0911.4430

[4] R. Hardt, P. Lambrechts, V. Turchin and I. Volic, Real homotopy theory of semi-algebraic sets, arXiv:0806.0476

[5] M. Kontsevich, Operads and motives in deformation quantization. Lett. Math. Phys. 48 (1999) 35–72.

[6] M. Kontsevich, Deformation Quantization of Poisson Manifolds. Lett. Math. Phys. 66 (2003) 157–216.

[7] P. Lambrechts and I. Volic, Formality of the little N-disks operad, arXiv:0808.0457

[8] P. Salvatore, Configuration spaces with summable labels. Cohomological methods in homotopy theory 375–395, Progr. Math., 196, Birkhäuser, Basel, 2001.

[9] P. Salvatore and N. Wahl, Framed discs operads and Batalin-Vilkovisky algebras, Q.J.Math. 54 (2003), 213-231.

[10] P. Severa, Formality of the chain operad of framed little disks, arXiv:math/0902.3576

[11] D. Tamarkin, Formality of chain operad of little discs, Lett. Math. Phys. 66 (2003), n.1-2, 65-72

[12] D. Tamarkin, Another proof of M. Kontsevich formality theorem, arXiv:math/9803025

MATHEMATICAL INSTITUTE, UNIVERSITY OF OXFORD, 24 - 29 ST. GILES, OXFORD, OX1 3LB, UNITED KINGDOM

E-mail address: giansira@maths.ox.ac.uk

DIPARTIMENTO DI MATEMATICA, UNIVERSITA' DI ROMA "TOR VERGATA", VIA DELLA RICERCA SCIENTIFICA, 00133 ROMA, ITALY

E-mail address: salvator@mat.uniroma2.it

Contemporary Mathematics
Volume **519**, 2010

James construction, Fox torus homotopy groups, and Hopf invariants

Marek Golasiński, Daciberg Gonçalves, and Peter Wong

ABSTRACT. For any pointed space Y, the James construction $J(Y)$ is a free monoid generated by Y and its suspension $\Sigma J(Y)$ has the homotopy type $\Sigma\bigvee_{n\geq 1}\Lambda^n Y$, where $\Lambda^n Y$ denotes the n-fold smash product of Y. For any path connected space X, the homotopy groups $\{\pi_i(X)\}$ can be encoded, as sets, in the homotopy group $[J(\mathbb{S}^1),\Omega X]$. In this paper, we make use of the Fox torus homotopy groups $\tau_n(X)$ and their generalizations to encode the homotopy groups of X by embedding $[J_n(\mathbb{S}^1),\Omega X]$ into $\tau_{n+1}(X)$ as a subgroup, where $J_n(Y)$ is the n-th stage of the James filtration of $J(Y)$. We generalize this embedding for generalized Fox groups $\tau_{Y^n}(X)$ and describe the image of $[J_n(Y),\Omega X]$ in $\tau_{Y^n}(X)$. Finally, we describe the generalized Hopf invariants in the context of generalized torus homotopy groups in connection with the James construction.

1. Introduction

In an unpublished manuscript [**2**], F. Cohen and T. Sato made a substantial contribution to classical homotopy theory. They studied, among other topics, the group $\prod[\Omega X]$ defined as the group of homotopy classes of maps from $J(\mathbb{S}^1)$ to ΩX, where $J(Y)$ denotes the James construction [**6**] of a pointed space Y. Since $\Sigma J(\mathbb{S}^1)$ has the homotopy type of $\Sigma\bigvee_{n\geq 1}\Lambda^n\mathbb{S}^1$, it follows that $\prod[\Omega X]$ is isomorphic, as *a set*, to $\prod_{n\geq 2}\pi_n(X)$. They described the group $[J_n(\mathbb{S}^1),\Omega X]$ as a central extension of $\pi_n(\Omega X)$ by $[J_{n-1}(\mathbb{S}^1),\Omega X]$ by showing that the 2-cocycle associated to this central extension is determined by James-Hopf invariants and Samelson products. On the other hand, an elegant construction of the so-called torus homotopy groups of a space X was introduced by R. Fox [**3**] in which the classical Whitehead products

---

2000 *Mathematics Subject Classification.* Primary: 55Q05, 55Q15, 55Q91; secondary: 55M20.

Key words and phrases. James construction, Fox torus homotopy groups, Hopf invariants, generalized Whitehead products.

This work was initiated during the first and second authors' visit to the Mathematics Department, Bates College, June 21 - 28, 2008, and was continued at the Faculty of Mathematics and Computer Science, Nicolaus Copernicus University, November 21 - 28, 2008 and at Bates College, June 2 - 8, 2009. The authors would like to thank Bates College and Copernicus University for their hospitality and support.

©2010 American Mathematical Society

are given by commutators. These groups have been generalized in [**4**] and furthered studied in [**5**].

The main objective of this paper is to provide an embedding of the groups $[J_n(\mathbb{S}^1), \Omega X]$ (Theorem 2.1) and $[J(\mathbb{S}^1), \Omega X]$ into a larger group and simultaneously give another *universal* approach to codify all homotopy groups than the approach used by F. Cohen and T. Sato in [**2**], via the James construction and its filtration. In particular, we conclude from Proposition 3.5 splitting central extensions:

$$1 \to \prod_{k>2}^{n+1} \pi_k(X) \to [J_n(\mathbb{S}^1), \Omega X] \leftrightarrows \mathbb{Z} \to 1$$

for $n \geq 2$ and

$$1 \to \prod_{k>2} \pi_k(X) \to [J(\mathbb{S}^1), \Omega X] \leftrightarrows \mathbb{Z} \to 1,$$

where $X = \mathbb{S}^2, \mathbb{R}P^2$.

Now, we review some facts about Fox groups [**3**] used in this work. Let X be a pointed space with a base-point x_0. For $n \geq 1$,

$$\tau_n(X, x_0) = \pi_1(X^{T^{n-1}}, \overline{x_0})$$

is called the n-th *Fox torus homotopy group* of X, where $X^{T^{n-1}}$ denotes the space of unbased maps from the $(n-1)$-torus T^{n-1} to X and $\overline{x_0}$ is the constant map at x_0. When $n = 1$, we get $\tau_1(X, x_0) = \pi_1(X, x_0)$.

To re-interpret Fox's result, we have shown in [**4**] that

$$\tau_n(X, x_0) \cong [\Sigma(T^{n-1} \sqcup *), X]$$

the group of homotopy classes of base-point preserving maps from the reduced suspension of T^{n-1} adjoined with a distinguished point to X. As a result, we call $F_n = \Sigma(T^{n-1} \sqcup *)$ the n-th *Fox space* with $F_1 = \mathbb{S}^1$, the circle.

One of the main results of [**3**] is the following split exact sequence:

$$0 \to \prod_{i=2}^{n} \pi_i(X, x_0)^{\alpha_i(n)} \to \tau_n(X, x_0) \leftrightarrows \tau_{n-1}(X, x_0) \to 1, \tag{1.1}$$

where $\alpha_i(n) = \binom{n-2}{i-2}$, the binomial coefficient.

With the isomorphism $\tau_{n-1}(\Omega X) \cong \prod_{i=2}^{n} \pi_i(X, x_0)^{\alpha_i(n)}$ shown in [**4**, Theorem 1.1], the sequence (1.1) becomes

$$0 \to \tau_{n-1}(\Omega X) \to \tau_n(X) \leftrightarrows \tau_{n-1}(X) \to 1 \tag{1.2}$$

or equivalently

$$0 \to [\Sigma F_{n-1}, X] \to [F_n, X] \leftrightarrows [F_{n-1}, X] \to 1.$$

Let X be a space and $x_0 \in X$. For any space V, the V-*Fox group* of X is defined to be

$$\tau_V(X, x_0) = [\Sigma(V \sqcup *), X].$$

Recall that the *n-th Fox space* of V is given by $F_n(V) = \Sigma(V^{n-1} \sqcup *)$. It is clear that τ_V reduces to τ_n when $V = T^{n-1}$.

Note that the obvious pointed projection $V \sqcup * \to \mathbb{S}^0$ leads to the inclusion $\pi_1(X) \subseteq \tau_V(X)$ for any nonempty topological space V and a pointed space X. In fact, it has been shown in [**4**] that $\tau_V(X) \cong [\Sigma V, X] \rtimes \pi_1(X)$ for any pointed spaces V and X. Furthermore, this can be deduced from the following split exact sequence of [**4**, Theorem 3.1]

$$1 \to [(V_1 \times V_2)/V_1, \Omega X] \to \tau_{V_1 \times V_2}(X) \overset{\leftarrow}{\to} \tau_{V_1}(X) \to 1 \tag{1.3}$$

which generalizes (1.1).

2. Central extensions and James filtration

In this section, we show how to embed $[J_n(Y), \Omega X]$ in the Fox homotopy group $\tau_{Y^n}(X)$ and derive from the splitting (1.3) of $\tau_{Y^n}(X)$ that $[\Lambda^n Y, \Omega X]$ is central not only in $[J_n(Y), \Omega X]$ but also in $\tau_{Y^n}(X)$. We also describe explicitly the images of elements of $[J_n(Y), \Omega X]$ in $\tau_{Y^n}(X)$. Write $\delta_n(Y) : Y^n/Y^{n-1} \to \Lambda^n Y$ for the canonical map.

THEOREM 2.1. *The following diagram of short exact sequences*

(2.1)

$$\begin{array}{ccccccccc} 1 & \longrightarrow & [\Lambda^n Y, \Omega X] & \longrightarrow & [J_n(Y), \Omega X] & \longrightarrow & [J_{n-1}(Y), \Omega X] & \longrightarrow & 1 \\ & & \downarrow{\scriptstyle \delta_n(Y)^*} & & \downarrow{\scriptstyle \varphi_n(Y)^*} & & \downarrow{\scriptstyle \varphi_{n-1}(Y)^*} & & \\ 1 & \longrightarrow & [Y^n/Y^{n-1}, \Omega X] & \longrightarrow & \tau_{Y^n}(X) & \overset{\leftarrow}{\longrightarrow} & \tau_{Y^{n-1}}(X) & \longrightarrow & 1 \end{array}$$

is commutative and the vertical arrows are injective homomorphisms. Moreover, the extension

(2.2)

$$1 \longrightarrow [\Lambda^n Y, \Omega X] \longrightarrow [J_n(Y), \Omega X] \longrightarrow [J_{n-1}(Y), \Omega X] \longrightarrow 1$$

is central if Y is a co-H-space.

PROOF. Let $e \in Y$ be a base-point. Consider the natural projection map $\varphi_n(Y) : Y^n \to J_n(Y)$ and the inclusion $j_{n-1}(Y) : J_{n-1}(Y) \to J_n(Y)$ given by $[y_1, \ldots, y_{n-1}] \mapsto [y_1, \ldots, y_{n-1}, e]$. Then, $\Sigma j_{n-1}(Y) : \Sigma J_{n-1}(Y) \to \Sigma J_n(Y)$ has a functorial retraction and consequently, the induced homomorphism $j_{n-1}(Y)^* : [J_n(Y), \Omega X] \to [J_{n-1}(Y), \Omega X]$ is onto. Together with the map $Y^n \sqcup * \to Y^n$ by sending $*$ to $(e, e, \ldots, e)$, the map $\varphi_n(Y)$ induces a homomorphism

$$\varphi_n(Y)^* : [J_n(Y), \Omega X] \to [Y^n \sqcup *, \Omega X].$$

Denote by $k_{n-1} : Y^{n-1} \hookrightarrow Y^n$ the inclusion given by $(y_1, \ldots, y_{n-1}) \mapsto (y_1, \ldots, y_{n-1}, e)$ for $(y_1, \ldots, y_{n-1}) \in Y^{n-1}$. Then, it follows that we have the following:

$$\begin{array}{ccccc} J_{n-1}(Y) & \xrightarrow{j_{n-1}(Y)} & J_n(Y) & \xrightarrow{\alpha} & J_n(Y)/J_{n-1}(Y) \\ \uparrow{\scriptstyle \varphi_{n-1}(Y)} & & \uparrow{\scriptstyle \varphi_n(Y)} & & \uparrow \\ Y^{n-1} & \xrightarrow{k_{n-1}(Y)} & Y^n & \xrightarrow{\beta} & Y^n/Y^{n-1} \end{array} \tag{2.3}$$

commutative diagram, where α and β are the canonical projections. Since the space $J_n(Y)/J_{n-1}(Y)$ has the homotopy type of the n-fold smash product $\Lambda^n Y$, the diagram (2.3) induces the commutative diagram (2.1).

To show that the vertical homomorphisms of (2.1) are injective, observe that there is a projection $F_{n+1}(Y) \to \Sigma Y^n$ and the composition $F_{n+1}(Y) \to \Sigma Y^n \overset{\varphi_n(Y)}{\to} \Sigma J_n(Y)$ has a functorial crossed-section.

It follows from the proof of [**4**, Theorem 3.1] that when Y is a co-H space, $\Sigma(Y^n/Y^{n-1})$ has the homotopy type of $\Sigma(Y \vee (Y^{n-1} \wedge Y))$ so that

$$[Y^n/Y^{n-1}, \Omega X] \cong [Y^{n-1} \wedge Y, \Omega X] \times [Y, \Omega X]$$

is abelian. One can then further decompose $\Sigma(Y^n/Y^{n-1})$ so that $\Sigma(Y^n/Y^{n-1}) \simeq \mathcal{Q} \vee \Sigma\Lambda^n Y$ for some space $\mathcal{Q}$. Given $a \in [\Lambda^n Y, \Omega X] \subseteq [Y^n/Y^{n-1}, \Omega X]$ and for any $b \in [Y^{n-1}, \Omega X]$, the generalized Whitehead product $a \circ b \in \tau_{Y^n}(X)$ is trivial for dimension reason. Since the generalized Whitehead product when regarded as an element in $\tau_{Y^n}(X)$ is a commutator, it follows, in view of (1.3), that a must be central in $\tau_{Y^n}(X)$. Thus, the image of $[\Lambda^n Y, \Omega X]$ in $\tau_{Y^n}(X)$ is central. In particular, using the commutative diagram of (2.1), we conclude that $[\Lambda^n Y, \Omega X]$ is central in $[J_n(Y), \Omega X]$. □

Next, we describe the image of $\varphi_n(Y) : [J_n(Y), \Omega X] \to \tau_{Y^n}(X)$.

THEOREM 2.2. *$\operatorname{Im} \varphi_n(Y) = \{a \in \tau_{Y^n}(X) | \iota_{1,n(Y)^*}(a) = \ldots = \iota_{n,n^*}(a)\}$, where $\iota_{j,n(Y)^*}$ is induced by the map $\iota_{j,n}(Y) : Y^{n-1} \to Y^n$ given by*

$$\iota_{j,n}(Y)(y_1, \ldots, y_{n-1}) = (y_1, \ldots, y_{j-1}, e, y_j, \ldots, y_{n-1})$$

for $j \leq n$ and $(y_1, \ldots, y_{n-1}) \in Y^{n-1}$, where $e \in Y$ is the base-point.

PROOF. We proceed by induction on n. For $n = 1$, the map $\varphi_1(Y) : Y \sqcup * \to Y$ is given by $\varphi_1(Y)(y) = y$ for $y \in Y$ and $\varphi_1(Y)(*) = e$. Moreover, $\iota_{1,1} : Y^0 = \{e\} \to Y$ is defined by $\iota_{1,1}(e) = e$ so the assertion follows. Suppose the assertion holds for $k < n$. Evidently, $\operatorname{Im} \varphi_n \subseteq \{a \in \tau_{Y^n}(X) | \iota_{1,n_*}(a) = \ldots = \iota_{n,n_*}(a)\}$. Note that for any $k, j \leq n$, the following diagram

(2.4)
$$\begin{array}{ccccccc}
Y^{n-1} & \xrightarrow{\iota_{j,n}(Y)} & Y^n & \longrightarrow & Y^n \sqcup * & \xrightarrow{\varphi_n(Y)} & J_n(Y) \\
\| & & & & & & \| \\
Y^{n-1} & \xrightarrow{\iota_{k,n}(Y)} & Y^n & \longrightarrow & Y^n \sqcup * & \xrightarrow{\varphi_n(Y)} & J_n(Y)
\end{array}$$

is commutative. Thus, $\iota_{j,n}(Y)^* \varphi_n(Y)^* = \iota_{k,n}(Y)^* \varphi_n(Y)^*$. Next, we will show that if $a \in \tau_{Y^n}(X)$ such that $\iota_{1,n(Y)^*}(a) = \ldots = \iota_{n,n(Y)^*}(a)$ then $a \in \operatorname{Im} \varphi_n(Y)^*$. The inclusions $j_{n-1}(Y)$ and $\iota_{j,n}(Y)$ induce surjective homomorphisms $j_{n-1}(Y)^* : [J_n(Y), \Omega X] \to [J_{n-1}(Y), \Omega X]$ and $\iota_{j,n}(Y)^* : \tau_{Y^n}(X) \to \tau_{Y^{n-1}}(X)$, respectively such that $\iota_{j,n}(Y)^* \circ \varphi_n(Y)^* = j_{n-1}(Y)^* \circ \varphi_{n-1}(Y)^*$. Moreover, it is straightforward to show that for $j < k-1 < n-1$, we have $\iota_{j,n}(Y) \circ \iota_{k,n-1}(Y) = \iota_{k+1,n}(Y) \circ \iota_{j,n-1}(Y)$. Thus, the assertion follows. □

REMARK 2.3. We thank the referee for pointing out that J. Wu obtained in [**11**, Lemma 2.9] similar results stated in Theorem 2.1 and Theorem 2.2. More precisely, J. Wu embedded the group $[J_n(Y), \Omega X]$ in the group $[Y^n, \Omega X]$ as a subgroup and described the image. Here, we embed the same group into the torus homotopy group $\tau_{Y^n}(X)$ which in turn is isomorphic to $[\Sigma Y^n, X] \rtimes \pi_1(X) \cong [Y^n, \Omega X] \rtimes \pi_1(X)$. The advantage of our approach is that $\tau_{Y^n}(X)$ contains all the (generalized) Whitehead products which appear as commutators. Thus, our approach immediately gives the centrality of the subgroup $[\Lambda^n Y, \Omega X]$ in $\tau_{Y^n}(X)$ and hence in $[J_n(Y), \Omega X]$ by showing the vanishing of the generalized Whitehead products.

In [**2**], F. Cohen and T. Sato determined the central extension (2.2) by computing the 2-cocycle that determines the class in the second cohomology group $H^2([J_{n-1}(\mathbb{S}^1), \Omega X], \pi_n(\Omega X)])$, using the James-Hopf invariants. By means of the Fox split exact sequence (1.1) for the torus homotopy groups, one can relate this cohomology class to an element in the first cohomology of $[J_{n-1}(\mathbb{S}^1), \Omega X]$ with coefficients in some quotient group of $\tau_n(\Omega X)$.

Using pushouts and pullbacks of extensions, the result below follows from [**7**, Exersices 5, p. 114].

LEMMA 2.4. *Let G, G' be groups and A, A' be G and G' modules respectively. Suppose Γ and Γ' are extensions such that the following diagram*

$$
\begin{array}{ccccccccc}
0 & \longrightarrow & A & \xrightarrow{i} & \Gamma & \xrightarrow{p} & G & \longrightarrow & 1 \\
 & & \alpha\big\downarrow & & \beta\big\downarrow & & \gamma\big\downarrow & & \\
0 & \longrightarrow & A' & \xrightarrow{i'} & \Gamma' & \xrightarrow{p'} & G' & \longrightarrow & 1
\end{array}
\tag{2.5}
$$

is commutative. If $\xi \in H^2(G, A)$ and $\eta \in H^2(G', A')$ are the cohomology classes corresponding to the extensions Γ and Γ' respectively, then $\gamma^*(\eta) = \alpha_*(\xi)$. Here, $\alpha_* : H^2(G, A) \to H^2(G, A')$ and $\gamma^* : H^2(G', A') \to H^2(G, A')$ are the respective induced homomorphisms.

In particular, we have the following:

COROLLARY 2.5. *Given the commutative diagram (2.5), if α is injective and $\eta = 0$ then there exists $\zeta \in H^1(G, A'/A)$ such that $\delta^*(\zeta) = \xi$, where*

$$\delta^* : H^1(G, A'/A) \to H^2(G, A)$$

is the connecting homomorphism.

PROOF. Since α is injective, we have a short exact sequence $0 \to A \to A' \to A'/A \to 0$ of coefficients. This induces a long exact sequence of cohomology of groups

$$\dots \longrightarrow H^1(G, A) \xrightarrow{\alpha_*} H^1(G, A') \longrightarrow H^1(G, A'/A) \xrightarrow{\delta^*} H^2(G, A) \xrightarrow{\alpha_*} H^2(G, A') \longrightarrow \dots$$

Since $\eta = 0$, by Lemma 2.4, it holds $\xi \in \operatorname{Ker} \alpha_* = \operatorname{Im} \delta^*$ so $\delta^*(\zeta) = \xi$ for some $\zeta \in H^1(G, A'/A)$. $\square$

It follows from (1.1), that $\tau_n(\Omega X)/\pi_n(\Omega X) \cong \bigoplus_{i=2}^n [\pi_i(X)]^{\alpha_i}$. If we apply Corollary 2.5 to the commutative diagram of (2.1) when $Y = \mathbb{S}^1$, then we obtain the following result.

THEOREM 2.6. *Suppose that $\xi \in H^2([J_{n-1}(\mathbb{S}^1), \Omega X], \pi_n(\Omega X)])$ represents the central extension of (2.2) then there exists an element*

$$\zeta \in H^1([J_{n-1}(\mathbb{S}^1), \Omega X], \bigoplus_{i=2}^{n} [\pi_i(X)]^{\alpha_i})$$

such that $\delta^*(\zeta) = \xi$, where $\alpha_i = \binom{n-1}{i-2}$ and δ^* is the connecting homomorphism.

3. Fox construction and Fox-Hopf maps

For any space V with a base-point, we write $V^\infty = \prod^* Y$ for its weak infinite power.

The *universal Fox space* of V is defined to be

$$F_\infty(V) := \Sigma(V^\infty \sqcup *).$$

One can show that

$$F_\infty(V) = \varinjlim F_n(V).$$

When $V = \mathbb{S}^1$, the unit circle, every map $\mathbb{S}^k \to (\mathbb{S}^1)^\infty$ factors through some torus $(\mathbb{S}^1)^t$. It follows that

$$\pi_k((\mathbb{S}^1)^\infty) = \varinjlim \pi_k((\mathbb{S}^1)^n) \quad = \begin{cases} 0, & \text{if } k \geq 2; \\ \prod^* \mathbb{Z}, & \text{if } k = 1 \end{cases}$$

and thus the Fox construction $(\mathbb{S}^1)^\infty$ of $\mathbb{S}^1$ is the Eilenberg-MacLane space $K(\prod^* \mathbb{Z}, 1)$. More generally, we have

$$(K(\pi, n))^\infty = K(\prod^* \pi, n)$$

for any group π and $n \geq 1$, where $\prod^*$ denotes the weak product when $n = 1$ or the direct sum $\bigoplus$ when $n \geq 2$.

We now define the *universal Fox homotopy group of X with respect to V* to be

$$\tau_{V^\infty}(X) := [F_\infty(V), X],$$

the group of homotopy classes of pointed maps from $F_\infty(V)$ to X.

It turns out that this universal Fox group is the same as the inverse limit of the system of Fox torus homotopy groups.

PROPOSITION 3.1. *Let V be a space with a base-point. Then, the induced map*

$$\tau_{V^\infty}(X) \to \varprojlim \tau_{V^n}(X)$$

is an isomorphism.

Using the commutativity of diagram (2.3), one can define a natural map $\varphi_\infty(V) : V^\infty \to J(V)$ by taking the colimit of $\{\varphi_n(V)\}$.

Recall the following universal property result due to I.M. James [**6**].

PROPOSITION 3.2. *Given any $f : Y \to \Omega X$, there is a unique map (up to homotopy) $\tilde{f} : J(Y) \to \Omega X$ such that $f = \tilde{f} \circ \mathrm{Ad}(1_{\Sigma Y})$, where* Ad *denotes the adjoint operator. The map $\tilde{f}$ is the looping of the adjoint of f. That is, if $\mathrm{Ad}(f) : \Sigma Y \to X$ denotes the adjoint of f, then $\tilde{f} = \Omega(\mathrm{Ad}(f))$ (provided $J(Y) = \Omega\Sigma Y$).*

Using the map $\varphi_\infty(Y)$ and the universal property of the James construction $J(Y)$ as in Proposition 3.2, we obtain a similar universal property for the Fox construction Y^∞ as follows.

PROPOSITION 3.3. *Let $n \geq 1$ and $j_n(Y) : Y \to Y^\infty$ be the inclusion to the n-th coordinate. Then for any map $f : Y \to \Omega X$, there exists a map $\hat{f}_n : Y^\infty \to \Omega X$ such that $f = \hat{f}_n \circ j_n(Y)$, up to homotopy.*

Similarly to the James-Hopf maps $h_n(Y) : J(Y) \to J(\Lambda^n(Y))$, one can also define similar maps $\lambda_n(Y) : Y^\infty \to \left(Y^n/Y^{n-1}\right)^\infty$ as follows. Given $(y_1, y_2, y_3, \ldots) \in Y^\infty$, consider all its subsequences of length n that are non-trivial in the sense that it is not equal to $(e, e, \ldots, e)$. Using the right lexicographical order, we obtain subsequences c_i of length n and thus points $\bar{c_i} \in Y^n/Y^{n-1}$. Then, we define the n-th Fox-Hopf map

$$\lambda_n(Y) : Y^\infty \to (Y^n/Y^{n-1})^\infty$$

to be given by $\lambda_n(y_1, y_2, \ldots) := (\bar{c_1}, \bar{c_2}, \ldots)$ for $(y_1, y_2, y_3, \ldots) \in Y^\infty$ which yields:

PROPOSITION 3.4. *The following diagram*

$$\begin{array}{ccccc} J(Y) & \xrightarrow{h_n(Y)} & J(\Lambda^n Y) & \xleftarrow{J(\delta_n(Y))} & J(Y^n/Y^{n-1}) \\ {\scriptstyle \varphi_\infty(Y)}\uparrow & & & & \| \\ Y^\infty & \xrightarrow[\lambda_n(Y)]{} & \left(Y^n/Y^{n-1}\right)^\infty & \xrightarrow[\varphi_\infty(Y^n/Y^{n-1})]{} & J(Y^n/Y^{n-1}) \end{array} \tag{3.1}$$

is commutative.

For every $f : Y^\infty \to \Omega X$ and every $n \geq 1$ there exists $\bar{f}_n : Y^n/Y^{n-1} \to \Omega X$ such that the following diagram

$$\begin{array}{ccccc} Y^\infty & \xrightarrow{\lambda_n(Y)} & \left(Y^n/Y^{n-1}\right)^\infty & \xleftarrow{j_n(Y^n/Y^{n-1}} & Y^n/Y^{n-1} \\ {\scriptstyle f}\downarrow & & {\scriptstyle \hat{f}_n}\downarrow & & {\scriptstyle \bar{f}_n}\downarrow \\ \Omega X & = & \Omega X & = & \Omega X \end{array} \tag{3.2}$$

is commutative up to homotopy.

While $[\Lambda^n Y, \Omega X]$ is a normal subgroup of $[J_n(Y), \Omega X]$, the group $[J_{n-1}(Y), \Omega X]$ is only a quotient of $[J_n(Y), \Omega X]$. On the other hand, we have both a directed and an inverse system

$$\tau_Y(X) \overset{-\to}{\longleftarrow} \tau_{Y^2}(X) \overset{-\to}{\longleftarrow} \tau_{Y^3}(X) \overset{-\to}{\longleftarrow} \ldots$$

so that we have both the direct limit $\varinjlim \tau_{Y^n}(X)$ and the inverse limit $\varprojlim \tau_{Y^n}(X) = \tau_{Y^\infty}(X)$. For any positive integer n, let $p_n : Y^\infty \to Y^n$ and $i_n : Y^n \to Y^\infty$ be the projection and inclusion of the first n coordinates, respectively. Then, $p_n \circ i_n = 1_{Y^n}$ and the induced homomorphism $p_{n*} : \tau_{Y^n}(X) \to \tau_{Y^\infty}(X)$ is injective so that $\tau_{Y^n}(X)$ can be embedded as a subgroup of $\tau_{Y^\infty}(X)$. In fact, there is an injective homomorphism $\varinjlim \tau_{Y^n}(X) \to \tau_{Y^\infty}(X)$. Further, functorial crossed-sections of $\Sigma F_{n+1}(Y) \to \Sigma Y^n \overset{\varphi_n(Y)}{\to} \Sigma J_n(Y)$ for $n \geq 0$ lead to such a section of $\Sigma F_\infty(Y) \to \Sigma J(Y)$. Hence, that there exists an injective homomorphism

$$[J(Y), \Omega X] \to \tau_{Y^\infty}(X).$$

When $Y = \mathbb{S}^1$, we write $\tau_\infty(X) := \tau_{(\mathbb{S}^1)^\infty}(X)$ so that $\tau_\infty(X) = \varprojlim \tau_n(X)$ is the inverse limit of the classical Fox torus homotopy groups of a pointed space X. In particular, we can embed $\prod[\Omega X]$ into $\tau_\infty(X)$.

Now, let X be such a pointed space that the Whitehead product $[\alpha, \beta] = 0$ for $\alpha \in \pi_k(X)$, $\beta \in \pi_l(X)$ provided $k > k_0$ and $l \geq 1$. Since $\tau_n(X)$ is determined by $\pi_k(X)$ for $k \leq n$ and its commutators by Whitehead products, it follows easily that $\pi_k(X)$ are in the center of $\tau_n(X)$ for $k_0 \leq n$. Proceeding inductively, in view of

1.1, we deduce that $\prod_{k>k_0}^{n+1}\pi_k(X)$ is in the center of $\tau_n(X)$ for $n \geq k_0$. Now, by embedding $[J_n(\mathbb{S}^1), \Omega X] \hookrightarrow \tau_{n+1}(X)$, in view of Theorem 2.1, we may state:

PROPOSITION 3.5. *Let $[\alpha, \beta] = 0$ for $\alpha \in \pi_k(X)$, $\beta \in \pi_l(X)$ provided $k > k_0$ and $l \geq 1$. Then, there are central extensions:*

$$1 \to \prod_{k>k_0}^{n+1} \pi_k(X) \to [J_n(\mathbb{S}^1), \Omega X] \to [J_{k_0-1}(\mathbb{S}^1), \Omega X] \to 1$$

for $n \geq k_0$
and

$$1 \to \prod_{k>k_0} \pi_k(X) \to [J(\mathbb{S}^1), \Omega X] \to [J_{k_0-1}(\mathbb{S}^1), \Omega X] \to 1.$$

Recall, that X is called a *W-space* if all of its Whitehead products vanish. For more details on W-spaces, see [**8**]. For instance, any H-space (or more generally, any T-space studied in [**1**]) is a W-space. Further, it is proved in [**9**] that $\mathbb{S}_p^{2n-1}$, the $(2n-1)$-sphere localized at a prime p, admits an A_{p-1}-form. In particular, $\mathbb{S}_p^{2n-1}$ is an H-space provided p is odd. Given a W-space X, the groups $\tau_n(X)$ must be abelian for all $n \geq 1$. Hence, we may deduce:

COROLLARY 3.6. *Let X be a W-space. For any positive integer n, we have*

$$\prod_{k=2}^{n+1} \pi_k(X) \cong [J_n(\mathbb{S}^1), \Omega X]$$

and hence

$$\prod_{n\geq 2} \pi_n(X) \cong \prod [\Omega X].$$

We end this section by presenting:

EXAMPLE 3.7. It is well-known that $[\alpha, \beta] = 0$ for $\alpha \in \pi_k(\mathbb{S}^2)$, $\beta \in \pi_l(\mathbb{S}^2)$ provided $k > 2$ and $l \geq 1$. For the real projective plane $\mathbb{R}P^2$, we believe that this relation is also a kind of the mathematical folklore. But, since we have no references, a sketch of proof is included.

If $k > 2$ and $l \geq 2$ then that follows by means of the universal covering $\mathbb{S}^2 \to \mathbb{R}P^2$ and the same relation on the Whitehead product for the sphere $\mathbb{S}^2$.

Now, suppose that $k = 3$ and $l = 1$. In this case, the result follows by means of the Hopf map $h : \mathbb{S}^3 \to \mathbb{S}^2$. Namely, the Whitehead product $[-,-] : \pi_1(\mathbb{R}P^2) \times \pi_3(\mathbb{R}P^2) \to \pi_3(\mathbb{R}P^2)$ corresponds to the composite of h with the antipodal map $\mathbb{S}^2 \to \mathbb{S}^2$ which is again (up to homotopy) h. So, the Whitehead product vanishes in that case.

For $k > 3$, any map $\mathbb{S}^k \to \mathbb{S}^2$ factors (up to homotopy) through the Hopf map $\mathbb{S}^3 \to \mathbb{S}^2$ and, in view of the above, the Whitehead product $[-,-] : \pi_1(\mathbb{R}P^2) \times \pi_k(\mathbb{R}P^2) \to \pi_k(\mathbb{R}P^2)$ vanishes as well.

Now, write $X = \mathbb{S}^2, \mathbb{R}P^2$. Because $[J_1(\mathbb{S}^1), \Omega X] = \pi_2(X) \cong \mathbb{Z}$, the infinite cyclic group, in light of Proposition 3.5, there are splitting central extensions:

$$1 \to \prod_{k>2}^{n+1} \pi_k(X) \to [J_n(\mathbb{S}^1), \Omega X] \leftrightarrows \mathbb{Z} \to 1$$

for $n \geq 2$ and

$$1 \to \prod_{k>2} \pi_k(X) \to [J(\mathbb{S}^1), \Omega X] \overset{\leftarrow}{\to} \mathbb{Z} \to 1.$$

4. Galois groups and torus homotopy groups

To prove the next result, we need to recall that a directed set $(\Lambda, \leq)$ is called *cofinite* if each of its elements has finitely many predecessors.

LEMMA 4.1. (1) *Let $\{G_\lambda\}_{\lambda\in\Lambda}$ be an inverse system of finitely generated groups with G_{λ_0} being abelian for any minimal element $\lambda_0 \in \Lambda$ and such that for each bonding map $G_{\lambda_2} \to G_{\lambda_1}$ whenever $\lambda_1 \leq \lambda_2$, there is a central extension*

$$1 \to A_{\lambda_1,\lambda_2} \to G_{\lambda_2} \to G_{\lambda_1} \to 1.$$

Suppose $G = \varprojlim H_\alpha$ is profinite and there is a topological isomorphism $\varprojlim H_\alpha \cong \varprojlim G_\lambda$. Then, G_λ is finite for all $\lambda \in \Lambda$.

(2) *Let $\{A_\lambda\}_{\lambda\in\Lambda}$ be an inverse system of finitely generated abelian groups such that the bonding maps $A_{\lambda_2} \to A_{\lambda_1}$ are surjective whenever $\lambda_1 \leq \lambda_2$. Suppose $A = \varprojlim B_\alpha$ and there is a topological isomorphism $A = \varprojlim B_\alpha \cong \varprojlim A_\lambda$. Then, G_λ is finite for all $\lambda \in \Lambda$.*

PROOF. (1) Because Λ is a cofinite ordered set, we may proceed by induction. First, observe that the projection maps $p_\lambda : G = \varprojlim G_\lambda \to G_\lambda$ are surjective because the bonding maps $G_{\lambda_2} \to G_{\lambda_1}$ are surjective whenever $\lambda_1 \leq \lambda_2$.

Suppose that G_λ is infinite for some initial element $\lambda \in \Lambda$. Because G_λ is abelian, there is an epimorphism $G_\lambda \to \mathbb{Z}$, where $\mathbb{Z}$ is the infinite cyclic group. Then, the continuous cohomology group $H^1(G)$ has no torsion, contrary to G being profinte.

Now, suppose that for a fixed (non-initial) element $\lambda \in \Lambda$ and for all $\lambda' < \lambda$ the groups $G_{\lambda'}$ are finite and we show that G_λ is finite as well. Suppose, G_λ is infinite. Then, by the Stallings-Stambach five-term exact sequence

$$H_2(G_\lambda) \to H_2(G_{\lambda'}) \to A_{\lambda',\lambda}/[G_\lambda, A_{\lambda',\lambda}] = A_{\lambda',\lambda} \to H_1(G_\lambda) \to H_1(G_{\lambda'}) \to 1$$

associated with the central extension $1 \to A_{\lambda',\lambda} \to G_\lambda \to G_{\lambda'} \to 1$, we derive that the abelainization $G_\lambda/[G_\lambda, G_\lambda]$ is an infinite abelian group. Consequently, there is an epimorphism $G_\lambda \to \mathbb{Z}$ and $H^1(G)$ has no torsion, contrary to G being profinte and the proof is complete.

(2) Because the bonding maps $p_{\lambda_1,\lambda_2} : A_{\lambda_2} \to A_{\lambda_1}$ are surjective whenever $\lambda_1 \leq \lambda_2$, we deduce that the projection maps $A = \varprojlim A_\lambda \to A_\lambda$ are also surjective for all $\lambda \in \Lambda$.

If A_λ is an infinite group for some $\lambda \in \Lambda$ then there is an epimorphism $A_\lambda \to \mathbb{Z}$ and consequently $H^1(A)$ is a torsion-free group, contrary to A being profinite and the proof is complete. □

THEOREM 4.2. *Let X be a CW-complex of finite type with an abelian finitely generated $\pi_1(X)$. Then, the following are equivalent:*

(i) *$\pi_i(X)$ is finite for all $i \geq 1$;*
(ii) *$\tau_i(X)$ is finite for all $i \geq 1$;*
(iii) *$\tau_\infty(X) = \varprojlim G_\lambda$ is a profinite group and there is a topological isomorphism $\tau_\infty(X) = \varprojlim G_\lambda \cong \varprojlim \tau_i(X)$.*

(iv) *the group* $[J(\mathbb{S}^1), \Omega X]$ *is a profinite group, there is a topological isomorphism* $\tau_\infty(X) = \varprojlim G_\lambda \cong \varprojlim [J_n(\mathbb{S}^1), \Omega(X)]$ *and* $\pi_1(X)$ *is finite;*

(v) *the group* $[J(\mathbb{S}^1), \Omega X]$ *is the Galois group for some choice of field extension* K/k.

Proof. (i)$\Rightarrow$(ii). If $\pi_i(X)$ are finite then by [**4**], the groups $\tau_i(X)$ are finite as well for all $i \geq 1$. The implication (ii)$\Rightarrow$(iii) follows directly from Theorem 3.1.

(iii)$\Rightarrow$(i). By [**5**], there is a spit extension $0 \to \tau_i(\Omega X) \to \tau_{i+1}(X) \to \tau_i(X) \to 1$ for all $i \geq 1$ and an isomorphism $\tau_n(\Omega X) \cong \Pi_{i=2}^{n+1} \pi_i(X)^{\alpha_i(n+1)}$ with the binomial coefficient $\alpha(n+1) = \binom{n-1}{i-2}$. Because X is a CW-complex of finite type with an abelian finitely generated $\pi_1(X)$, the groups $\tau_i(X)$ are finitely generated. Thus, we may apply Lemma 4.1 and the result follows.

Because $[J_n(\mathbb{S}^1), \Omega X]$ is isomorphic (as a set) to the direct product of $\pi_i(X)$ for $i \leq n$, the implication (i)$\Rightarrow$(iv) follows.

(iv)$\Rightarrow$(i). By Theorem 2.1 (or [**2**, Theorem 4.1]), there is a central extension

$$1 \to \pi_{n+1}(\Omega X) \to [J_{n+1}(\mathbb{S}^1), \Omega X] \to [J_n \mathbb{S}^1), \Omega X] \to 1.$$

Because X is a CW-complex of finite type with an abelian finitely generated $\pi_1(X)$, the groups $[J_n(\mathbb{S}^1), \Omega X]$ are finitely generated. Thus, we may apply Lemma 4.1 and the implication follows. By [**2**, Theorem 8.2], it follows that (i)$\Leftrightarrow$(v) and the proof is complete. $\square$

References

[1] Aguadé, J., Decomposable free loop spaces, *Canad. J. Math.* **39** (1987), no. 4, 938–955.

[2] Cohen, F. and Sato, T., On groups of homotopy groups, and braid-like groups, preprint (1997).

[3] Fox, R., Homotopy groups and torus homotopy groups, *Ann. of Math.* **49** (1948), 471–510.

[4] Golasiński, M., Gonçalves D., and Wong, P., Generalizations of Fox homotopy groups, Whitehead products, and Gottlieb groups, *Ukrain. Math. Zh.* **57** (2005), no. 3, 320–328 (translated in Ukrainian Math J. **57** (2005) no. 3, 382-393).

[5] ———, On Fox spaces and Jacobi identities, *Math. J. Okayama U.* **50** (2008), 161–176.

[6] James, I.M., Reduced product spaces, *Ann. of Math.* **62** (1955), 170–197.

[7] MacLane, S., "Homology," Springer-Verlag, Berlin-Heidelberg, 1963.

[8] Siegel, J., *G*-spaces, *H*-spaces and *W*-spaces, *Pacific J. Math.* **31** (1969), 209–214.

[9] Stasheff, J.D., Homotopy assiciativity of *H*-spaces, I, *Trans. Amer. Math. Soc.* **108** (1963), 275-292.

[10] Whitehead, G., "Elements of Homotopy Theory," Springer-Verlag, New York, 1978.

[11] Wu, J., "Homotopy theory of the suspensions of the projective plane," Mem. Amer. Math. Soc. 162 (2003), no. 769.

Faculty of Mathematics and Computer Science, Nicolaus Copernicus University, Chopina 12/18, 87-100, Toruń, Poland

E-mail address: marek@mat.uni.torun.pl

Dept. de Matemática - IME - USP, Caixa Postal 66.281 - CEP 05314-970, São Paulo - SP, Brasil

E-mail address: dlgoncal@ime.usp.br

Department of Mathematics, Bates College, Lewiston, ME 04240, U.S.A.

E-mail address: pwong@bates.edu

Contemporary Mathematics
Volume **519**, 2010

Notes on the Triviality of Adjoint Bundles

Akira Kono and Shuichi Tsukuda

ABSTRACT. Let G be a connected compact Lie group and B a finite CW-complex. We show that, if p is a prime large enough, then, for every principal G bundle $P \to B$, the adjoint bundle $\mathrm{Ad}P$ is fibrewise p-local trivial and in particular, the gauge group has the p-local homotopy type of $\mathrm{Map}(B, G_{(p)})$.

1. Introduction

Let G be a connected compact Lie group, $P \to B$ a principal G bundle over a connected finite CW-complex B and $\mathrm{Ad}P = P \times_{\mathrm{Ad}} G$ its adjoint bundle where G acts on itself by the adjoint action. The space of sections of the adjoint bundle $\mathcal{G}(P) = \Gamma(\mathrm{Ad}P)$ has a natural group structure and is called the gauge group of P.

It is well known that the classifying space of the gauge group has the homotopy type of the path component of the mapping space including f, $\mathrm{Map}(B, BG; f)$ where $f\colon B \to BG$ is a map classifying the bundle P (see for example [**2**],[**5**]).

M. Crabb and W. Sutherland [**5**] showed that, for fixed G and B, the number of H-types (homotopy types as H-spaces) of $\mathcal{G}(P)$ is finite as P ranges over all principal G bundles over B. In the proof, they showed that the adjoint bundle is H-trivial i.e., trivial as a fibrewise H-space after fibrewise rationalisation (see §3 for an alternative proof).

In fact it is H-trivial after fibrewise localisation at a prime p if p is large enough and the main purpose of this note is to show this.

THEOREM 1.1. *Let G be a connected compact Lie group and B a finite CW-complex. If p is a prime large enough, then, for every principal G bundle $P \to B$, $\mathrm{Ad}P$ is fibrewise p-local H-trivial.*

In particular, we have the following. Let $-_{(p)}$ denote a localisation at p.

COROLLARY 1.2. *Under the same assumptions as above, for every principal G bundle $P \to B$, there is a p-equivalence $\mathcal{G}(P) \to \mathrm{Map}(B, G)$ which induces H-equivalence $\mathcal{G}(P)_{(p)} \simeq \mathrm{Map}(B, G_{(p)})$.*

1991 *Mathematics Subject Classification.* Primary 55P15, Secondary 55R05.

The first author was partially supported by JSPS KAKENHI (18340016).
The second author was partially supported by JSPS KAKENHI (19540095).

©2010 American Mathematical Society

Recall that a Lie group G is said to be p-regular if it is p-equivalent to a product of odd dimensional spheres

$$G \underset{p}{\simeq} S^{2n_1-1} \times \cdots \times S^{2n_l-1}$$

where $n_1 \leq \cdots \leq n_l$. The list of integers $(2n_1 - 1, \ldots, 2n_l - 1)$ is called the *type* of G. If G is p-regular, then we can describe the relation between the prime p for the triviality (not as a fibrewise H-space) and the dimension of B using obstruction theory:

THEOREM 1.3. *Let p be an odd prime. If G is p-regular of type $(2n_1 - 1, \ldots, 2n_l - 1)$ and $\dim B < 2(p - n_l + 1)$, then the adjoint bundle $\mathrm{Ad}P$ is fibrewise p-local trivial.*

In particular, we have the following:

COROLLARY 1.4. *Under the same assumptions as above, there is a homotopy equivalence $\mathrm{Map}(B, G_{(p)}) \simeq \mathcal{G}(P)_{(p)}$.*

EXAMPLE 1.5. Let X be a closed 4 manifold and $p > 3$ an odd prime. Then, $\mathrm{Map}(X, \mathrm{SU}(2)_{(p)}) \simeq \mathcal{G}(P)_{(p)}$ for any principal $\mathrm{SU}(2)$ bundle P over X.

In the case when G is a classical group, the estimate given in Theorem 1.3 is the best possible, namely, we have the following:

THEOREM 1.6. *Let p be an odd prime, $EG \to BG$ the universal bundle, $B^{(k)}$ the k skeleton of BG and $E_k \to B^{(k)}$ the restriction of the universal bundle. If G is a noncommutative classical group which is p-regular of type $(2n_1 - 1, \ldots, 2n_l - 1)$ except $\mathrm{SU}(p)$, then $\mathrm{Ad}E_k$ is fibrewise p-local trivial if and only if $k < 2(p - n_l + 1)$.*

The paper is organised as follows: In §2, we recall some fundamental definitions and facts concerning fibrewise homotopy theory. In §3, we give a proof of the rational triviality of adjoint bundles. We show the fibrewise p-local triviality of adjoint bundles in §4 and consider the p-regular case in §5. We study the classical group case in §6.

The second author is grateful to the Aberdeen University Department of Mathematical Sciences for the warm hospitality while part of this work being carried out.

2. Fibrewise homotopy theory

In this section, for the reader's convenience, we recall some fundamental definitions and facts concerning fibrewise homotopy theory. See [**4**] for details.

Fix a space B. A space over B is a space X equipped with a map called the projection $p\colon X \to B$. A space over B is said to be pointed if a section $s\colon B \to X$ is fixed and the section is called a basepoint. A map $f\colon X \to Y$ over B is called a fibrewise map. When X and Y are pointed, f is said to be pointed if it preserves basepoint, i.e., it commutes with the sections.

2.1. Fibrewise homotopy. For a space T and a space X over B, the Cartesian product $T \times X$ is a space over B by the projection $p \circ p_X$ where p_X is the canonical projection to X.

A fibrewise map $H\colon I \times X \to Y$ is called a fibrewise homotopy, where $I = [0, 1]$ is the unit interval. If $H(t, -)\colon X \to Y$ is pointed for all $t \in I$, H is called a pointed homotopy.

If X and Y are (pointed) fibrewise homotopy equivalent, then the space of sections $\Gamma(X)$ and $\Gamma(Y)$ are (pointed) homotopy equivalent.

2.2. Fibrewise H-space. If X is pointed with basepoint s, then so is $X \times_B X$ with the basepoint $s \times_B s$. A pointed fibrewise map $\mu\colon X \times_B X \to X$ is called a fibrewise multiplication. A pair (X, μ) is called a fibrewise H-space if the following diagram is commutative up to fibrewise pointed homotopy

$$\begin{array}{ccc} X & & \\ {\scriptstyle (1,\eta)}\downarrow & \searrow^{1} & \\ X \times_B X & \xrightarrow[\mu]{} & X \\ {\scriptstyle (\eta,1)}\uparrow & \nearrow_{1} & \\ X & & \end{array}$$

where η is the constant map at the basepoint given by $\eta = s \circ p$. If (X, μ) is a fibrewise H-space, then $\Gamma(X)$ has an H-space structure induced by μ. A pointed fibrewise map $f\colon X \to Y$ between fibrewise H-spaces is called a fibrewise H-map if it commutes with the multiplication up to fibrewise pointed homotopy. A fibrewise H-map is called a fibrewise H-equivalence if it is a pointed fibrewise homotopy equivalence. If X and Y are fibrewise H-equivalent, then $\Gamma(X)$ and $\Gamma(Y)$ are H-equivalent, that is, there exists an H-map between them, which is also a pointed homotopy equivalence.

The constant map at the unit $e \in G$ induces a section s_e of $\mathrm{Ad}P$, thus the bundle $\mathrm{Ad}P \to B$ is pointed. The group multiplication of G makes $\mathrm{Ad}P$ a fibrewise H-space.

2.3. Fibrewise localisation. Let $F \to E \to B$ be a fibration with nilpotent fibre F and p a prime number. There exists a fibrewise p-localisation ([**3**],[**8**],[**9**]), that is, there is a fibration $F_{(p)} \to E^f_{(p)} \to B$ and a map of fibration

$$\begin{array}{ccccc} F & \longrightarrow & E & \longrightarrow & B \\ \downarrow & & \downarrow & & \| \\ F_{(p)} & \longrightarrow & E^f_{(p)} & \longrightarrow & B \end{array}$$

where $F \to F_{(p)}$ is a p-localisation.

In the case when E and B are nilpotent (then so is the fibre F [**7**]), we have a fibration of p-local spaces $F_{(p)} \to E_{(p)} \to B_{(p)}$ and the induced fibration

$$\begin{array}{ccccc} F & \longrightarrow & E & \longrightarrow & B \\ \downarrow & & \downarrow & & \| \\ F_{(p)} & \longrightarrow & l^* E_{(p)} & \longrightarrow & B \end{array}$$

gives a fibrewise localisation where $l\colon B \to B_{(p)}$ is a localisation.

Fibrewise localisation has the usual universality.

The induced map $\Gamma(E) \to \Gamma(E^f_{(p)})$ gives a p-localisation ([**13**]).

We say that a fibration $F \to E \to B$ is fibrewise p-local trivial if its fibrewise p-localisation is fibrewise homotopy equivalent to the trivial fibration $B \times F_{(p)} \to B$.

3. Milnor's universal bundle and the rational triviality of adjoint bundles

In [**5**], [**1**], it was shown that adjoint bundles are rationally trivial as fibrewise H-spaces. The proof is based on the "triviality" of the adjoint action of G on BG and the fact that the rationalisation $G_{(0)}$ is an (abelian) group. Since the adjoint action of an abelian group is trivial, it is natural to expect that the fibrewise rationalisation of adjoint bundles are trivial. We give an alternative proof of the triviality along these lines using Milnor's universal bundle [**10**].

PROPOSITION 3.1. *There is a fibrewise (loop) map*

$$\begin{array}{ccc} G & \xrightarrow{l_0} & G_{(0)} \\ \downarrow & & \downarrow \\ EG \times_{\mathrm{Ad}} G & \longrightarrow & BG \times G_{(0)} \\ \downarrow & & \downarrow \\ BG & = & BG \end{array}$$

such that l_0 is a rationalisation.

PROOF. Let X be a countable connected simplicial complex. Milnor constructed a principal bundle $\tilde{E} \to X$ with the structure group $\tilde{G}$ enjoying the following property: $\tilde{E}$ and $\tilde{G}$ are CW-complexes and $\tilde{E}$ is contractible. Moreover for any topological group H and principal H-bundle $P \to X$, there is a continuous homomorphism $h\colon \tilde{G} \to H$ such that P is associated to $\tilde{E}$ by h.

Take BG as the X and let $\tilde{G} \to \tilde{E} \to BG$ be the universal bundle as above. Let $G \to EG \to BG$ be the (usual) universal bundle. There is a continuous homomorphism $h\colon \tilde{G} \to G$ with $EG = \tilde{E} \times_h G$. Note that h is a homotopy equivalence.

Recall that the rationalisation $G_{(0)}$ is an abelian group and let $E_0 \to B(G_{(0)})$ be the universal $G_{(0)}$ bundle. Note that $B(G_{(0)}) \simeq (BG)_{(0)}$. Let $l\colon BG \to B(G_{(0)})$ be the rationalisation and $G_{(0)} \to l^*E_0 \to BG$ the induced $G_{(0)}$ bundle. By the universality, there is a continuous homomorphism $h_0\colon \tilde{G} \to G_{(0)}$ with $l^*E_0 = \tilde{E} \times_{h_0} G_{(0)}$. Note that l^*E_0 is rationally contractible and h_0 is a rational equivalence. Thus we obtain equivariant maps

$$\tilde{E} \to \tilde{E} \times_h G = EG$$

$$\tilde{E} \to \tilde{E} \times_{h_0} G_{(0)} = l^*E_0$$

which induce fibrewise group homomorphisms $\bar{h}$ and $\bar{h}_0$ over BG:

$$\begin{array}{ccccc} G & \xleftarrow{h} & \tilde{G} & \xrightarrow{h_0} & G_{(0)} \\ \downarrow & & \downarrow & & \downarrow \\ EG \times_{\mathrm{Ad}} G & \xleftarrow{\bar{h}} & \tilde{E} \times_{\mathrm{Ad}} \tilde{G} & \xrightarrow{\bar{h}_0} & BG \times G_{(0)} \\ \downarrow & & \downarrow & & \downarrow \\ BG & = & BG & = & BG. \end{array}$$

Since Bh can be taken to be equivariant with respect to the adjoint action, we may think of $\bar{h}$ as a fibrewise loop of the fibrewise equivalence $\overline{Bh}$:

$$\begin{array}{ccc} BG & \xleftarrow{Bh} & B\tilde{G} \\ \downarrow & & \downarrow \\ EG \times_{\mathrm{Ad}} BG & \xleftarrow{\overline{Bh}} & \tilde{E} \times_{\mathrm{Ad}} B\tilde{G} \\ \downarrow & & \downarrow \\ BG & = & BG. \end{array}$$

By a theorem of Dold [**6**, Theorem 6.3], $\bar{h}$ is a fibrewise (loop) equivalence which completes the proof. □

4. Fibrewise p-local triviality of adjoint bundles

Recall the telescope construction of the localisation ([**12**]). For a prime p, let $\rho\colon \mathbb{N} \to \mathbb{N}$ be the increasing function such that the image of ρ is all the numbers prime to p and f_n the self map of G defined by $f_n = \psi_{\rho(n)}$ where ψ_k is the power map sending g to g^k. Let $\mathrm{Tel}^p_n(G)$ (resp. $\mathrm{Tel}^p_\infty(G)$) be the telescope associated to the sequence

$$G_0 \xrightarrow{f_1} G_1 \xrightarrow{f_2} \cdots \xrightarrow{f_n} G_n \quad (resp. \to \cdots)$$

where each G_i is a copy of G and $i_n\colon G = G_0 \to \mathrm{Tel}^p_n(G)$ the inclusion. The inclusion $i_\infty\colon G \to \mathrm{Tel}^p_\infty(G)$ gives a localisation at p. We write $i_\infty = l_p$ and $\mathrm{Tel}^p_\infty(G) = G_{(p)}$. Note that $\mathrm{Tel}^p_n(G) \simeq G_n = G$ for each $n < \infty$ and the composition $G \xrightarrow{i_n} \mathrm{Tel}^p_n(G) \xrightarrow{\simeq} G$ is a p-equivalence given by the power map ψ_N for a certain $N \in \mathbb{N}$. Let $j_n\colon G \simeq \mathrm{Tel}^p_n(G) \to \mathrm{Tel}^p_\infty(G)$ be the composition of the homotopy equivalence and the inclusion. Thus we have a factorisation $l_p \simeq j_n \circ \psi_N\colon G \to G \to G_{(p)}$. Note also that if $p > n$, $\mathrm{Tel}^p_n(G) = \mathrm{Tel}^0_n(G)$, whence l_p factors through $\mathrm{Tel}^0_n(G)$.

LEMMA 4.1. *If B is a finite complex, then there is a fibrewise map $h\colon \mathrm{Ad}P \to B \times G$ over B such that the composition $h_0 = (1 \times l_0) \circ h$ is a fibrewise H-map and a fibrewise rationalisation.*

PROOF. As we have shown, there is a fibrewise H-map which is also a fibrewise rationalisation $\mathrm{Ad}P \to B \times G_{(0)}$. Since $\mathrm{Ad}P$ is compact, this map factors through $B \times \mathrm{Tel}^0_n(G)$ for some $n \in \mathbb{N}$. Hence we obtain a factorisation of the map through $B \times G$ and let h be the factorisation:

$$\mathrm{Ad}P \xrightarrow{h} B \times G \xrightarrow{1\times j_n} B \times G_{(0)}.$$

Since $l_0 \simeq j_n \circ \psi_N$ and ψ_N is a rational equivalence, $\psi_{N(0)} \circ j_n \simeq l_0$, whence $(1 \times l_0) \circ h \simeq (1 \times \psi_N) \circ (1 \times j_n) \circ h$. Since $G_{(0)}$ is homotopy abelian, we have the following diagram which is commutative up to homotopy over B

$$\begin{array}{ccccccc} \mathrm{Ad}P \times_B \mathrm{Ad}P & \xrightarrow{h\times_B h} & B \times G \times G & \xrightarrow{1\times j_n \times j_n} & B \times G_{(0)} \times G_{(0)} & \xrightarrow{1\times\psi_N\times\psi_N} & B \times G_{(0)} \times G_{(0)} \\ \downarrow & & & & \downarrow & & \downarrow \\ \mathrm{Ad}P & \xrightarrow[h]{} & B \times G & \xrightarrow[1\times j_n]{} & B \times G_{(0)} & \xrightarrow[1\times\psi_N]{} & B \times G_{(0)} \end{array}$$

which completes the proof. □

The next proposition and Remark 4.3 imply Theorem 1.1 and Corollary 1.2.

PROPOSITION 4.2. *If B is a finite complex and p is a prime large enough, then there is a fibrewise p-equivalence $h\colon \mathrm{Ad}P \to B\times G$ over B such that the composition $(1\times l_p)\circ h\colon \mathrm{Ad}P \to B\times G_{(p)}$ is a fibrewise H-map.*

PROOF. Note that the map h constructed in the above lemma is a fibrewise p-equivalence if p is large ($p > N$). We have to show that $(1\times l_p)\circ h\colon \mathrm{Ad}P \to B\times G_{(p)}$ is a fibrewise H-map. Since $(1\times l_0)\circ h$ is a fibrewise H-map, we have homotopies $(1\times l_0)\circ(1\times\mu)\circ(h\times_B h)\simeq(1\times\mu)\circ(1\times l_0\times l_0)\circ(h\times_B h)\simeq(1\times l_0)\circ h\circ\mu_P$ over B:

$$\begin{array}{ccccc}
\mathrm{Ad}P\times_B \mathrm{Ad}P & \xrightarrow{h\times_B h} & B\times G\times G & & \\
\Big\downarrow{\scriptstyle \mu_P} & & \Big\downarrow{\scriptstyle 1\times\mu} & & \\
\mathrm{Ad}P & \xrightarrow[h]{} & B\times G & \xrightarrow[1\times l_0]{} & B\times G_{(0)}.
\end{array}$$

Since $\mathrm{Ad}P\times_B \mathrm{Ad}P$ is compact, the homotopy factors through $B\times \mathrm{Tel}^0_n(G)$ for a certain $n\in\mathbb{N}$. If $p > n$, the localisation l_p factors through $\mathrm{Tel}^0_n(G)$. Hence we obtain a homotopy $(1\times l_p)\circ(1\times\mu)\circ(h\times_B h)\simeq(1\times l_p)\circ h\circ\mu_P$ over B, which completes the proof. □

REMARK 4.3. By [**5**, Theorem 2.1], the number of fibrewise H-equivalence classes of $\mathrm{Ad}P$ is finite, whence we can take the prime p in Proposition 4.2 independent of P.

As observed in the arguments above, fibrewise p-local trivialization of adjoint bundles over finite complexes are given by integral maps. Moreover, the map can be chosen to be pointed. We record this as a lemma:

LEMMA 4.4. *If B is a finite complex, then the adjoint bundle $\mathrm{Ad}P$ is fibrewise p-local trivial, namely, there exists a fibrewise homotopy equivalence*

$$\begin{array}{ccc}
B\times G_{(p)} & \longrightarrow & \mathrm{Ad}P^f_{(p)} \\
\Big\downarrow & & \Big\downarrow \\
B & = & B
\end{array}$$

if and only if there exists a fibrewise pointed map

$$\begin{array}{ccc}
B\times G & \longrightarrow & \mathrm{Ad}P \\
\Big\downarrow & & \Big\downarrow \\
B & = & B
\end{array}$$

such that the map over the base point $G\to G$ is a p-equivalence where $B\times G\to B$ is pointed by the unit of G.

PROOF. The if part is clear by a theorem of Dold [**6**] and the universality of fibrewise localisations.

Assume that $\mathrm{Ad}P$ is fibrewise p-local trivial. Note that the telescope construction is equivariant under the adjoint action, whence, $\mathrm{Ad}P^f_{(p)}$ is given by $P\times_{\mathrm{Ad}} G_{(p)}$.

Since $B \times G$ is compact, by a similar argument to the previous lemma, we have a factorisation

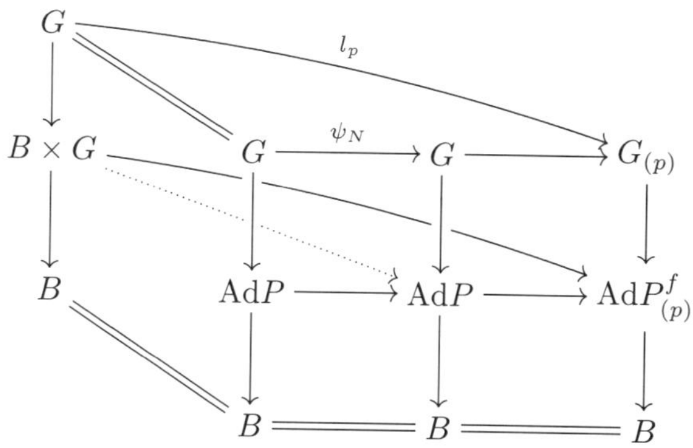

where N is an integer which is prime to p and ψ_N is a p-equivalence. Using the fibrewise group structure of $\mathrm{Ad}P$, we can modify the map to be pointed and obtain a desired fibrewise pointed map:

$$\begin{array}{ccc} G & \xrightarrow{\psi_N} & G \\ \downarrow & & \downarrow \\ B \times G & \longrightarrow & \mathrm{Ad}P \\ \downarrow & & \downarrow \\ B & = & B. \end{array}$$

□

5. The p-regular case

Let p be an odd prime and G a compact connected Lie group which is p-regular, then there exist integers $1 \leq n_1 \leq \cdots \leq n_l$ and maps $\theta_j \colon S^{2n_j-1} \to G$ such that the composite

$$\theta \colon S = \prod S^{2n_j-1} \xrightarrow{\prod \theta_j} \prod G \xrightarrow{\mu} G$$

is a p-equivalence, where μ is the multiplication of G. The next lemma implies Theorem 1.3.

LEMMA 5.1. *If* $\dim B < 2(p - n_j + 1)$, *then there exists a fibrewise pointed map:*

$$\begin{array}{ccc} S^{2n_j-1} & \xrightarrow{l_p \theta_j} & G_{(p)} \\ \downarrow & & \downarrow \\ B \times S^{2n_j-1} & \longrightarrow & \mathrm{Ad}P^f_{(p)} \\ \downarrow & & \downarrow \\ B & = & B. \end{array}$$

PROOF OF THEOREM 1.3. Taking fibrewise product and fibrewise multiplication, we obtain a fibrewise p-equivalence:

$$\begin{array}{ccc}
\prod S^{2n_j-1} & \xrightarrow{l_p\theta} & G_{(p)} \\
\downarrow & & \downarrow \\
B\times \prod S^{2n_j-1} & \longrightarrow & \mathrm{Ad}P^f_{(p)} \\
\downarrow & & \downarrow \\
B & = & B.
\end{array}$$

□

PROOF OF LEMMA 5.1. Note that the usual adjunction

$$\mathrm{Map}\left(B\times S^{2n_j-1}, \mathrm{Ad}P^f_{(p)}\right) \cong \mathrm{Map}\left(S^{2n_j-1}, \mathrm{Map}\left(B, \mathrm{Ad}P^f_{(p)}\right)\right)$$

gives rise to a one to one correspondence between the fibrewise pointed maps and pointed sections of the evaluation map:

$$\begin{array}{ccc}
S^{2n_j-1} & \xrightarrow{l_p\theta_j} & G_{(p)} \\
\downarrow & & \downarrow \\
B\times S^{2n_j-1} & \underset{ptd.}{\dashrightarrow} & \mathrm{Ad}P^f_{(p)} \\
\downarrow & & \downarrow \\
B & \xrightarrow[f]{} & B
\end{array}
\quad\Longleftrightarrow\quad
\begin{array}{ccc}
 & & \Gamma\left(\mathrm{Ad}P^f_{(p)}\right) \\
 & \overset{ptd.}{\nearrow} & \downarrow ev \\
S^{2n_j-1} & \xrightarrow[l_p\theta_j]{} & G_{(p)}.
\end{array}$$

Recall that the space of sections $\Gamma\left(\mathrm{Ad}P^f_{(p)}\right)$ is a localisation of the gauge group $\mathcal{G}(P) = \Gamma(\mathrm{Ad}P)$ whose classifying space is $\mathrm{Map}(B, BG; f)$, where $f\colon B \to BG$ is a classifying map of the bundle P. Therefore, by noting that the evaluation map is a fibration, we see that the existence of the fibrewise pointed map is equivalent to the existence of a pointed map which makes the following diagram homotopy commutative

$$\begin{array}{ccc}
 & & \Omega\mathrm{Map}(B, BG_{(p)}; l_pf) \\
 & \overset{ptd.}{\nearrow} & \downarrow \Omega ev \\
S^{2n_j-1} & \xrightarrow[l_p\theta_j]{} & \Omega BG_{(p)}
\end{array}$$

and, by taking adjoint, is equivalent to the existence of a map which makes the following diagram homotopy commutative:

$$\begin{array}{ccc}
S^{2n_j}\vee B & \xrightarrow{\varphi_j\vee l_pf} & BG_{(p)} \\
\downarrow & \nearrow & \\
S^{2n_j}\times B & &
\end{array}$$

where $\varphi_j\colon S^{2n_j} \to BG_{(p)}$ denotes the adjoint of the map $l_p\theta_j$.

Since G is p-regular, there exist cohomology classes $x_1, \cdots, x_l \in H^*(BG; \mathbb{Z}_{(p)})$, $|x_i| = 2n_i$ such that $H^*(BG; \mathbb{Z}_{(p)}) = \mathbb{Z}_{(p)}[x_1, \cdots, x_l]$ and $\mathrm{ad}(\theta_i)^*(x_i) = \alpha_i$ where $\alpha_i \in H^{2n_i}(S^{2n_i}; \mathbb{Z}_{(p)})$ is a generator. Let K_i denote the Eilenberg-MacLane space $K(\mathbb{Z}_{(p)}, 2n_i)$ and $K = \prod_{i=1}^{l} K_i$. Thus we obtain a homotopy commutative diagram

$$\begin{array}{ccccc} S^{2n_j} \vee B & \xrightarrow{\varphi_j \vee l_p f} & BG_{(p)} & \xrightarrow{x_p} & K \\ \downarrow & & & & \uparrow \mu \\ S^{2n_j} \times B & & \xrightarrow[\alpha_j \times xf]{} & & K \times K \end{array}$$

where $x = \prod x_i \colon BG \to K$ and x_p is the factorisation of x.

Since $(x_p)_* \colon \pi_i(BG_{(p)}) \to \pi_i(K)$ is surjective for $i \leq 2p+1$ and bijective for $i < 2p+1$ ([**14**]), by a theorem of J. H. C. Whitehead, if $\dim B < 2(p - n_j + 1)$, we see that the map $\varphi_j \vee l_p f$ extends to $S^{2n_j} \times BG_{(p)}$:

$$\begin{array}{ccc} [S^{2n_j} \vee B, BG_{(p)}] & \longleftarrow & [S^{2n_j} \times B, BG_{(p)}] \\ (x_p)_* \downarrow \cong & & \downarrow\!\!\downarrow (x_p)_* \\ [S^{2n_j} \vee B, K] & \longleftarrow & [S^{2n_j} \times B, K]. \end{array}$$

□

6. Classical groups

The estimate given in Theorem 1.3 is the best possible in the case when G is a classical group. We prove Theorem 1.6 in this section.

Let p be an odd prime, $EG \to BG$ the universal bundle, $B^{(k)}$ the k skeleton of BG, $E_k \to B^{(k)}$ the restriction of the universal bundle. Recall Serre's mod p decomposition:

$$\begin{array}{rlccc} \mathrm{U}(n) & \underset{p}{\simeq} S^1 \times S^3 \times \cdots \times S^{2n-1} & \Leftrightarrow & p \geq n, & n_l = n \\ \mathrm{SU}(n) & \underset{p}{\simeq} S^3 \times \cdots \times S^{2n-1} & \Leftrightarrow & p \geq n, & n_l = n \\ \mathrm{Sp}(n) & \underset{p}{\simeq} S^3 \times \cdots \times S^{4n-1} & \Leftrightarrow & p \geq 2n, & n_l = 2n \\ \mathrm{Spin}(2n+1) & \underset{p}{\simeq} S^3 \times \cdots \times S^{4n-1} & \Leftrightarrow & p \geq 2n, & n_l = 2n \\ \mathrm{Spin}(2n+2) & \underset{p}{\simeq} \mathrm{Spin}(2n+1) \times S^{2n+1} & & & \\ & \underset{p}{\simeq} S^3 \times \cdots \times S^{4n-1} \times S^{2n+1} & \Leftrightarrow & p \geq 2n+1, & n_l = 2n. \end{array}$$

PROOF OF THEOREM 1.6. The coefficients of the cohomology are taken to be $\mathbb{Z}/p$. We set $k = 2(p - n_l + 1)$. Assume that the adjoint bundle $\mathrm{Ad}E_k$ is fibrewise p-local trivial. Then, as in the proof of Lemma 5.1 (see also Lemma 4.4), we obtain a homotopy commutative diagram

$$\begin{array}{ccc} B^{(k)} \vee S^{2n_l} & \xrightarrow{i \vee \varphi} & BG_{(p)} \vee BG_{(p)} \\ \downarrow & & \downarrow \nabla \\ B^{(k)} \times S^{2n_l} & \xrightarrow[\tilde{\varphi}]{} & BG_{(p)} \end{array}$$

where φ represents a generator of $\pi_{2n_l}(BG_{(p)}) = \mathbb{Z}_{(p)}$. (When $G = \mathrm{Spin}(4) = S^3 \times S^3$, φ represents a generator of, say, the first factor.) We show that the existence of the map $\tilde{\varphi}$ implies a contradiction.

We treat the low rank groups $\mathrm{SU}(2) = \mathrm{Sp}(1) = \mathrm{Spin}(3)$, $\mathrm{SU}(3)$ and $\mathrm{Spin}(4)$ separately.

(1) $G = \mathrm{SU}(n)$, $n \geq 4$, $n \neq p$.

In this case, we have $p > n$, $n_l = n$ and $k = 2p + 2 - 2n$.

Let T^n be the standard maximal torus of $\mathrm{U}(n)$ and $\lambda \colon BT^n \to B\mathrm{U}(n)$ the natural map. The induced homomorphism

$$\lambda^* \colon H^*(B\mathrm{U}(n)) = \mathbb{Z}/p[c_1, \dots, c_n] \to \mathbb{Z}/p[t_1, \dots, t_n] = H^*(BT^n)$$

is monic and $\lambda^*(c_j) = \sigma_j(t_1, \dots, t_n)$, where $|t_j| = 2$. Let $s_q \in H^{2q}(B\mathrm{U}(n))$ be the element satisfying $\lambda^*(s_q) = t_1^q + \cdots + t_n^q$, then $s_1 = c_1$, $s_2 = c_1^2 - 2c_2$ and so on. We denote the image of s_q in $H^*(B\mathrm{SU}(n))$ by the same symbol s_q.

We see that the coefficient of c_n of $\wp^1(s_2) \in H^{2p+2}(B\mathrm{SU}(n)) \subset \mathbb{Z}/p[c_2, \dots, c_n]$ as a polynomial in c_n with coefficients in $\mathbb{Z}/p[c_2, \dots, c_{n-1}]$ is not zero , namely, if we write $\wp^1(s_2) = \sum_{j=0}^d h_j c_n^j$, $h_j \in \mathbb{Z}/p[c_2, \dots, c_{n-1}]$, then $h_1 \neq 0$ as follows: We have

$$\wp^1(s_2) = 2s_{p+1}$$

and by Girard's formula (see [**11**]),

$$= 2(p+1) \sum_{2i_2+3i_3+\cdots+ni_n=p+1} (-1)^{i_2+\cdots+i_n} \frac{(i_2 + \cdots + i_n - 1)!}{i_2! \cdots i_n!} c_2^{i_2} \cdots c_n^{i_n}.$$

Note that $1 < i_2 + \cdots + i_n \leq (p+1)/2 < p$, whence each term in the expression above is not zero. Since

$$p+1 = \begin{cases} 2 \cdot \frac{p-n+1}{2} + n \cdot 1, & n\text{: even,} \\ 2 \cdot \frac{p-n-2}{2} + 3 \cdot 1 + n \cdot 1 & n\text{: odd} \end{cases}$$

(note that $p > n$ and $n \geq 4$), we see that

$$\wp^1(s_2) = \begin{cases} \pm 2c_2^{\frac{p-n+1}{2}} c_n + \cdots, & n\text{: even,} \\ \pm(p-n)c_2^{\frac{p-n-2}{2}} c_3 c_n + \cdots, & n\text{: odd,} \end{cases}$$

whence $h_1 \neq 0$. Moreover we see that $i^*(h_1) \neq 0$ since $B^{(k)}$ is the k-skeleton of $B\mathrm{SU}(n)$ where $k = 2p + 2 - 2n \geq 4$ and if n is odd, $k \geq 6$.

Note that $a = \varphi^*(c_n) \in H^{2n}(S^{2n})$ is a generator. We have

$$\begin{aligned} \tilde{\varphi}^*(c_j) &= i^*(c_j) \otimes 1, \quad 2 \leq j \leq n-1, \\ \tilde{\varphi}^*(c_n) &= i^*(c_n) \otimes 1 + 1 \otimes a, \\ \tilde{\varphi}^*(h_j) &= i^*(h_j) \otimes 1, \\ \tilde{\varphi}^*(s_2) &= i^*(s_2) \otimes 1. \end{aligned}$$

Therefore we have $\wp^1(\tilde{\varphi}^*(s_2)) = 0$. Since

$$\tilde{\varphi}^*(c_n)^j = i^*(c_n)^j \otimes 1 + j i^*(c_n)^{j-1} \otimes a,$$

we have

$$\begin{aligned}0 &= \wp^1(\tilde{\varphi}^*(s_2))\\ &= \tilde{\varphi}^*(\wp^1(s_2))\\ &= \sum ji^*(c_n^{j-1}h_j)\otimes a.\end{aligned}$$

In particular, $i^*(h_1)=0$, which is a contradiction.
(2) $G=\mathrm{U}(n)$, $n\geq 2$.

We can prove this case similarly to the $\mathrm{SU}(n)$ case. In this case, we have $p\geq n$, $n_l=n$, $k=2p+2-2n\geq 2$ and

$$\wp^1(s_2)=(-1)^{p-n}2c_1^{p-n+1}c_n+\cdots$$

whence the proof works for the case $p=n$.
(3) $G=\mathrm{Spin}(n)$, $n\geq 5$.

We can prove this case similarly to the $\mathrm{U}(n)$ case using $\sum t_j^{2q}$ and Pontrjagin classes in the place of $\sum t_j^q$ and Chern classes respectively.

We consider the $\mathrm{Spin}(2n+2)$ case. In this case, we have $p\geq 2n+1$, $n_l=2n$ and $k=2p+2-4n\geq 4$. The inclusion of a maximal torus induces monomorphism

$$\lambda^*\colon H^*(B\mathrm{Spin}(2n+2))=\mathbb{Z}/p[p_1,\ldots,p_n,\chi]\to\mathbb{Z}/p[t_1,\ldots,t_{n+1}]=H^*(BT^{n+1})$$

such that $\lambda^*(p_j)=\sigma_j(t_1^2,\ldots,t_{n+1}^2)$ and $\lambda^*(\chi)=\sigma_{n+1}(t_1,\ldots,t_{n+1})=t_1\cdots t_{n+1}$. Let $\bar{s}_q=s_{2q}\in H^{4q}(B\mathrm{Spin}(2n+2))$, then $s_2=\bar{s}_1=p_1$. Note that $\frac{p+1}{2}\geq n+1$ and $|\chi|=2n+2<4n$. We have

$$\begin{aligned}\wp^1(p_1)&=\wp^1(s_2)=2s_{p+1}=2\bar{s}_{\frac{p+1}{2}}\\ &=(-1)^{\frac{p+1}{2}}(p+1)\sum_{i_1+\cdots+(n+1)i_{n+1}=\frac{p+1}{2}}(-1)^{i_1+\cdots+i_{n+1}}\frac{(i_1+\cdots+i_{n+1}-1)!}{i_1!\cdots i_{n+1}!}p_1^{i_1}\cdots p_n^{i_n}\chi^{2i_{n+1}}\\ &=\pm p_1^{\frac{p+1}{2}-n}p_n+\cdots\end{aligned}$$

and get a contradiction similarly to the $\mathrm{U}(n)$ case.

The case for $\mathrm{Spin}(2n+1)$ is a little bit simpler.
(4) $G=\mathrm{Sp}(n)$, $n\geq 2$.

We can prove this case similarly to the $\mathrm{Spin}(2n+1)$ case using symplectic Pontrjagin classes in the place of Pontrjagin classes.
(5) Low rank groups.

These cases are shown by similar but straightforward calculations: See [**15**] for the case $\mathrm{SU}(2)=\mathrm{Sp}(1)=\mathrm{Spin}(3)$, the $\mathrm{SU}(3)$ case is shown similarly and the case of $\mathrm{Spin}(4)$ reduces to the $\mathrm{SU}(2)$ case.

□

Note that if G is commutative, then the adjoint bundle is always trivial. In the case when $G=\mathrm{SU}(p)$, $2(p-n_l+1)=2$ and $B^{(2)}=B^{(3)}=*$. In this case, we have the following:

Proposition 6.1. *Let p be an odd prime and $G=\mathrm{SU}(p)$. Then, $\mathrm{Ad}E_k$ is not fibrewise p-local trivial if $k\geq 4$.*

Proof. We can show this similarly to Theorem 1.6 by noting that $p+1=2\cdot 1+(p-1)\cdot 1$, with a little care about $\mathrm{SU}(3)$, $p=3$. □

References

1. S. Bauer, M. C. Crabb and M. Spreafico, The classifying space of the gauge group of an SO(3)-bundle over S^2, Proc. Roy. Soc. Edinburgh **A 131** (2001), 767–783
2. R. Bencivenga, Approximating groups of bundle automorphisms by loop spaces, Trans. Amer. Math. Soc. **285** (1984), 703–715
3. A. K. Bousfield and D. M. Kan, Homotopy limits, completions and localizations, Lecture Notes in Math., 304, Springer, Berlin, (1972)
4. M. Crabb and I. James, Fibrewise homotopy theory, Mathematical Monographs, Springer, Berlin (1998)
5. M. C. Crabb and W. A. Sutherland, Counting homotopy types of gauge groups, Proc. London Math. Soc. **81** (2000), 747–768.
6. A. Dold, Partitions of unity in the theory of fibrations, Ann. of Math. (2) **78** (1963), 223–255.
7. P. Hilton, G. Mislin and J. Roitberg, Localization of nilpotent groups and spaces, North Holland (1975)
8. I. Llerena, Localization of fibrations with nilpotent fibre, Math. Z. **188** (1985), 397–410.
9. J. P. May, Fibrewise localization and completion, Trans. Amer. Math. Soc. **258** (1980), 127–146
10. J. Milnor, Construction of universal bundles. I, Ann. of Math. **63** (1956), 272–284.
11. J. Milnor and J. D. Stasheff, Characteristic classes, Ann. of Math. Stud. **76**, Princeton University Press, Princeton, N. J. (1974)
12. M. Mimura, G. Nishida and H. Toda, Localization of CW-complexes and its applications, J. Math. Soc. Japan **23** (1971), 593–624.
13. J. M. Møller, Nilpotent spaces of sections, Trans. Amer. Math. Soc. **303** (1987), 733–741.
14. H. Toda, Composition methods in homotopy groups of spheres, Ann. of Math. Stud. **49**, Princeton University Press, Princeton, N.J. (1962)
15. S. Tsukuda, A remark on the homotopy type of the classifying space of certain gauge groups, J. Math. Kyoto Univ. **36** (1996), 123–128.

Department of Mathematics, Kyoto University, Kyoto 606-8502, JAPAN

Department of Mathematical Sciences, University of the Ryukyus, Okinawa 903-0213, JAPAN

Contemporary Mathematics
Volume **519**, 2010

Spaces of algebraic maps from real projective spaces into complex projective spaces

Andrzej Kozlowski and Kohhei Yamaguchi

ABSTRACT. We study the homotopy types of spaces of algebraic (rational) maps from real projective spaces into complex projective spaces. We showed in [**1**] that in this setting the inclusion of the space of rational maps into the space of all continuous maps is a homotopy equivalence. In this paper we prove that the homotopy types of the terms of the natural 'degree' filtration approximate closer and closer the homotopy type of the space of continuous maps and obtain bounds that describe the closeness of the approximation in terms of the degree. Moreover, we compute low dimensional homotopy groups of these spaces. These results combined with those of [**1**] can be formulated as a single statement about $\mathbb{Z}/2$-equivariant homotopy equivalence between these spaces, where the $\mathbb{Z}/2$-action is induced by the complex conjugation. This generalizes a theorem of [**8**].

1. Introduction.

1.1. Summary of the contents. Let M and N be manifolds with some additional structure, e.g holomorphic, symplectic, real algebraic etc. The relation between the topology of the space of continuous maps preserving this structure and that of the space of all continuous maps has long been an object of study in several areas of topology and geometry. Early examples were provided by Gromov's h-principle for holomorphic maps [**7**]. In these cases the manifolds are complex, the structure preserving maps are the holomorphic ones, and the spaces of holomorphic and continuous maps turn out to be homotopy equivalent. However, in many other cases, the space of structure preserving maps approximates, in some sense, the space of all continuous ones and becomes homotopy equivalent to it only after some kind of stabilization. A paradigmatic example of this type was given in a seminal paper of Segal [**14**], where the space of rational (or holomorphic) maps of a fixed degree from the Riemann sphere to a complex projective space was shown to approximate the space of all continuous maps in homotopy, with the approximation becoming better as the degree increases. Segal's result was extended to a variety of other target spaces by various authors (e.g. [**2**]). Although it has been sometimes stated that these phenomena are inherently related to complex or at least symplectic structures, real analogs of Segal's result were given in [**14**], [**11**], [**8**], [**16**]. In fact, Segal formulated the complex and real approximation theorems which he had

2000 *Mathematics Subject Classification.* Primary 55P35, 55P10; Secondly 55P91.
Key words and phrases. Spaces of algebraic maps, equivariant homotopy equivalence.

©2010 American Mathematical Society

proved as a single statement involving equivariant equivalence, with respect to complex conjugation (see the remark after Proposition 1.4 of [**14**]). A similar idea was used in [**8**, Theorem 3.7]. This theorem amounts to two equivariant ones, one of which is equivalent to (a stable version of) Segal's equivariant one, while the other one is related to the 'real version' of Segal's theorem proved in [**11**].

All the results mentioned above (except the ones involving Gromov's h-principle), assume that the domain of the mappings is one dimensional (complex or real). It is natural to try to generalize them to the situation where the domain is higher dimensional. That such generalizations might be possible was first suggested by Segal (see the remark under Proposition 1.3 of [**14**]). A large step in this direction appeared to have been made when Mostovoy [**12**] showed that the homotopy types of spaces of holomorphic maps from $\mathbb{C}\mathrm{P}^m$ to $\mathbb{C}\mathrm{P}^n$ (for $m \leq n$) approximate the homotopy types of the spaces of continuous maps, with the approximation becoming better as the degree increases. Unfortunately Mostovoy's published argument contains several gaps. A new version of the paper, currently only available from the author, appears to correct all the mistakes, with the main results remaining essentially unchanged. There are two major changes in the proofs. One is that the space $\mathrm{Rat}_f(p,q)$ of (p,q) maps from $\mathbb{C}\mathrm{P}^m$ to $\mathbb{C}\mathrm{P}^n$ that restrict to a fixed map f on a fixed hyperplane, used in section 2 of the published article is replaced by the space $\overline{\mathrm{Rat}}_f(p,q)$ of pairs of $n+1$-tuples of polynomials in m variables that produce these maps. In the published version of the article it is assumed that these two spaces are homotopy equivalent, which is clearly not the case. However, they are homotopy equivalent after stabilization, both being equivalent to $\Omega^{2m}\mathbb{C}\mathrm{P}^n$ - the space of continuous maps that restrict to f on a fixed hyperplane. The second important change is the introduction of a new filtration on the simplicial resolution $X^{\Delta} \subset \mathbb{R}^N \times Y$ of a map $h : X \to Y$ and an embedding $i : X \to \mathbb{R}^N = \mathbb{C}^{N/2}$. This filtration is defined by means of complex skeleta (where the complex k-skeleton of a simplex in a complex affine space is the union of all its faces that are contained in complex affine subspaces of dimension at most k) and replaces the analogous "real" filtration in the arguments in [**12**, section 4].

In [**1**], a variant of Mostovoy's idea was applied to the case of algebraic maps from $\mathbb{R}\mathrm{P}^m$ to $\mathbb{R}\mathrm{P}^n$ for $m < n$. This leads naturally to the question whether one can generalize the $\mathbb{Z}/2$-equivariant Theorem 3.7 of [**8**] to an analogous equivariant equivalence of the spaces of algebraic maps and continuous maps between projective spaces, in which the domain is either a real or a complex projective space of dimension $m > 1$, the range is a complex projective space of dimension greater or equal to that of the domain. (Here, the $\mathbb{Z}/2$-action is induced by complex conjugation.)

Note that Theorem 3.7 of [**8**] has two parts, in the first the domain being complex and in the second real. Clearly, to prove the first part we need a Mostovoy's complex theorem. We plan to consider this problem in a future paper. Here we concentrate on the second part, concerning equivariant algebraic maps from real projective spaces to complex projective ones. Our main theorem is new, but it uses the main result of [**1**] and where our arguments are very similar to those used in that paper we omit their details and refer the reader to [**1**].

In the remainder of this section we introduce our notation and state the main definitions and theorems.

1.2. Notation and Main Results. We first introduce notation which is analogous to the one used in [**1**], the presence of $\mathbb{C}$ indicating that the complex case is being considered (i.e. maps take values in $\mathbb{C}\mathrm{P}^n$ or polynomials have coefficients in $\mathbb{C}$).

Let m and n be positive integers such that $1 \leq m < 2\cdot(n+1)-1$. We choose $\mathbf{e}_m = [1:0:\cdots:0] \in \mathbb{R}\mathrm{P}^m$ and $\mathbf{e}'_n = [1:0:\cdots:0] \in \mathbb{C}\mathrm{P}^n$ as the base points of $\mathbb{R}\mathrm{P}^m$ and $\mathbb{C}\mathrm{P}^n$, respectively. Let $\mathrm{Map}^*(\mathbb{R}\mathrm{P}^m, \mathbb{C}\mathrm{P}^n)$ denote the space consisting of all based maps $f : (\mathbb{R}\mathrm{P}^m, \mathbf{e}_m) \to (\mathbb{C}\mathrm{P}^n, \mathbf{e}'_n)$. When $m \geq 2$, we denote by $\mathrm{Map}^*_\epsilon(\mathbb{R}\mathrm{P}^m, \mathbb{C}\mathrm{P}^n)$ the corresponding path component of $\mathrm{Map}^*(\mathbb{R}\mathrm{P}^m, \mathbb{C}\mathrm{P}^n)$ for each $\epsilon \in \mathbb{Z}/2 = \{0,1\} = \pi_0(\mathrm{Map}^*(\mathbb{R}\mathrm{P}^m, \mathbb{C}\mathrm{P}^n))$ ([**5**]). Similarly, let $\mathrm{Map}(\mathbb{R}\mathrm{P}^m, \mathbb{C}\mathrm{P}^n)$ denote the space of all free maps $f : \mathbb{R}\mathrm{P}^m \to \mathbb{C}\mathrm{P}^n$ and $\mathrm{Map}_\epsilon(\mathbb{R}\mathrm{P}^m, \mathbb{C}\mathrm{P}^n)$ the corresponding path component of $\mathrm{Map}(\mathbb{R}\mathrm{P}^m, \mathbb{C}\mathrm{P}^n)$.

We shall use the symbols z_i when we refer to complex valued coordinates or variables or when we refer to complex and real valued ones at the same time while the notation x_i will be restricted to the purely real case.

A map $f : \mathbb{R}\mathrm{P}^m \to \mathbb{C}\mathrm{P}^n$ is called a *algebraic map of the degree* d if it can be represented as a rational map of the form $f = [f_0 : \cdots : f_n]$ such that $f_0, \cdots, f_n \in \mathbb{C}[z_0, \cdots, z_m]$ are homogeneous polynomials of the same degree d with no common *real* roots except $\mathbf{0}_{m+1} = (0, \cdots, 0) \in \mathbb{R}^{m+1}$. We denote by $\mathrm{Alg}_d(\mathbb{R}\mathrm{P}^m, \mathbb{C}\mathrm{P}^n)$ (resp. $\mathrm{Alg}^*_d(\mathbb{R}\mathrm{P}^m, \mathbb{C}\mathrm{P}^n)$) the space consisting of all (resp. based) algebraic maps $f : \mathbb{R}\mathrm{P}^m \to \mathbb{C}\mathrm{P}^n$ of degree d. It is easy to see that there are inclusions $\mathrm{Alg}_d(\mathbb{R}\mathrm{P}^m, \mathbb{C}\mathrm{P}^n) \subset \mathrm{Map}_{[d]_2}(\mathbb{R}\mathrm{P}^m, \mathbb{C}\mathrm{P}^n)$ and $\mathrm{Alg}^*_d(\mathbb{R}\mathrm{P}^m, \mathbb{C}\mathrm{P}^n) \subset \mathrm{Map}^*_{[d]_2}(\mathbb{R}\mathrm{P}^m, \mathbb{C}\mathrm{P}^n)$, where $[d]_2 \in \mathbb{Z}/2 = \{0,1\}$ denotes the integer d mod 2.

Let $A_d(m,n)(\mathbb{C})$ denote the space consisting of all $(n+1)$-tuples $(f_0, \cdots, f_n) \in \mathbb{C}[z_0, \cdots, z_m]^{n+1}$ of homogeneous polynomials of degree d with coefficients in $\mathbb{C}$ and without non-trivial common real roots (but possibly with non-trivial common *non-real* ones).

Let $A^{\mathbb{C}}_d(m,n) \subset A_d(m,n)(\mathbb{C})$ be the subspace consisting of $(n+1)$-tuples $(f_0, \cdots, f_n) \in A_d(m,n)(\mathbb{C})$ such that the coefficient of z_0^d in f_0 is 1 and 0 in the other f_k's ($k \neq 0$). Then there is a natural projection map

$$\Psi^{\mathbb{C}}_d : A^{\mathbb{C}}_d(m,n) \to \mathrm{Alg}^*_d(\mathbb{R}\mathrm{P}^m, \mathbb{C}\mathrm{P}^n). \tag{1.1}$$

If $d = 2d^*$ is an even positive integer, we also have a natural projection map

$$j^{\mathbb{C}}_d : A^{\mathbb{C}}_d(m,n) \to \mathrm{Map}^*(\mathbb{R}\mathrm{P}^m, \mathbb{C}^{n+1} \setminus \{\mathbf{0}\}) \simeq \mathrm{Map}^*(\mathbb{R}\mathrm{P}^m, S^{2n+1}) \tag{1.2}$$

defined by

$$j^{\mathbb{C}}_d(f)([x_0 : \cdots : x_m]) = \Big(\frac{f_0(x_0, \cdots, x_m)}{(\sum_{k=0}^m x_k^2)^{d^*}}, \cdots, \frac{f_n(x_0, \cdots, x_m)}{(\sum_{k=0}^m x_k^2)^{d^*}}\Big)$$

for $f = (f_0, \cdots, f_n) \in A^{\mathbb{C}}_d(m,n)$. Note that the map $j^{\mathbb{C}}_d$ is well defined only if $d \geq 2$ is an even integer.

For $m \geq 2$ and $g \in \mathrm{Alg}^*_d(\mathbb{R}\mathrm{P}^{m-1}, \mathbb{C}\mathrm{P}^n)$ a fixed algebraic map, we denote by $\mathrm{Alg}^{\mathbb{C}}_d(m,n;g)$ and $F(m,n;g)$ the spaces defined by

$$\begin{cases} \mathrm{Alg}^{\mathbb{C}}_d(m,n;g) &= \{f \in \mathrm{Alg}^*_d(\mathbb{R}\mathrm{P}^m, \mathbb{C}\mathrm{P}^n) : f|\mathbb{R}\mathrm{P}^{m-1} = g\}, \\ F(m,n;g) &= \{f \in \mathrm{Map}^*_d(\mathbb{R}\mathrm{P}^m, \mathbb{C}\mathrm{P}^n) : f|\mathbb{R}\mathrm{P}^{m-1} = g\}. \end{cases}$$

It is well-known that there is a homotopy equivalence $F(m,n;g)\simeq\Omega^m\mathbb{C}\mathrm{P}^n$ ([**13**]). Let $A_d^{\mathbb{C}}(m,n;g)\subset A_d^{\mathbb{C}}(m,n)$ denote the subspace given by

$$A_d^{\mathbb{C}}(m,n;g)=(\Psi_d^{\mathbb{C}})^{-1}(\mathrm{Alg}_d^{\mathbb{C}}(m,n;g)).$$

Observe that if an algebraic map $f\in\mathrm{Alg}_d^*(\mathbb{R}\mathrm{P}^m,\mathbb{C}\mathrm{P}^n)$ can be represented as $f=[f_0:\cdots:f_n]$ for some $(f_0,\cdots,f_n)\in A_d^{\mathbb{C}}(m,n)$ then the same map can also be represented as $f=[\tilde{g}_mf_0:\cdots:\tilde{g}_mf_n]$, where $\tilde{g}_m=\sum_{k=0}^m z_k^2$. So there is an inclusion $\mathrm{Alg}_d^*(\mathbb{R}\mathrm{P}^m,\mathbb{C}\mathrm{P}^n)\subset\mathrm{Alg}_{d+2}^*(\mathbb{R}\mathrm{P}^m,\mathbb{C}\mathrm{P}^n)$ and we can define *the stabilization map* $s_d:A_d^{\mathbb{C}}(m,n)\to A_{d+2}^{\mathbb{C}}(m,n)$ by $s_d(f_0,\cdots,f_n)=(\tilde{g}_mf_0,\cdots,\tilde{g}_mf_n)$.

A map $f\in\mathrm{Alg}_d^*(\mathbb{R}\mathrm{P}^m,\mathbb{C}\mathrm{P}^n)$ is called an algebraic map of *minimal degree* d if $f\in\mathrm{Alg}_d^*(\mathbb{R}\mathrm{P}^m,\mathbb{C}\mathrm{P}^n)\setminus\mathrm{Alg}_{d-2}^*(\mathbb{R}\mathrm{P}^m,\mathbb{C}\mathrm{P}^n)$. It is easy to see that if $g\in\mathrm{Alg}_d^*(\mathbb{R}\mathrm{P}^{m-1},\mathbb{C}\mathrm{P}^n)$ is an algebraic map of minimal degree d, then the restriction

$$\Psi_d^{\mathbb{C}}|A_d^{\mathbb{C}}(m,n;g):A_d^{\mathbb{C}}(m,n;g)\xrightarrow{\cong}\mathrm{Alg}_d^{\mathbb{C}}(m,n;g)\tag{1.3}$$

is a homeomorphism. Let

$$\begin{cases}i_{d,\mathbb{C}}:\mathrm{Alg}_d^*(\mathbb{R}\mathrm{P}^m,\mathbb{C}\mathrm{P}^n)\xrightarrow{\subsetneq}\mathrm{Map}_{[d]_2}^*(\mathbb{R}\mathrm{P}^m,\mathbb{C}\mathrm{P}^n)\\ i'_{d,\mathbb{C}}:\mathrm{Alg}_d^{\mathbb{C}}(m,n;g)\xrightarrow{\subsetneq}F(m,n;g)\simeq\Omega^m\mathbb{C}\mathrm{P}^n\end{cases}\tag{1.4}$$

denote the inclusions and let

$$i_d^{\mathbb{C}}=i_{d,\mathbb{C}}\circ\Psi_d^{\mathbb{C}}:A_d^{\mathbb{C}}(m,n)\to\mathrm{Map}_{[d]_2}^*(\mathbb{R}\mathrm{P}^m,\mathbb{C}\mathrm{P}^n)\tag{1.5}$$

be the natural map. For a connected space X, let $F(X,r)$ denote the configuration space of distinct r points in X. The symmetric group S_r of r letters acts on $F(X,r)$ freely by permuting coordinates, and let $C_r(X)$ be the configuration space of unordered r-distinct points in X given by $C_r(X)=F(X,r)/S_r$. Note that there is a stable homotopy equivalence $\Omega^mS^{m+l}\simeq_s\bigvee_{r=1}^\infty F(\mathbb{R}^m,r)_+\wedge_{S_r}(\bigwedge^rS^l)$ ([**15**]), and it is known that there is an isomorphism $H_k(F(\mathbb{R}^m,r)_+\wedge_{S_r}(\bigwedge^rS^l),\mathbb{Z})\cong H_{k-rl}(C_r(\mathbb{R}^m),(\pm\mathbb{Z})^{\otimes r})$ for $k,l\geq 1$ ([**4**], [**16**]), where $\bigwedge^rX=X\wedge\cdots\wedge X$ (r times). Let $G_{m,N}^d$ denote the abelian group

$$G_{m,N}^d=\bigoplus_{r=1}^{\lfloor\frac{d+1}{2}\rfloor}H_{k-(N-m)r}(C_r(\mathbb{R}^m),(\pm\mathbb{Z})^{\otimes(N-m)}),\tag{1.6}$$

where the meaning of $(\pm\mathbb{Z})^{\otimes(N-m)}$ is the same as in [**16**]. Let $D_{\mathbb{K}}(d;m,n)$ be the positive integer defined by

$$D_{\mathbb{K}}(d;m,n)=\begin{cases}(n-m)(\lfloor\frac{d+1}{2}\rfloor+1)-1 & \text{if } \mathbb{K}=\mathbb{R},\\ (2n-m+1)(\lfloor\frac{d+1}{2}\rfloor+1)-1 & \text{if } \mathbb{K}=\mathbb{C},\end{cases}\tag{1.7}$$

and $\lfloor x\rfloor$ is the integer part of a real number x. Note that $D_{\mathbb{C}}(d;m,n)=D_{\mathbb{R}}(d;m,2n+1)$. Now, recall the following 3 results, where we use the notations used in [**1**].

THEOREM 1.1 ([**9**], [**17**]). *If $n\geq 2$, then the natural map $i_d:A_d(1,n)\to\mathrm{Map}_{[d]_2}^*(\mathbb{R}\mathrm{P}^1,\mathbb{R}\mathrm{P}^n)\simeq\Omega S^n$ is a homotopy equivalence up to dimension $D_1(d,n)=(d+1)(n-1)-1$.*

THEOREM 1.2 ([**1**]). *Let $2\leq m<n$ be integers and let $g\in\mathrm{Alg}_d^*(\mathbb{R}\mathrm{P}^{m-1},\mathbb{R}\mathrm{P}^n)$ be an algebraic map of minimal degree d.*

(i) *The inclusion* $i'_d : \mathrm{Alg}_d(m,n;g) \to F^{\mathbb{R}}(m,n;g) \simeq \Omega^m S^n$ *is a homotopy equivalence through dimension* $D_{\mathbb{R}}(d;m,n)$ *if* $m+2 \leq n$ *and a homology equivalence through dimension* $D_{\mathbb{R}}(d;m,n)$ *if* $m+1=n$. *Here,* $F^{\mathbb{R}}(m,n;g)$ *denotes the space defined by*

$$F^{\mathbb{R}}(m,n;g) = \{f \in \mathrm{Map}^*_{[d]_2}(\mathbb{R}\mathrm{P}^m, \mathbb{R}\mathrm{P}^n) : f|\mathbb{R}\mathrm{P}^{m-1} = g\}.$$

(ii) *For any* $k \geq 1$, $H_k(\mathrm{Alg}_d(m,n;g),\mathbb{Z})$ *contains the subgroup* $G^d_{m,n}$ *as a direct summand.*

THEOREM 1.3 ([**1**]). *If* $2 \leq m < n$, *the map* $i_d : A_d(m,n) \to \mathrm{Map}^*_{[d]_2}(\mathbb{R}\mathrm{P}^m, \mathbb{R}\mathrm{P}^n)$ *is a homotopy equivalence through dimension* $D_{\mathbb{R}}(d;m,n)$ *if* $m+2 \leq n$ *and a homology equivalence through dimension* $D_{\mathbb{R}}(d;m,n)$ *if* $m+1=n$.

Remark. A map $f : X \to Y$ is called *a homotopy* (resp. *a homology*) *equivalence up to dimension* D if $f_* : \pi_k(X) \to \pi_k(Y)$ (resp. $f_* : H_k(X,\mathbb{Z}) \to H_k(Y,\mathbb{Z})$) is an isomorphism for any $k < D$ and an epimorphism for $k = D$. Similarly, it is called *a homotopy* (resp. *a homology*) *equivalence through dimension* D if $f_* : \pi_k(X) \to \pi_k(Y)$ (resp. $f_* : H_k(X,\mathbb{Z}) \to H_k(Y,\mathbb{Z})$) is an isomorphism for any $k \leq D$.

Our main results of this paper are as follows.

THEOREM 1.4. *Let* $2 \leq m \leq 2n$, *and let* $g \in \mathrm{Alg}^*_d(\mathbb{R}\mathrm{P}^{m-1}, \mathbb{C}\mathrm{P}^n)$ *be an algebraic map of minimal degree* d.

(i) *The inclusion* $i'_{d,\mathbb{C}} : \mathrm{Alg}^{\mathbb{C}}_d(m,n;g) \to F(m,n;g) \simeq \Omega^m S^{2n+1}$ *is a homotopy equivalence through dimension* $D_{\mathbb{C}}(d;m,n)$ *if* $m < 2n$ *and a homology equivalence through dimension* $D_{\mathbb{C}}(d;m,n)$ *if* $m = 2n$.

(ii) *For any* $k \geq 1$, $H_k(\mathrm{Alg}^{\mathbb{C}}_d(m,n;g),\mathbb{Z})$ *contains the subgroup* $G^d_{m,2n+1}$ *as a direct summand.*

THEOREM 1.5. *If* $2 \leq m \leq 2n$, *the map* $i^{\mathbb{C}}_d : A^{\mathbb{C}}_d(m,n) \to \mathrm{Map}^*_{[d]_2}(\mathbb{R}\mathrm{P}^m, \mathbb{C}\mathrm{P}^n)$ *is a homotopy equivalence through dimension* $D_{\mathbb{C}}(d;m,n)$ *if* $m < 2n$ *and a homology equivalence through dimension* $D_{\mathbb{C}}(d;m,n)$ *if* $m = 2n$.

COROLLARY 1.6. *If* $2 \leq m \leq 2n$, *the stabilization map* $s_d : A^{\mathbb{C}}_d(m,n) \to A^{\mathbb{C}}_{d+2}(m,n)$ *is a homotopy equivalence through dimension* $D_{\mathbb{C}}(d;m,n)$ *if* $m < 2n$ *and a homology equivalence through dimension* $D_{\mathbb{C}}(d;m,n)$ *if* $m = 2n$.

Note that the complex conjugation on $\mathbb{C}$ naturally induces $\mathbb{Z}/2$-actions on the spaces $\mathrm{Alg}^{\mathbb{C}}_d(m,n;g)$ and $A^{\mathbb{C}}_d(m,n)$. In the same way it also induces a $\mathbb{Z}/2$-action on $\mathbb{C}\mathrm{P}^n$ and this action extends to actions on the spaces $\mathrm{Map}^*(\mathbb{R}\mathrm{P}^m, S^{2n+1})$ and $\mathrm{Map}^*_\epsilon(\mathbb{R}\mathrm{P}^m, \mathbb{C}\mathrm{P}^n)$, where we identify $S^{2n+1} = \{(w_0,\cdots,w_n) \in \mathbb{C}^{n+1} : \sum_{k=0}^n |w_k|^2 = 1\}$ and regard $\mathbb{R}\mathrm{P}^m$ as a $\mathbb{Z}/2$-space with the trivial $\mathbb{Z}/2$-action. Since the maps $i'_{d,\mathbb{C}}$ and $i^{\mathbb{C}}_d$ are $\mathbb{Z}/2$-equivariant and $(i'_{d,\mathbb{C}})^{\mathbb{Z}/2} = i'_d$, $(i^{\mathbb{C}}_d)^{\mathbb{Z}/2} = i_d$, we easily obtain the following result.

COROLLARY 1.7. *If* $2 \leq m \leq 2n$ *and* $d \geq 1$ *be positive integers, two maps*

$$\begin{cases} i'_{d,\mathbb{C}} : \mathrm{Alg}^{\mathbb{C}}_d(m,n;g) \to F(m,n;g) \simeq \Omega^m S^{2n+1} \\ i^{\mathbb{C}}_d : A^{\mathbb{C}}_d(m,n) \to \mathrm{Map}^*_{[d]_2}(\mathbb{R}\mathrm{P}^m, \mathbb{C}\mathrm{P}^n) \end{cases}$$

are $\mathbb{Z}/2$-*equivariant homotopy equivalences through dimension* $D_{\mathbb{R}}(d;m,n)$ *if* $m < 2n$ *and* $\mathbb{Z}/2$-*equivariant homology equivalence through dimension* $D_{\mathbb{R}}(d;m,n)$ *if* $m = 2n$.

Remark. Let G be a finite group and let $f : X \to Y$ be a G-equivariant map. Then a map $f : X \to Y$ is called *a G-equivariant homotopy* (resp. *homology*) *equivalence through dimension D* if the induced homomorphism $f_*^H : \pi_k(X^H) \xrightarrow{\cong} \pi_k(Y^H)$ (resp. $f_*^H : H_k(X^H, \mathbb{Z}) \xrightarrow{\cong} H_k(Y^H, \mathbb{Z})$) are isomorphisms for any $k \leq D$ and any subgroup $H \subset G$.

Since there is a homotopy equivalence $\mathrm{Map}_0^*(\mathbb{R}\mathrm{P}^m, \mathbb{C}\mathrm{P}^n) \simeq \mathrm{Map}^*(\mathbb{R}\mathrm{P}^m, S^{2n+1})$ if $m \geq 2$ (cf. Lemma 4.5), it is not so difficult to investigate the homotopy type of the zero component of $\mathrm{Map}^*(\mathbb{R}\mathrm{P}^m, \mathbb{C}\mathrm{P}^n)$. So we would also like to study the cases $d \equiv 1 \pmod 2$. However, in this case the homotopy types of $\mathrm{Alg}_d^*(\mathbb{R}\mathrm{P}^m, \mathbb{C}\mathrm{P}^n)$ and $\mathrm{Map}_1^*(\mathbb{R}\mathrm{P}^m, \mathbb{R}\mathrm{P}^n)$ appear hard to investigate in general. However, for $d = 1$, $\Psi_1^{\mathbb{K}} : A_1^{\mathbb{C}}(m, n) \xrightarrow{\cong} \mathrm{Alg}_1^*(\mathbb{R}\mathrm{P}^m, \mathbb{C}\mathrm{P}^n)$ is a homeomorphism and we can prove the following:

THEOREM 1.8. (i) *If $2 \leq m < 2n$, the inclusion*

$$i_{1,\mathbb{C}} : \mathrm{Alg}_1^*(\mathbb{R}\mathrm{P}^m, \mathbb{C}\mathrm{P}^n) \to \mathrm{Map}_1^*(\mathbb{R}\mathrm{P}^m, \mathbb{C}\mathrm{P}^n)$$

is a homotopy equivalence up to dimension $D_{\mathbb{C}}(1; m, n) = 4n - 2m + 1$.

(ii) *If $m = 2n \geq 4$, the inclusion $i_{1,\mathbb{C}}$ induces an isomorphism*

$$(i_{1,\mathbb{C}})_* : \pi_1(\mathrm{Alg}_1^*(\mathbb{R}\mathrm{P}^{2n}, \mathbb{C}\mathrm{P}^n)) \xrightarrow{\cong} \pi_1(\mathrm{Map}_1^*(\mathbb{R}\mathrm{P}^{2n}, \mathbb{C}\mathrm{P}^n)) \cong \mathbb{Z}/2.$$

COROLLARY 1.9. (i) *If $2 \leq m < 2n$ and $d \equiv 0 \pmod 2$ are positive integers, the space $\mathrm{Alg}_d^*(\mathbb{R}\mathrm{P}^m, \mathbb{C}\mathrm{P}^n)$ is $(2n - m)$-connected and*

$$\pi_{2n-m+1}(\mathrm{Alg}_d^*(\mathbb{R}\mathrm{P}^m, \mathbb{C}\mathrm{P}^n)) \cong \begin{cases} \mathbb{Z} & \text{if } m \equiv 1 \pmod 2, \\ \mathbb{Z}/2 & \text{if } m \equiv 0 \pmod 2. \end{cases}$$

(ii) *If $2 \leq m \leq 2n$ and $\epsilon \in \{0, 1\}$, the two spaces $\mathrm{Alg}_1^*(\mathbb{R}\mathrm{P}^m, \mathbb{C}\mathrm{P}^n)$ and $\mathrm{Map}_\epsilon^*(\mathbb{R}\mathrm{P}^m, \mathbb{C}\mathrm{P}^n)$ are $(2n - m)$-connected, and*

$$\begin{aligned} \pi_{2n-m+1}(\mathrm{Alg}_1^*(\mathbb{R}\mathrm{P}^m, \mathbb{C}\mathrm{P}^n)) &\cong \pi_{2n-m+1}(\mathrm{Map}_\epsilon^*(\mathbb{R}\mathrm{P}^m, \mathbb{C}\mathrm{P}^n)) \\ &\cong \begin{cases} \mathbb{Z} & \text{if } m \equiv 1 \pmod 2, \\ \mathbb{Z}/2 & \text{if } m \equiv 0 \pmod 2. \end{cases} \end{aligned}$$

Remark. We conjecture that $\pi_1(\mathrm{Alg}_d^*(\mathbb{R}\mathrm{P}^m, \mathbb{C}\mathrm{P}^n)) = \mathbb{Z}/2$ if $m = 2n \geq 4$ and $d \equiv 0 \pmod 2$, but at this time we cannot prove this.

This paper is organized as follows. In section 2, we consider the space of algebraic maps $\mathbb{R}\mathrm{P}^m \to \mathbb{C}\mathrm{P}^n$ and recall the stable Theorem obtained in [**1**]. In section 3 we study simplicial resolutions and the spectral sequences induced from them and prove Theorem 1.4. In section 4 we prove Theorem 1.5 and Corollary 1.6 by using Theorem 1.3, and in section 5 we study the homotopy type of the stabilized space $A_{\infty+\epsilon}^{\mathbb{C}}(m, n)$. Finally in section 6, we investigate $A_d^{\mathbb{C}}(m, n)$ for the case $d = 1$ and prove Theorem 1.8 and Corollary 1.9.

2. Spaces of algebraic maps.

An algebraic map $f : \mathbb{R}\mathrm{P}^m \to \mathbb{C}\mathrm{P}^n$ can always be represented as $f = [f_0 : f_1 : \cdots : f_n]$, where $f_0, \cdots, f_n \in \mathbb{C}[z_0, z_1, \cdots, z_m]$ are homogeneous polynomials of the same degree d with no common *real* root other than $\mathbf{0}_{m+1} = (0, \cdots, 0) \in \mathbb{R}^{m+1}$ (but possibly with common *non-real* roots).

Clearly, an element of $\mathrm{Alg}_d^*(\mathbb{R}\mathrm{P}^m, \mathbb{C}\mathrm{P}^n)$ can always be represented in the form $f = [f_0 : f_1 : \cdots : f_n]$, such that the coefficient of z_0^d in f_0 is 1 and in the other polynomials f_i $(i \neq 0)$ 0. In general, such a representation is also not unique. For example, if we multiply all polynomials f_i by two different homogeneous polynomials in $z_0, z_1, \cdots, z_m$, which contain a power of z_0 with coefficient 1 and are always positive on $\mathbb{R}\mathrm{P}^m$, we will obtain two distinct representations of the same algebraic map. So the map $\Psi_d^{\mathbb{C}} : A_d^{\mathbb{C}}(m,n) \to \mathrm{Alg}_d^*(\mathbb{R}\mathrm{P}^m, \mathbb{C}\mathrm{P}^n)$ is a surjective projection map. It is easy to see that any fiber of $\Psi_d^{\mathbb{C}}$ is homeomorphic to the space consisting of non-negative positive homogeneous polynomial functions of some fixed even degree. So it is convex and contractible. Hence, it seems plausible to expect the following may be true.

CONJECTURE 2.1. *The map* $\Psi_d^{\mathbb{C}} : A_d^{\mathbb{C}}(m,n) \to \mathrm{Alg}_d^*(\mathbb{R}\mathrm{P}^m, \mathbb{C}\mathrm{P}^n)$ *is a homotopy equivalence.*

Although we cannot prove this conjecture, we show in section 5 that it is true if $d \to \infty$.

We always have $\mathrm{Alg}_d^*(\mathbb{R}\mathrm{P}^m, \mathbb{C}\mathrm{P}^n) \subset \mathrm{Alg}_{d+2}^*(\mathbb{R}\mathrm{P}^m, \mathbb{C}\mathrm{P}^n)$ and $\mathrm{Alg}_d(\mathbb{R}\mathrm{P}^m, \mathbb{C}\mathrm{P}^n) \subset \mathrm{Alg}_{d+2}(\mathbb{R}\mathrm{P}^m, \mathbb{C}\mathrm{P}^n)$, because $[f_0 : f_1 : \cdots : f_n] = [\tilde{g}_m f_0 : \tilde{g}_m f_1 : \cdots : \tilde{g}_m f_n]$.

DEFINITION 2.2. For $\epsilon \in \{0,1\}$, define subspaces $\mathrm{Alg}_\epsilon^*(m,n) \subset \mathrm{Map}_\epsilon^*(\mathbb{R}\mathrm{P}^m, \mathbb{C}\mathrm{P}^n)$ and $\mathrm{Alg}_\epsilon(m,n) \subset \mathrm{Map}_\epsilon(\mathbb{R}\mathrm{P}^m, \mathbb{C}\mathrm{P}^n)$ by

$$\begin{cases} \mathrm{Alg}_\epsilon^*(m,n) &= \bigcup_{k=1}^\infty \mathrm{Alg}_{\epsilon+2k}^*(\mathbb{R}\mathrm{P}^m, \mathbb{C}\mathrm{P}^n), \\ \mathrm{Alg}_\epsilon(m,n) &= \bigcup_{k=1}^\infty \mathrm{Alg}_{\epsilon+2k}(\mathbb{R}\mathrm{P}^m, \mathbb{C}\mathrm{P}^n). \end{cases}$$

THEOREM 2.3. *If* $1 \leq m \leq 2n$ *and* $\epsilon = 0$ *or* 1, *the inclusion maps*

$$\begin{cases} i : \mathrm{Alg}_\epsilon^*(m,n) \xrightarrow{\simeq} \mathrm{Map}_\epsilon^*(\mathbb{R}\mathrm{P}^m, \mathbb{C}\mathrm{P}^n) \\ j : \mathrm{Alg}_\epsilon(m,n) \xrightarrow{\simeq} \mathrm{Map}_\epsilon(\mathbb{R}\mathrm{P}^m, \mathbb{C}\mathrm{P}^n) \end{cases}$$

are homotopy equivalences.

PROOF. It follows from [**1**, Theorem 2.3] that j is a homotopy equivalence. The statement and the proof in it are valid also for spaces of based maps, and we can show that i is also a homotopy equivalence. □

3. Spectral sequences of the Vassiliev type.

From now on, we assume $2 \leq m \leq 2n$ and let $g \in \mathrm{Alg}_d^*(\mathbb{R}\mathrm{P}^{m-1}, \mathbb{C}\mathrm{P}^n)$ be a fixed algebraic map of minimal degree d, such that $g = [g_0 : \cdots : g_n]$ with $(g_0, \cdots, g_n) \in A_d^{\mathbb{C}}(m-1, n)$. Note that $(g_0, \cdots, g_n)$ is uniquely determined by g (because of the minimal degree condition).

Let $\mathcal{H}_d \subset \mathbb{C}[z_0, \cdots, z_m]$ denote the subspace consisting of all homogeneous polynomials of degree d. For $\epsilon \in \{0,1\}$, let $H_d^\epsilon \subset \mathcal{H}_d$ be the subspace consisting of all homogeneous polynomials $f \in \mathcal{H}_d$ such that the coefficient of $(z_0)^d$ of f is ϵ.

Since $A_d^{\mathbb{C}}(m,n)$ is the space consisting of all $(n+1)$-tuples $(f_0, \cdots, f_n) \in A_d(m,n)(\mathbb{C})$ such that the coefficient of z_0^d in f_0 is 1 and those of other f_k's are all zero, $A_d^{\mathbb{C}}(m,n) \subset \mathcal{H}_d^0 \times (\mathcal{H}_d^1)^n$. Note that $\mathcal{H}_d^0 \times (\mathcal{H}_d^1)^n$ is an affine space of real dimension of $N_d = 2(n+1)\Big(\binom{m+d}{m} - 1\Big)$.

Next, we set $B_k = \{g_k + z_m h : h \in \mathcal{H}_{d-1}\}$ $(k = 0, 1, \cdots, n)$ and define the subspace $A_d^* \subset \mathcal{H}_d^0 \times (\mathcal{H}_d^1)^n$ by $A_d^* = B_0 \times B_1 \times \cdots \times B_n$. Note that A_d^* is an affine space of real dimension $N_d^* = 2(n+1)\binom{m+d-1}{m}$.

DEFINITION 3.1. Let $A_d^{\mathbb{C}}(m,n;g) \subset A_d^*$ be the subspace $A_d^{\mathbb{C}}(m,n;g) = A_d^* \cap A_d^{\mathbb{C}}(m,n)$. Let $\Sigma_d \subset A_d$ denote the discriminant of $A_d^{\mathbb{C}}(m,n;g)$ in A_d^* defined by $\Sigma_d = A_d^* \setminus A_d^{\mathbb{C}}(m,n;g)$.

Since $g \in \mathrm{Alg}_d^*(\mathbb{R}\mathrm{P}^{m-1}, \mathbb{R}\mathrm{P}^n)$ has minimal degree d, clearly the restriction

$$\Psi_d^{\mathbb{C}}|_{A_d^{\mathbb{C}}(m,n;g)} : A_d^{\mathbb{C}}(m,n;g) \xrightarrow{\cong} \mathrm{Alg}_d^{\mathbb{C}}(m,n;g) \tag{3.1}$$

is a homeomorphism.

LEMMA 3.2. (i) *If* $(f_0, \cdots, f_n) \in \Sigma_d$ *and* $\mathbf{x} = (x_0, \cdots, x_m) \in \mathbb{R}^{m+1}$ *is a non-trivial common root of* $f_0, \cdots, f_n$, *then* $x_m \neq 0$.

(ii) $A_d^{\mathbb{C}}(m,n;g)$ *and* $A_d^{\mathbb{C}}(m,n)$ *are simply connected if* $m < 2n$.

PROOF. The proof is completely analogous to that of [**1**, Lemma 4.1]. □

DEFINITION 3.3. (i) For a finite set $\mathbf{x} = \{x_1, \cdots, x_l\} \subset \mathbb{R}^N$, let $\sigma(\mathbf{x})$ denote the convex hull spanned by $\mathbf{x}$. Note that $\sigma(\mathbf{x})$ is an $(l-1)$-dimensional simplex if and only if vectors $\{x_k - x_1\}_{k=2}^{l}$ are linearly independent. In particular, it is in general position if $x_1, \cdots, x_l$ are linearly independent over $\mathbb{R}$.

(ii) Let $h : X \to Y$ be a surjective map such that $h^{-1}(y)$ is a finite set for any $y \in Y$, and let $i : X \to \mathbb{R}^n$ be an embedding. Let $\mathcal{X}^\Delta$ and $h^\Delta : \mathcal{X}^\Delta \to Y$ denote the space and the map defined by

$$\mathcal{X}^\Delta = \big\{(y,w) \in Y \times \mathbb{R}^N : w \in \sigma(i(h^{-1}(y)))\big\} \subset Y \times \mathbb{R}^N,\ h^\Delta(y,w) = y.$$

The pair $(\mathcal{X}^\Delta, h^\Delta)$ is called *a simplicial resolution of* (h,i). In particular, $(\mathcal{X}^\Delta, h^\Delta)$ is called *a non-degenerate simplicial resolution* if for each $y \in Y$ and any k points of $i(h^{-1}(y))$ span $(k-1)$-dimensional simplex of $\mathbb{R}^N$.

(iii) For each $k \geq 0$, let $\mathcal{X}_k^\Delta \subset \mathcal{X}^\Delta$ be the subspace given by

$$\mathcal{X}_k^\Delta = \big\{(y,\omega) \in \mathcal{X}^\Delta : \omega \in \sigma(\mathbf{u}), \mathbf{u} = \{u_1, \cdots, u_l\} \subset i(h^{-1}(y)), l \leq k\big\}.$$

We make identification $X = \mathcal{X}_1^\Delta$ by identifying the point $x \in X$ with the pair $(h(x), i(x)) \in \mathcal{X}_1^\Delta$, and we note that there is an increasing filtration

$$\emptyset = \mathcal{X}_0^\Delta \subset X = \mathcal{X}_1^\Delta \subset \mathcal{X}_2^\Delta \subset \cdots \subset \mathcal{X}_k^\Delta \subset \mathcal{X}_{k+1}^\Delta \subset \cdots \subset \bigcup_{k=0}^{\infty} \mathcal{X}_k^\Delta = \mathcal{X}^\Delta.$$

LEMMA 3.4 ([**12**], [**16**]). *Let* $h : X \to Y$ *be a surjective map such that* $h^{-1}(y)$ *is a finite set for any* $y \in Y$, *and let* $i : X \to \mathbb{R}^N$ *be an embedding.*

(i) *If* X *and* Y *are closed semi-algebraic spaces and the two maps* h, i *are polynomial maps, then* $h^\Delta : \mathcal{X}^\Delta \xrightarrow{\simeq} Y$ *is a homotopy equivalence.*

(ii) *There is an embedding* $j : X \to \mathbb{R}^M$ *such that the associated simplicial resolution* $(\tilde{\mathcal{X}}^\Delta, \tilde{h}^\Delta)$ *of* (h,j) *is non-degenerate, and the space* $\tilde{\mathcal{X}}^\Delta$ *is uniquely determined up to homeomorphism. Moreover, there is a filtration preserving homotopy equivalence* $q^\Delta : \tilde{\mathcal{X}}^\Delta \xrightarrow{\simeq} \mathcal{X}^\Delta$ *such that* $q^\Delta|X = id_X$. □

Remark. Even when h is not finite to one, it is still possible to define its simplicial resolution and associated non-degenerate one. We omit the details of this construction and refer the reader to [**12**].

DEFINITION 3.5. Let $Z_d \subset \Sigma_d \times \mathbb{R}^m$ denote the tautological normalization of Σ_d consisting of all pairs $(\mathbf{f},\mathbf{x}) = ((f_0,\cdots,f_n),(x_0,\cdots,x_{m-1})) \in \Sigma_d \times \mathbb{R}^m$ such that the polynomials $f_0,\cdots,f_n$ have a non-trivial common real root $(\mathbf{x},1) = (x_0,\cdots,x_{m-1},1)$. Projection on the first factor gives a surjective map $\pi'_d : Z_d \to \Sigma_d$.

Let $\phi_d : A^*_d \overset{\cong}{\to} \mathbb{R}^{N^*_d}$ be any fixed homeomorphism, and let H_d be the set consisting of all monomials $\varphi_I = z^I = z_0^{i_0} z_1^{i_1} \cdots z_m^{i_m}$ of degree d ($I = (i_0, i_1, \cdots, i_m) \in \mathbb{Z}^{m+1}_{\geq 0}$, $|I| = \sum_{k=0}^m i_k = d$). Next, we define the Veronese embedding, which will play a key role in our argument. Let $\psi^*_d : \mathbb{R}^m \to \mathbb{R}^{M_d}$ be the map given by $\psi^*_d(x_0,\cdots,x_{m-1}) = \Big(\varphi_I(x_0,\cdots,x_{m-1},1)\Big)_{\varphi_I\in \mathrm{H}_d}$, where $M_d := \binom{d+m}{m}$. Now define the embedding $\Phi^*_d : Z^*_d \to \mathbb{R}^{N^*_d + M_d}$ by

$$\Phi^*_d((f_0,\cdots,f_n),\mathbf{x}) = (\phi^*_d(f_0,\cdots,f_n),\psi^*_d(\mathbf{x})).$$

DEFINITION 3.6. Let $(\mathcal{Z}^\Delta(d), {}'\pi^\Delta_d : \mathcal{Z}^\Delta(d) \to \Sigma_d)$ and $(\tilde{\mathcal{Z}}^\Delta(d), {}'\tilde{\pi}^\Delta_d : \tilde{\mathcal{Z}}^\Delta(d) \to \Sigma_d)$ denote the simplicial resolution of (π'_d, Φ^*_d) and the corresponding non-degenerate simplicial resolution with the natural increasing filtrations

$$\begin{cases} \mathcal{Z}^\Delta(d)_0 = \emptyset \subset \mathcal{Z}^\Delta(d)_1 \subset \mathcal{Z}^\Delta(d)_2 \subset \cdots \subset \mathcal{Z}^\Delta(d) = \displaystyle\bigcup_{k=0}^\infty \mathcal{Z}^\Delta(d)_k, \\ \tilde{\mathcal{Z}}^\Delta(d)_0 = \emptyset \subset \tilde{\mathcal{Z}}^\Delta(d)_1 \subset \tilde{\mathcal{Z}}^\Delta(d)_2 \subset \cdots \subset \tilde{\mathcal{Z}}^\Delta(d) = \displaystyle\bigcup_{k=0}^\infty \tilde{\mathcal{Z}}^\Delta(d)_k. \end{cases}$$

By Lemma 3.4, the map ${}'\pi^\Delta_d : \mathcal{Z}^\Delta(d) \overset{\simeq}{\to} \Sigma_d$ is a homotopy equivalence. It is easy to see that it extends to a homotopy equivalence ${}'\pi^\Delta_{d+} : \mathcal{Z}^\Delta(d)_+ \overset{\simeq}{\to} \Sigma_{d+}$, where X_+ denotes the one-point compactification of a locally compact space X.

Since $\mathcal{Z}^\Delta(d)_{r_+}/\mathcal{Z}^\Delta(d)_{r-1_+} \cong (\mathcal{Z}^\Delta(d)_r \setminus \mathcal{Z}^\Delta(d)_{r-1})_+$, we have the Vassiliev type spectral sequence

$$\{E^{r,s}_t(d), d_t : E^{r,s}_t(d) \to E^{r+t,s+1-t}_t(d)\} \Rightarrow H^{r+s}_c(\Sigma_d, \mathbb{Z}),$$

where $H^k_c(X,\mathbb{Z})$ denotes the cohomology group with compact supports given by $H^k_c(X,\mathbb{Z}) := H^k(X_+,\mathbb{Z})$ and $E^{r,s}_1(d) := H^{r+s}_c(\mathcal{Z}^\Delta(d)_r \setminus \mathcal{Z}^\Delta(d)_{r-1}, \mathbb{Z})$.

It follows from the Alexander duality that there is a natural isomorphism

$$H_k(A^{\mathbb{C}}_d(m,n;g),\mathbb{Z}) \cong H^{N^*_d - k - 1}_c(\Sigma_d,\mathbb{Z}) \quad \text{for } 1 \leq k \leq N^*_d - 2. \tag{3.2}$$

Using (3.2) and reindexing we obtain a spectral sequence

$$\{\tilde{E}^t_{r,s}(d), \tilde{d}^t : \tilde{E}^t_{r,s}(d) \to \tilde{E}^t_{r+t,s+t-1}(d)\} \Rightarrow H_{s-r}(A^{\mathbb{C}}_d(m,n;g),\mathbb{Z}) \tag{3.3}$$

if $s - r \leq N^*_d - 2$, where $\tilde{E}^1_{r,s}(d) = H^{N^*_d + r - s - 1}_c(\mathcal{Z}^\Delta(d)_r \setminus \mathcal{Z}^\Delta(d)_{r-1}, \mathbb{Z})$.

LEMMA 3.7. (i) *If $\{y_1,\cdots,y_r\} \in C_r(\mathbb{R}^m)$ is any set of r distinct points in $\mathbb{R}^m$ and $r \leq d+1$, then the r vectors $\{\psi^*_d(y_k) : 1 \leq k \leq r\}$ are linearly independent over $\mathbb{R}$ and span an $(r-1)$-dimensional simplex in $\mathbb{R}^{M_d}$.*

(ii) *If $1 \leq r \leq d+1$, there is a homeomorphism*

$$\mathcal{Z}^\Delta(d)_r \setminus \mathcal{Z}^\Delta(d)_{r-1} \cong \tilde{\mathcal{Z}}^\Delta(d)_r \setminus \tilde{\mathcal{Z}}^\Delta(d)_{r-1}$$

PROOF. The proof is completely analogous to that of [**1**, Lemma 4.3]. □

LEMMA 3.8. *If $1 \leq r \leq \lfloor \frac{d+1}{2} \rfloor$, $\mathcal{Z}^\Delta(d)_r \setminus \mathcal{Z}^\Delta(d)_{r-1}$ is homeomorphic to the total space of a real vector bundle $\xi_{d,r}$ over $C_r(\mathbb{R}^m)$ with rank $l^*_{d,r} = N^*_d - (2n+1)r - 1$.*

PROOF. The proof is completely analogous to that of [**1**, Lemma 4.4]. □

LEMMA 3.9. *All non-zero entries of* $\tilde{E}^1_{r,s}(d)$ *are situated in the range* $s \geq r(2n+2-m)$.

PROOF. The proof is completely analogous to that of [**1**, Lemma 4.5]. □

LEMMA 3.10. *If* $1 \leq r \leq \lfloor \frac{d+1}{2} \rfloor$, *there is a natural isomorphism*

$$\tilde{E}^1_{r,s}(d) \cong H_{s-(2n-m+2)r}(C_r(\mathbb{R}^m), (\pm\mathbb{Z})^{\otimes(2n-m+1)}).$$

PROOF. By using the Thom isomorphism and Poincaré duality, we obtain the desired isomorphism. □

Now we recall the spectral sequence constructed by V. Vassiliev [**16**]. From now on, we will assume that $m \leq 2n$, and we identify $\mathrm{Map}(S^m, S^{2n+1})$ with the space $\mathrm{Map}(S^m, \mathbb{R}^{2n+2} \setminus \{\mathbf{0}_{2n+2}\})$. We also choose a map $\varphi : S^m \to \mathbb{R}^{2n+2} \setminus \{\mathbf{0}_{2n+2}\}$ and fix it. Observe that $\mathrm{Map}(S^m, \mathbb{R}^{2n+2})$ is a linear space and consider the complements $\mathfrak{A}^n_m = \mathrm{Map}(S^m, \mathbb{R}^{2n+2}) \setminus \mathrm{Map}(S^m, S^{2n+1})$ and $\tilde{\mathfrak{A}}^n_m = \mathrm{Map}^*(S^m, \mathbb{R}^{2n+2}) \setminus \mathrm{Map}^*(S^m, S^{2n+1})$.

Note that $\mathfrak{A}^n_m$ consists of all continuous maps $f : S^m \to \mathbb{R}^{2n+2}$ passing through $\mathbf{0}_{2n+2}$. We will denote by $\Theta^k_\varphi \subset \mathrm{Map}(S^m, \mathbb{R}^{2n+2})$ the subspace consisting of all maps f of the forms $f = \varphi + p$, where p is the restriction to S^m of a polynomial map $\mathbb{R}^{m+1} \to \mathbb{R}^{2n+2}$ of degree $\leq k$. Let $\Theta^k \subset \Theta^k_\varphi$ denote the subspace consisting of all $f \in \Theta^k_\varphi$ passing through $\mathbf{0}_{n+1}$. In [**16**, page 111-112], Vassiliev uses the space Θ^k as a finite dimensional approximation of $\mathfrak{A}^n_m$.

Let $\tilde{\Theta}^k$ denote the subspace of Θ^k consisting of all maps $f \in \Theta^k$ which preserve the base points. By a variation of the preceding argument, Vassiliev also shows that $\tilde{\Theta}^k$ can be used as a finite dimensional approximation of $\tilde{\mathfrak{A}}^n_m$.

Let $\mathcal{X}_k \subset \tilde{\Theta}^k \times \mathbb{R}^{m+1}$ denote the subspace consisting of all pairs $(f, \alpha) \in \tilde{\Theta}^k \times \mathbb{R}^{m+1}$ such that $f(\alpha) = \mathbf{0}_{2n+2}$, and let $p_k : \mathcal{X}_k \to \tilde{\Theta}^k$ be the projection onto the first factor. Then, by making use of simplicial resolutions of the surjective maps $\{p_k : k \geq 1\}$, one can construct an associated geometric resolution $\{\tilde{\mathfrak{A}}^n_m\}$ of $\tilde{\mathfrak{A}}^n_m$, whose cohomology approximates the homology of $\mathrm{Map}^*(S^m, S^{2n+1}) = \Omega^m S^{2n+1}$ to any desired dimension. From the natural filtration on the approximating space $F_1 \subset F_2 \subset F_3 \subset \cdots \subset \bigcup_{k=1}^{\infty} F_k = \{\tilde{\mathfrak{A}}^n_m\}$, we obtain the associated spectral sequence:

$$\{E^t_{r,s}, d^t : E^t_{r,s} \to E^t_{r+t,s+t-1}\} \Rightarrow H_{s-r}(\Omega^m S^{2n+1}, \mathbb{Z}). \tag{3.4}$$

The following result follows easily from [**16**, Theorem 2 (page 112) and (32)].

LEMMA 3.11 ([**16**]). *Let* $2 \leq m \leq 2n$ *be integers and let* X *be a finite* m*-dimensional simplicial complex with a fixed base point* $x_0 \in X$.

(i) $E^1_{r,s} = H_{s-(2n-m+2)r}(C_r(\mathbb{R}^m), (\pm\mathbb{Z})^{\otimes(2n-m+1)})$ *if* $r \geq 1$, *and* $E^1_{r,s} = 0$ *if* $r < 0$ *or* $s < 0$ *or* $s < (2n-m+2)r$.

(ii) *For any* $t \geq 1$, $d^t = 0 : E^t_{r,s} \to E^t_{r+t,s+t-1}$ *for all* (r, s), *and* $E^1_{r,s} = E^\infty_{r,s}$ *Moreover, for any* $k \geq 1$, *the extension problem for the graded group* $Gr(H_k(\Omega^m S^{2n+1}, \mathbb{Z})) = \bigoplus_{r=1}^\infty E^\infty_{r,r+k} = \bigoplus_{r=1}^\infty E^1_{r,r+k}$ *is trivial and there is an isomorphism*

$$H_k(\Omega^m S^{2n+1},\mathbb{Z}) \cong \bigoplus_{r=1}^{\infty} H_{k-(2n-m+1)r}(C_r(\mathbb{R}^m),(\pm\mathbb{Z})^{\otimes(2n-m+1)}). \quad \square$$

DEFINITION 3.12. We identify $\Omega^m S^{2n+1} = \Omega^m(\mathbb{C}^{n+1}\setminus\{\mathbf{0}\})$ and define the map $j'_d : A^{\mathbb{C}}_d(m,n;g)\to \Omega^m S^{2n+1}$ by

$$j'_d(f_0,\cdots,f_n)(x_0,\cdots,x_m) = (f_0(x_0,\cdots,x_m),\cdots\cdots,f_n(x_0,\cdots,x_m))$$

for $((f_0,\cdots,f_n),(x_0,\cdots,x_m))\in A^{\mathbb{C}}_d(m,n;g)\times S^m$.

Now, by applying the spectral sequence (3.4), we prove the following result.

THEOREM 3.13. *Let $m,n\geq 2$ be positive integers such that $2\leq m\leq 2n$, and let $g\in \mathrm{Alg}^*_d(\mathbb{RP}^{m-1},\mathbb{CP}^n)$ be an algebraic map of minimal degree d.*

(i) *The map $j'_d : A^{\mathbb{C}}_d(m,n;g)\to\Omega^m S^{2n+1}$ is a homotopy equivalence through dimension $D_{\mathbb{C}}(d;m,n)$ if $m<2n$ and a homology equivalence through dimension $D_{\mathbb{C}}(d;m,n)$ if $m=2n$.*
(ii) *For any $k\geq 1$, $H_k(A^{\mathbb{C}}_d(m,n;g),\mathbb{Z})$ contains the subgroup $G^d_{m,2n+1}$ as a direct summand.*

PROOF. Consider the spectral sequence (3.4). First, note that, by Lemma 3.4, there is a filtration preserving homotopy equivalence $'q^\Delta : \tilde{\mathcal{Z}}^\Delta(d)\xrightarrow{\simeq}\mathcal{Z}^\Delta(d)$. Note also that the image of the map j'_d lies in a space of polynomial mappings, which approximates the space of continuous mappings $S^m\to S^{2n+1}$. Since $\tilde{\mathcal{Z}}^\Delta(d)$ is non-degenerate, the mapj'_d naturally extends to a filtration preserving map $'\tilde{\pi}$: $\tilde{\mathcal{Z}}^\Delta(d)\to\{\tilde{\mathfrak{A}}^n_m\}$ between resolutions. Thus the filtration preserving maps

$$\mathcal{Z}^\Delta(d) \xleftarrow[\simeq]{'q^\Delta} \tilde{\mathcal{Z}}^\Delta(d) \xrightarrow{'\tilde{\pi}} \{\tilde{\mathfrak{A}}^n_m\}$$

induce a homomorphism of spectral sequences $\{\tilde{\theta}^t_{r,s} : \tilde{E}^t_{r,s}(d)\to E^t_{r,s}\}$, where $\{E^t_{r,s},d^t\}\Rightarrow H_{s-r}(\Omega^m S^n,\mathbb{Z})$.

Observe that, by Lemma 3.10, Lemma 3.11 and the naturality of Thom isomorphism, for $r\leq\lfloor\frac{d+1}{2}\rfloor$ there is a commutative diagram

$$\begin{array}{ccc}
\tilde{E}^1_{r,s}(d) & \xrightarrow[\cong]{T} & H_{s-r(2n-m+2)}(C_r(\mathbb{R}^m),(\pm\mathbb{Z})^{\otimes(2n-m+1)}) \\
\tilde{\theta}^1_{r,s}\downarrow & & \| \\
E^1_{r,s} & \xrightarrow[\cong]{T} & H_{s-r(2n-m+2)}(C_r(\mathbb{R}^m),(\pm\mathbb{Z})^{\otimes(2n-m+1)})
\end{array} \tag{3.5}$$

Hence, if $r\leq\lfloor\frac{d+1}{2}\rfloor$, $\tilde{\theta}^1_{r,s}:\tilde{E}^1_{r,s}(d)\xrightarrow{\cong}E^1_{r,s}$ and thus so is $\tilde{\theta}^\infty_{r,s}:\tilde{E}^\infty_{r,s}(d)\xrightarrow{\cong}E^\infty_{r,s}$.

Next, we will compute the number

$$D_{min} = \min\{N|\ N\geq s-r,\ s\geq(2n+2-m)r,\ 1\leq r<\lfloor\frac{d+1}{2}\rfloor+1\}.$$

It is easy to see that D_{min} is the largest integer N which satisfies the inequality $(n+1-m)r>r+N$ for $r=\lfloor\frac{d+1}{2}\rfloor+1$, hence

$$D_{min} = (2n-m+1)(\lfloor\frac{d+1}{2}\rfloor+1)-1 = D_{\mathbb{C}}(d;m,n). \tag{3.6}$$

We note that, for dimensional reasons, $\tilde{\theta}^\infty_{r,s}:\tilde{E}^\infty_{r,s}(d)\xrightarrow{\cong}E^\infty_{r,s}$ is always an isomorphism if $r\leq\lfloor\frac{d+1}{2}\rfloor$ and $s-r\leq D_{\mathbb{C}}(d;m,n)$.

On the other hand, from Lemma 3.9 it easily follows that $\tilde{E}^1_{r,s}(d) = E^1_{r,s} = 0$ if $s - r \leq D_{\mathbb{C}}(d;m,n)$ and $r > \lfloor \frac{d+1}{2} \rfloor$. Hence, we have:

(3.6.1) If $s \leq r + D_{\mathbb{C}}(d;m,n)$, then $\tilde{\theta}^{\infty}_{r,s} : \tilde{E}^{\infty}_{r,s}(d) \xrightarrow{\cong} E^{\infty}_{r,s}$ is always an isomorphism.

Hence, by using the comparison Theorem of spectral sequences, we have that j'_d is a homology equivalence through dimension $D_{\mathbb{C}}(d;m,n)$. Since $A^{\mathbb{C}}_d(m,n;g)$ and $\Omega^m S^{2n+1}$ are simply connected if $m < 2n$, j'_d is a homotopy equivalence through dimension $D_{\mathbb{C}}(d;m,n)$ if $m < 2n$. Hence, (i) is proved.

It remains to show (ii). Since $d^t = 0$ for any $t \geq 1$, from the equality $d^t \circ \tilde{\theta}^t_{r,s} = \tilde{\theta}^t_{r+t,s+t-1} \circ \tilde{d}^t$ and some diagram chasing, we obtain $\tilde{E}^1_{r,s}(d) = \tilde{E}^{\infty}_{r,s}(d)$ for all $r \leq \lfloor \frac{d+1}{2} \rfloor$. Moreover, since the extension problem for the graded group

$$Gr(H_k(\Omega^m S^{2n+1}, \mathbb{Z})) = \bigoplus_{r=1}^{\infty} E^{\infty}_{r,k+r} = \bigoplus_{r=1}^{\infty} E^1_{r,k+r}$$

is trivial, by using (3.6.1) we can prove that the associated graded group

$$Gr(H_k(A^{\mathbb{C}}_d(m,n;g), \mathbb{Z})) = \bigoplus_{r=1}^{\infty} \tilde{E}^{\infty}_{r,k+r}(d) = \bigoplus_{r=1}^{\infty} \tilde{E}^1_{r,k+r}(d)$$

is also trivial until the $\lfloor \frac{d+1}{2} \rfloor$-th term of the filtration. Hence, $H_k(A^{\mathbb{C}}_d(m,n;g), \mathbb{Z})$ contains the subgroup

$$\bigoplus_{r=1}^{\lfloor \frac{d+1}{2} \rfloor} \tilde{E}^1_{r,k+r}(d) = \bigoplus_{r=1}^{\lfloor \frac{d+1}{2} \rfloor} \tilde{E}^{\infty}_{r,k+r}(d) \cong \bigoplus_{r=1}^{\lfloor \frac{d+1}{2} \rfloor} H_{k-r(2n-m+1)}(C_r(\mathbb{R}^m), (\pm\mathbb{Z})^{\otimes(2n-m+1)})$$

as a direct summand, which proves the assertion (ii). □

COROLLARY 3.14. *Let $m, n \geq 2$ be positive integers such that $2 \leq m \leq 2n$, let $g \in \mathrm{Alg}^*_d(\mathbb{R}\mathrm{P}^{m-1}, \mathbb{C}\mathrm{P}^n)$ be an algebraic map of minimal degree d, and let $\mathbb{F} = \mathbb{Z}/p$ (p: prime) or $\mathbb{F} = \mathbb{Q}$. Then the map $j'_d : A^{\mathbb{C}}_d(m,n;g) \to \Omega^m S^{2n+1}$ induces an isomorphism on the homology group $H_k(\ ,\mathbb{F})$ for any $1 \leq k \leq D_{\mathbb{C}}(d;m,n)$.*

PROOF. In the proof of Theorem 3.13 replace the homology groups $H_k(\ ,\mathbb{Z})$ and $H_k(\ ,(\pm\mathbb{Z})^{\otimes k})$ by $H_k(\ ,\mathbb{F})$ and $H_k(\ ,(\pm\mathbb{F})^{\otimes k})$ and use the same argument. □

Let $\gamma_m : S^m \to \mathbb{R}\mathrm{P}^m$ and $\gamma^{\mathbb{C}}_n : S^{2n+1} \to \mathbb{C}\mathrm{P}^n$ denote the usual double covering and the Hopf fibration map, respectively. Let $\gamma^{\#}_m : \mathrm{Map}^*_{\epsilon}(\mathbb{R}\mathrm{P}^m, \mathbb{C}\mathrm{P}^n) \to \Omega^m \mathbb{C}\mathrm{P}^n$ be given by $\gamma^{\#}_m(h) = h \circ \gamma_m$. It is easy to verify that the following diagram

$$(3.7)\qquad \begin{array}{ccccc} A^{\mathbb{C}}_d(m,n;g) & \xrightarrow[\cong]{\Psi'_d} & \mathrm{Alg}^{\mathbb{C}}_d(m,n;g) & \xrightarrow[\subset]{i'_d} & F(m,n;g) \\ \Big\downarrow{\scriptstyle j'_d} & & & & \Big\downarrow{\scriptstyle i'}\ \cap \\ \Omega^m S^{2n+1} & \xrightarrow[\simeq]{\Omega^m \gamma^{\mathbb{C}}_n} & \Omega^m \mathbb{C}\mathrm{P}^n & \xleftarrow{\gamma^{\#}_m} & \mathrm{Map}^*_{[d]_2}(\mathbb{R}\mathrm{P}^m, \mathbb{C}\mathrm{P}^n) \end{array}$$

is commutative, where $i' : F(m,n;g) \xrightarrow{\subsetneq} \mathrm{Map}^*_{[d]_2}(\mathbb{R}\mathrm{P}^m, \mathbb{C}\mathrm{P}^n)$ and Ψ'_d denote the inclusion and the restriction $\Psi'_d = \Psi^{\mathbb{C}}_d | A^{\mathbb{C}}_d(m,n;g)$, respectively.

LEMMA 3.15. *If $2 \leq m < 2n$ and $g \in \mathrm{Alg}^*_d(\mathbb{R}\mathrm{P}^{m-1}, \mathbb{R}\mathrm{P}^n)$ a fixed map of minimal degree d, then the map $\gamma^{\#}_m \circ i' : F(m,n;g) \to \Omega^m \mathbb{C}\mathrm{P}^n$ is a homotopy equivalence through dimension $D_{\mathbb{C}}(d;m,n)$.*

PROOF. Since there is a homotopy equivalence $F(m,n;g) \simeq \Omega^m\mathbb{C}\mathrm{P}^n$, the two spaces $F(m,n;g)$ and $\Omega^m\mathbb{C}\mathrm{P}^n$ are simple. So it suffices to show that the map $\gamma_m^{\#}\circ i'$ is a homology equivalence through dimension $D_{\mathbb{C}}(d;m,n)$.

Let $\mathbb{F} = \mathbb{Z}/p$ (p: prime) or $\mathbb{F} = \mathbb{Q}$, and consider the induced homomorphism $(\gamma_m^{\#}\circ i')_* = H_k(\gamma_m^{\#}\circ i', \mathbb{F}) : H_k(F(m,n;g),\mathbb{F}) \to H_k(\Omega^m\mathbb{C}\mathrm{P}^n,\mathbb{F})$. Since $\Omega^m\gamma_n^{\mathbb{C}}$ is a homotopy equivalence by Corollary 3.14 and the commutativity of the diagram (3.7) $(\gamma_m^{\#}\circ i')_*$ is an epimorphism for any $1 \leq k \leq D_{\mathbb{C}}(d;m,n)$.

However, since there is a homotopy equivalence $F(m,n;g) \simeq \Omega^m\mathbb{C}\mathrm{P}^n$, we have $\dim_{\mathbb{F}} H_k(F(m,n;g),\mathbb{F}) = \dim_{\mathbb{F}} H_k(\Omega^m\mathbb{C}\mathrm{P}^n,\mathbb{F}) < \infty$ for any k. Hence, $H_k(\gamma_m^{\#}\circ i',\mathbb{F})$ is an isomorphism for any $1 \leq k \leq D_{\mathbb{C}}(d;m,n)$. By the Universal Coefficient Theorem $\gamma_m^{\#}\circ i'$ is a homology equivalence through dimension $D_{\mathbb{C}}(d;m,n)$. □

PROOF OF THEOREM 1.4. Since $\Psi_d' : A_d^{\mathbb{C}}(m,n;g) \xrightarrow{\cong} \mathrm{Alg}_d^{\mathbb{C}}(m,n;g)$ is a homeomorphism, the assertion (ii) follows from Theorem 3.13. Because $\gamma_m^{\#}\circ i'$ a homotopy equivalence through dimension $D_{\mathbb{C}}(d;m,n)$, by using the diagram (3.7) and Theorem 3.13 we easily obtain the assertion (i). □

4. The space $A_d^{\mathbb{C}}(m,n)$.

In this section we shall consider the unstable problem for the space $A_d^{\mathbb{C}}(m,n)$, where $d = 2d^* \geq 2$ is even.

DEFINITION 4.1. Define $\psi_{m,n} : A_d^{\mathbb{R}}(m,2n+1) \to A_d^{\mathbb{C}}(m,n)$ by

$$\psi_{m,n}(f_0,\cdots,f_{2n+1}) = (f_0+\sqrt{-1}f_1, f_2+\sqrt{-1}f_3,\cdots,f_{2n}+\sqrt{-1}f_{2n+1}).$$

It is easy to see that

LEMMA 4.2. *$\psi_{m,n} : A_d^{\mathbb{R}}(m,2n+1) \xrightarrow{\cong} A_d^{\mathbb{C}}(m,n)$ is a homeomorphism.* □

LEMMA 4.3. (i) *$\mathrm{Map}^*(\mathbb{R}\mathrm{P}^m, S^{2n+1})$ is $(2n-m)$-connected.*

(ii) $\pi_{2n-m+1}(\mathrm{Map}^*(\mathbb{R}\mathrm{P}^m,S^{2n+1})) \cong \begin{cases}\mathbb{Z} & \text{if } m\equiv 1 \pmod 2,\\ \mathbb{Z}/2 & \text{if } m \equiv 0 \pmod 2.\end{cases}$

PROOF. (i) We argue by induction on m. For $m = 1$ the result follows from the homotopy equivalence $\mathrm{Map}^*(\mathbb{R}\mathrm{P}^1,S^{2n+1}) \simeq \Omega S^{2n+1}$. Suppose that the space $\mathrm{Map}^*(\mathbb{R}\mathrm{P}^{m-1},S^{2n+1})$ is $(2n-m+1)$-connected for some $m \geq 2$. Since $\Omega^m S^{2n+1}$ is $(2n-m)$-connected, from the restriction fibration sequence $\Omega^m S^{2n+1} \to \mathrm{Map}^*(\mathbb{R}\mathrm{P}^m,S^{2n+1}) \to \mathrm{Map}^*(\mathbb{R}\mathrm{P}^{m-1},S^{2n+1})$, we deduce that $\mathrm{Map}^*(\mathbb{R}\mathrm{P}^m,S^{2n+1})$ is $(2n-m)$-connected. Hence, (i) has been proved.

(ii) First, consider the case $m = 1$. Since $\mathrm{Map}^*(\mathbb{R}\mathrm{P}^1,S^{2n+1}) \simeq \Omega S^{2n+1}$, (ii) clearly holds for $m = 1$. Next, consider the case $m = 2$. If we consider the fibration sequence $\mathrm{Map}^*(\mathbb{R}\mathrm{P}^2,S^{2n+1}) \to \Omega S^{2n+1} \xrightarrow{2} \Omega S^{2n+1}$ induced from the cofibration sequence $S^1 \xrightarrow{2\iota_1} S^1 \to \mathbb{R}\mathrm{P}^2$, an easy computation shows that $\pi_{2n-1}(\mathrm{Map}^*(\mathbb{R}\mathrm{P}^2,S^{2n+1})) \cong \mathbb{Z}/2$. Hence, (ii) holds for $m = 2$, too.

Now we assume that $m \geq 3$ and consider the fibration sequence

$$\mathrm{Map}^*(\mathbb{R}\mathrm{P}^m/\mathbb{R}\mathrm{P}^{m-2},S^{2n+1}) \to \mathrm{Map}^*(\mathbb{R}\mathrm{P}^m,S^{2n+1}) \xrightarrow{r} \mathrm{Map}^*(\mathbb{R}\mathrm{P}^{m-2},S^{2n+1}).$$

Since $\mathrm{Map}^*(\mathbb{R}\mathrm{P}^{m-2},S^{2n+1})$ is $(2n-m+2)$-connected, there is an isomorphism $\pi_{2n-m+1}(\mathrm{Map}^*(\mathbb{R}\mathrm{P}^m,S^{2n+1})) \cong \pi_{2n-m+1}(\mathrm{Map}^*(\mathbb{R}\mathrm{P}^m/\mathbb{R}\mathrm{P}^{m-2},S^{2n+1}))$. Thus it remains to show the following:

$$\pi_{2n-m+1}(\mathrm{Map}^*(\mathbb{R}\mathrm{P}^m/\mathbb{R}\mathrm{P}^{m-2}, S^{2n+1})) \cong \begin{cases} \mathbb{Z} & \text{if } m \equiv 1 \pmod 2, \\ \mathbb{Z}/2 & \text{if } m \equiv 0 \pmod 2. \end{cases} \tag{4.1}$$

If $m \equiv 1 \pmod 2$, $\mathbb{R}\mathrm{P}^m/\mathbb{R}\mathrm{P}^{m-2} = S^{m-1} \vee S^m$ and there is an isomorphism

$$\pi_{2n-m+1}(\mathrm{Map}^*(\mathbb{R}\mathrm{P}^m/\mathbb{R}\mathrm{P}^{m-2}, S^{2n+1})) \cong \pi_{2n-m+1}(\Omega^{m-1}S^{2n+1} \times \Omega^m S^{2n+1}) \cong \mathbb{Z}.$$

Hence (4.1) holds for $m \equiv 1 \pmod 2$. Finally suppose that $m \equiv 0 \pmod 2$. Then, because $\mathbb{R}\mathrm{P}^m/\mathbb{R}\mathrm{P}^{m-2} = S^{m-1} \cup_2 e^m$, there is a fibration sequence

$$\mathrm{Map}^*(\mathbb{R}\mathrm{P}^m/\mathbb{R}\mathrm{P}^{m-2}, S^{2n+1}) \to \Omega^{m-1}S^{2n+1} \xrightarrow{2} \Omega^{m-1}S^{2n+1}.$$

From the homotopy exact sequence induced by the above sequence, we deduce that $\pi_{2n-m+1}(\mathrm{Map}^*(\mathbb{R}\mathrm{P}^m/\mathbb{R}\mathrm{P}^{m-2}, S^{2n+1})) \cong \mathbb{Z}/2$. □

DEFINITION 4.4. (i) Let $\gamma_n : S^n \to \mathbb{R}\mathrm{P}^n$ and $\gamma_n^{\mathbb{C}} : S^{2n+1} \to \mathbb{C}\mathrm{P}^n$ denote the usual double covering and the Hopf fibering as before. Define the following two maps

$$\begin{cases} \gamma_{n\#} : \mathrm{Map}^*(\mathbb{R}\mathrm{P}^m, S^n) \to \mathrm{Map}^*(\mathbb{R}\mathrm{P}^m, \mathbb{R}\mathrm{P}^n) \\ \gamma_{n\#}^{\mathbb{C}} : \mathrm{Map}^*(\mathbb{R}\mathrm{P}^m, S^{2n+1}) \to \mathrm{Map}^*(\mathbb{R}\mathrm{P}^m, \mathbb{C}\mathrm{P}^n) \end{cases}$$

by $\gamma_{n\#}(h) = \gamma_n \circ h$ and $\gamma_{n\#}^{\mathbb{C}}(h') = \gamma_n^{\mathbb{C}} \circ h'$.

Since $\mathrm{Map}^*(\mathbb{R}\mathrm{P}^m, S^n)$ and $\mathrm{Map}^*(\mathbb{R}\mathrm{P}^m, S^{2n+1}))$ contain the subspace of constant maps, the images of $\gamma_{n\#}$ and that of $\gamma_{n\#}^{\mathbb{C}}$ are contained in $\mathrm{Map}_0^*(\mathbb{R}\mathrm{P}^m, \mathbb{R}\mathrm{P}^n)$ and $\mathrm{Map}_0^*(\mathbb{R}\mathrm{P}^m, \mathbb{C}\mathrm{P}^n)$, respectively. Thus we obtain two maps

$$\begin{cases} \gamma_{n\#} : \mathrm{Map}^*(\mathbb{R}\mathrm{P}^m, S^n) \to \mathrm{Map}_0^*(\mathbb{R}\mathrm{P}^m, \mathbb{R}\mathrm{P}^n), \\ \gamma_{n\#}^{\mathbb{C}} : \mathrm{Map}^*(\mathbb{R}\mathrm{P}^m, S^{2n+1}) \to \mathrm{Map}_0^*(\mathbb{R}\mathrm{P}^m, \mathbb{C}\mathrm{P}^n). \end{cases}$$

(ii) Let $\mu_n : \mathbb{R}\mathrm{P}^{2n+1} \to \mathbb{C}\mathrm{P}^n$ denote the usual projection given by

$$\mu_n([x_0 : x_1 : \cdots : x_{2n+1}] = [x_0 + \sqrt{-1}x_1 : \cdots : x_{2n} + \sqrt{-1}x_{2n+1}],$$

and define the map $\mu_{n\#} : \mathrm{Map}_0^*(\mathbb{R}\mathrm{P}^m, \mathbb{R}\mathrm{P}^{2n+1}) \to \mathrm{Map}_0^*(\mathbb{R}\mathrm{P}^m, \mathbb{C}\mathrm{P}^n)$ by $\mu_{n\#}(h) = \mu_n \circ h$.

LEMMA 4.5. (i) *If* $1 \leq m < n$, $\gamma_{n\#} : \mathrm{Map}^*(\mathbb{R}\mathrm{P}^m, S^n) \xrightarrow{\simeq} \mathrm{Map}_0^*(\mathbb{R}\mathrm{P}^m, \mathbb{R}\mathrm{P}^n)$ *is a homotopy equivalence.*

(ii) *If* $2 \leq m \leq 2n$, $\gamma_{n\#}^{\mathbb{C}} : \mathrm{Map}^*(\mathbb{R}\mathrm{P}^m, S^{2n+1}) \xrightarrow{\simeq} \mathrm{Map}_0^*(\mathbb{R}\mathrm{P}^m, \mathbb{C}\mathrm{P}^n)$ *is a homotopy equivalence.*

Remark. Since $\pi_1(\mathrm{Map}^*(\mathbb{R}\mathrm{P}^1, S^{2n+1})) = 0$ and $\pi_1(\mathrm{Map}_0^*(\mathbb{R}\mathrm{P}^1, \mathbb{C}\mathrm{P}^n)) = \mathbb{Z}$, (ii) of Theorem 4.5 does not hold for $m = 1$.

PROOF. Since the proof of (i) is analogous to that of (ii), we only prove the assertion (ii). We prove it by induction on m. First, assume that $m = 2$, and consider the following commutative diagram of restriction fibration sequences

$$\begin{array}{ccccc} \Omega^2 S^{2n+1} & \longrightarrow & \mathrm{Map}^*(\mathbb{R}\mathrm{P}^2, S^{2n+1}) & \xrightarrow{r} & \Omega S^{2n+1} \\ \Omega^2\gamma_n^{\mathbb{C}} \downarrow \simeq & & \gamma_{n\#} \downarrow & & \Omega\gamma_n^{\mathbb{C}} \downarrow \\ \Omega^2\mathbb{C}\mathrm{P}^n & \longrightarrow & \mathrm{Map}_0^*(\mathbb{R}\mathrm{P}^2, \mathbb{C}\mathrm{P}^n) & \xrightarrow{r} & \Omega\mathbb{C}\mathrm{P}^n \end{array} \tag{4.2}$$

Since $\Omega^2\gamma_n^{\mathbb{C}}$ is a homotopy equivalence and $\Omega\gamma_{n*}^{\mathbb{C}} : \pi_k(\Omega S^{2n+1}) \xrightarrow{\cong} \pi_k(\Omega\mathbb{C}\mathrm{P}^n)$ is an isomorphism for any $k \geq 2$, by the Five Lemma the induced homomorphism ${\gamma_{n\#}^{\mathbb{C}}}_* : \pi_k(\mathrm{Map}^*(\mathbb{R}\mathrm{P}^2, S^{2n+1})) \xrightarrow{\cong} \pi_k(\mathrm{Map}_0^*(\mathbb{R}\mathrm{P}^2, \mathbb{C}\mathrm{P}^n))$ is an isomorphism for any $k \geq 2$. On the other hand, because the homotopy exact sequence of the lower row of (4.2) is

$$0 \to \pi_1(\mathrm{Map}_0^*(\mathbb{R}\mathrm{P}^2, \mathbb{C}\mathrm{P}^n)) \to \mathbb{Z} = \pi_1(\Omega\mathbb{C}\mathrm{P}^n) \xrightarrow{\partial} \pi_0(\Omega^2\mathbb{C}\mathrm{P}^n) = \mathbb{Z} \to 0,$$

we see that $\pi_1(\mathrm{Map}_0^*(\mathbb{R}\mathrm{P}^2, \mathbb{C}\mathrm{P}^n)) = 0$. Thus, since $\mathrm{Map}^*(\mathbb{R}\mathrm{P}^2, S^{2n+1})$ is $(2n-2)$-connected (by Lemma 4.3), the induced homomorphism $\Omega\gamma_{n*}^{\mathbb{C}} : \pi_1(\Omega S^{2n+1}) \xrightarrow{\cong} \pi_1(\Omega\mathbb{C}\mathrm{P}^n)$ is an isomorphism. Hence, $\pi_k(\gamma_{n\#}^{\mathbb{C}})$ is an isomorphism for any $k \geq 1$ and the assertion is true for $m = 2$.

Now suppose that $\gamma_{n\#}' : \mathrm{Map}^*(\mathbb{R}\mathrm{P}^{m-1}, S^{2n+1}) \xrightarrow{\simeq} \mathrm{Map}_0^*(\mathbb{R}\mathrm{P}^{m-1}, \mathbb{C}\mathrm{P}^n)$ is a homotopy equivalence for some $m \geq 3$, and consider the following commutative diagram of restriction fibration sequences

$$\begin{array}{ccccc}
\Omega^m S^{2n+1} & \longrightarrow & \mathrm{Map}^*(\mathbb{R}\mathrm{P}^m, S^{2n+1}) & \xrightarrow{r} & \mathrm{Map}^*(\mathbb{R}\mathrm{P}^{m-1}, S^{2n+1}) \\
{\scriptstyle \Omega^m\gamma_n^{\mathbb{C}}}\downarrow\simeq & & {\scriptstyle \gamma_{n\#}}\downarrow & & {\scriptstyle \gamma_{n\#}'}\downarrow\simeq \\
\Omega^m\mathbb{C}\mathrm{P}^n & \longrightarrow & \mathrm{Map}_0^*(\mathbb{R}\mathrm{P}^m, \mathbb{C}\mathrm{P}^n) & \xrightarrow{r} & \mathrm{Map}_0^*(\mathbb{R}\mathrm{P}^{m-1}, \mathbb{C}\mathrm{P}^n)
\end{array}$$

Since $\Omega^m\gamma_n^{\mathbb{C}}$ and $\gamma_{n\#}'$ are homotopy equivalences, by the Five Lemma, we see that $\gamma_{n\#} : \mathrm{Map}^*(\mathbb{R}\mathrm{P}^m, S^{2n+1}) \xrightarrow{\simeq} \mathrm{Map}_0^*(\mathbb{R}\mathrm{P}^m, \mathbb{C}\mathrm{P}^n)$ is also a homotopy equivalence. □

COROLLARY 4.6. *If* $2 \leq m \leq 2n$, $\mathrm{Map}_0^*(\mathbb{R}\mathrm{P}^m, \mathbb{C}\mathrm{P}^n)$ *is* $(2n-m)$*-connected and*

$$\pi_{2n-m+1}(\mathrm{Map}_0^*(\mathbb{R}\mathrm{P}^m, \mathbb{C}\mathrm{P}^n)) \cong \begin{cases} \mathbb{Z} & \text{if } m \equiv 1 \pmod 2, \\ \mathbb{Z}/2 & \text{if } m \equiv 0 \pmod 2. \end{cases}$$

PROOF. This follows from Lemma 4.3 and Lemma 4.5. □

COROLLARY 4.7. *If* $2 \leq m \leq 2n$, *the map* $\mu_{n\#} : \mathrm{Map}_0^*(\mathbb{R}\mathrm{P}^m, \mathbb{R}\mathrm{P}^{2n+1}) \xrightarrow{\simeq} \mathrm{Map}_0^*(\mathbb{R}\mathrm{P}^m, \mathbb{C}\mathrm{P}^n)$ *is a homotopy equivalence.*

PROOF. Consider the commutative diagram

$$\begin{array}{ccc}
\mathrm{Map}^*(\mathbb{R}\mathrm{P}^m, S^{2n+1}) & \xrightarrow[\simeq]{\gamma_{n\#}^{\mathbb{C}}} & \mathrm{Map}_0^*(\mathbb{R}\mathrm{P}^m, \mathbb{C}\mathrm{P}^n) \\
{\scriptstyle \gamma_{2n+1\#}}\downarrow\simeq & & \| \\
\mathrm{Map}_0^*(\mathbb{R}\mathrm{P}^m, \mathbb{R}\mathrm{P}^{2n+1}) & \xrightarrow{\mu_{n\#}} & \mathrm{Map}_0^*(\mathbb{R}\mathrm{P}^m, \mathbb{C}\mathrm{P}^n)
\end{array}$$

Since $\gamma_{2n+1\#}$ and $\gamma_{n\#}^{\mathbb{C}}$ are homotopy equivalences by Lemma 4.5, the assertion easily follows from the above commutative diagram. □

PROOF OF THEOREM 1.5. We can prove Theorem 1.5 by using the same method as in the proof of Theorem 3.5 of [**1**]. However, that argument depends on difficult results from the theory of stratified spaces [**6**]. To provide an alternative here we shall give a much more elementary proof (which we used in an early preprint of [**1**]), which however only works in the case $d \equiv 0 \pmod 2$ because the map $j_d^{\mathbb{C}}$ (defined in (1.2)) is not well-defined if $d \equiv 1 \pmod 2$.

Assume that $d \equiv 0 \pmod 2$, and consider the commutative diagram

$$(4.3)\qquad \begin{array}{ccc} A_d^{\mathbb{R}}(m,2n+1) & \xrightarrow{j_d^{\mathbb{R}}} & \mathrm{Map}^*(\mathbb{RP}^m, S^{2n+1}) \\ \psi_{m,n}\downarrow\cong & & \gamma^{\mathbb{C}}_{n\#}\downarrow\simeq \\ A_d^{\mathbb{C}}(m,n) & \xrightarrow{j_d^{\mathbb{C}}} & \mathrm{Map}_0^*(\mathbb{RP}^m,\mathbb{CP}^n) \end{array}$$

where $j_d^{\mathbb{R}}$ is defined in a similar way as the map $j_d^{\mathbb{C}}$. Since the map $j_d^{\mathbb{R}}$ is a homotopy equivalence through dimension $D_{\mathbb{R}}(d;m,2n+1)$ if $m<2n$ and a homology equivalence through dimension $D_{\mathbb{R}}(d;m,2n+1)$ if $m=2n$ [**1**, Theorem 4.10], so is the map $j_d^{\mathbb{C}}$. Because $D_{\mathbb{C}}(d;m,n)=D_{\mathbb{R}}(d;m,2n+1)$, we see that the map $j_d^{\mathbb{C}}$ is a homotopy equivalence through dimension $D_{\mathbb{C}}(d;m,n)$ if $m<2n$ and a homology equivalence through dimension $D_{\mathbb{C}}(d;m,n)$ if $m=2n$. It remains to show that the same holds for map $i_d^{\mathbb{C}}$. However, this follows easily from the facts that $i_d^{\mathbb{C}}=\gamma^{\mathbb{C}}_{n\#}\circ j_d^{\mathbb{C}}$ and $\gamma^{\mathbb{C}}_{n\#}$ is a homotopy equivalence. □

PROOF OF COROLLARY 1.6. This easily follows from Theorem 1.5. □

5. The stabilized space $A^{\mathbb{C}}_{\infty+\epsilon}(m,n)$.

Although we cannot prove Conjecture 2.1, we can prove the following stabilized version.

DEFINITION 5.1. For $\epsilon=0$ or 1, let $A^{\mathbb{C}}_{\infty+\epsilon}(m,n)$ denote the stabilized space $A^{\mathbb{C}}_{\infty+\epsilon}(m,n)=\lim_{k\to\infty}A^{\mathbb{C}}_{2k+\epsilon}(m,n)$, where the limit is taken over the stabilization maps, $A^{\mathbb{C}}_{\epsilon}(m,n)\xrightarrow{s_\epsilon}A^{\mathbb{C}}_{2+\epsilon}(m,n)\xrightarrow{s_{2+\epsilon}}A^{\mathbb{C}}_{4+\epsilon}(m,n)\xrightarrow{s_{4+\epsilon}}A^{\mathbb{C}}_{6+\epsilon}(m,n)\xrightarrow{s_{6+\epsilon}}\cdots$.
From the commutative diagram

$$\begin{array}{ccccccc} \longrightarrow & A^{\mathbb{C}}_{2k+\epsilon}(m,n) & \xrightarrow{s_{2k+\epsilon}} & A^{\mathbb{C}}_{2k+2+\epsilon}(m,n) & \xrightarrow{s_{2k+2+\epsilon}} & \cdots \\ & \Psi^{\mathbb{C}}_{2k+\epsilon}\downarrow & & \Psi^{\mathbb{C}}_{2k+2+\epsilon}\downarrow & & \\ \longrightarrow & \mathrm{Alg}^*_{2k+\epsilon}(\mathbb{RP}^m,\mathbb{CP}^n) & \xrightarrow{\subset} & \mathrm{Alg}^*_{2k+2+\epsilon}(\mathbb{RP}^m,\mathbb{CP}^n) & \xrightarrow{\subset} & \cdots \end{array}$$

we obtain a stabilized map $\Psi^{\mathbb{C}}_{\infty+\epsilon}=\lim_{k\to\infty}\Psi^{\mathbb{C}}_{2k+\epsilon}:A^{\mathbb{C}}_{\infty+\epsilon}(m,n)\to\mathrm{Alg}^*_{\epsilon}(m,n)$.

PROPOSITION 5.2. *If $2\le m\le 2n$, the map*

$$\Psi^{\mathbb{C}}_{\infty+\epsilon}:A^{\mathbb{C}}_{\infty+\epsilon}(m,n)\xrightarrow{\simeq}\mathrm{Alg}^*_{\epsilon}(m,n)\simeq\mathrm{Map}^*_{\epsilon}(\mathbb{RP}^m,\mathbb{CP}^n)$$

is a homotopy equivalence if $m<2n$ and a homology equivalence if $m=2n$.

PROOF. The assertion easily follows from Theorem 2.3, Corollary 1.7 and the equality $i\circ\Psi^{\mathbb{C}}_{\infty+\epsilon}=\lim_k i^{\mathbb{C}}_{2k+\epsilon}$. □

6. The space $\mathrm{Alg}_1^*(\mathbb{RP}^m,\mathbb{CP}^n)$.

In this section we investigate the homotopy of $A_1^{\mathbb{C}}(m,n)\cong\mathrm{Alg}_1^*(\mathbb{RP}^m,\mathbb{RP}^n)$.

DEFINITION 6.1. For integers $1\le m\le 2n$, let $V_{2n+1,m}$ denote the real Stiefel manifold of all orthogonal m-frames in $\mathbb{R}^{2n+1}$ and $(\mathbf{b}_1,\cdots,\mathbf{b}_m)\in V_{2n+1,m}$ any element such that $\mathbf{b}_k={}^t(b_{k,1},\cdots,b_{k,2n+1})\in\mathbb{R}^{2n+1}$ $(1\le k\le m)$. Consider the $(n+1)$-tuple of polynomial defined by

$$(f_0,\cdots,f_n)=(z_0,\cdots,z_m)\begin{pmatrix}1 & 0 & 0 & \cdots & 0 & 0\\ \sqrt{-1}\mathbf{b} & \mathbf{c}_1 & \mathbf{c}_2 & \cdots & \mathbf{c}_{n-1} & \mathbf{c}_n\end{pmatrix},$$

where $\mathbf{b} \in \mathbb{R}^m$ and $\mathbf{c}_k \in \mathbb{C}^m$ $(k = 1, 2, \cdots, n)$ are given by

$$\begin{cases} \mathbf{b} = & {}^t(b_{1,1}, b_{1,2}, b_{1,3}, \cdots . b_{1,m}), \\ \mathbf{c}_k = & {}^t(b_{2k,1} + \sqrt{-1}b_{2k+1,1}, b_{2k,2} + \sqrt{-1}b_{2k+1,2}, \cdots, b_{2k,m} + \sqrt{-1}b_{2k+1,m}) \end{cases}$$

Since it is easy to see that $(f_0, \cdots, f_n) \in A_1^{\mathbb{C}}(m,n)$, one can define the map $\varphi_{m,n}$: $V_{2n+1,m} \to A_1^{\mathbb{C}}(m,n)$ by $\varphi_{m,n}(\mathbf{b}_1, \cdots, \mathbf{b}_m) = (f_0, \cdots, f_n)$.

LEMMA 6.2. *If $1 \leq m \leq 2n$, the map $\varphi_{m,n} : V_{2n+1,m} \xrightarrow{\simeq} A_1^{\mathbb{C}}(m,n)$ is a homotopy equivalence.*

PROOF. Let us consider the element $(f_0, \cdots, f_n) \in \mathbb{C}[z_0, \cdots, z_m]^{n+1}$ of the form

$$(6.1) \qquad (f_0, \cdots, f_n) = (z_0, \cdots, z_m) \begin{pmatrix} 1 & 0 & \cdots & 0 \\ a_{1,0} & a_{1,1} & \cdots & a_{1,n} \\ \vdots & \vdots & \vdots & \vdots \\ a_{m,0} & a_{m,1} & \cdots & a_{m,n} \end{pmatrix},$$

where $a_{k,j} = b_{k,j} + \sqrt{-1}c_{k,j}$ $(b_{k,j}, c_{k,j} \in \mathbb{R})$ and we write

$$\begin{cases} \mathbf{a} = {}^t(b_{1,0}, b_{2,0}, \cdots, b_{m,0}) \in \mathbb{R}^m \\ B = \big(b_{k,j}\big)_{1 \leq k \leq m, 1 \leq j \leq n}, \ C = \big(c_{k,j}\big)_{1 \leq k \leq m, 0 \leq j \leq n}. \end{cases}$$

It is easy to see that the polynomials $f_0, \cdots, f_n$ have no common *real* root beside zero if and only if the equation

$$(z_0, z_1, \cdots, z_m) \begin{pmatrix} 1 & 0 & 0 & 0 & \cdots & 0 & 0 \\ b_{1,0} & c_{1,0} & b_{1,1} & c_{1,1} & \cdots & b_{1,n} & c_{1,n} \\ \vdots & \vdots & \vdots & \vdots & \vdots & \vdots & \vdots \\ b_{m,0} & c_{m,0} & b_{m,1} & c_{m,1} & \cdots & b_{m,n} & c_{m,n} \end{pmatrix} = \begin{pmatrix} 0 \\ 0 \\ \vdots \\ 0 \end{pmatrix}$$

has no non-zero solution. So we see that $(f_0, \cdots, f_n) \in A_1^{\mathbb{C}}(m,n)$ if and only if the $(m \times (2n+1))$-matrix (B, C) has rank m. Thus the map $\Phi : V_{2n+1,m} \times \mathbb{R}^m \to A_1^{\mathbb{C}}(m,n)$ given by $(B, C, \mathbf{a}) \mapsto (f_0, \cdots, f_n)$ is clearly a homeomorphism, where $(f_0, \cdots, f_n)$ is defined by (6.1). Since $\varphi_{m,n} = \Phi | V_{2n+1,m} \times \{\mathbf{0}_m\}$, the map $\varphi_{m,n}$ is a homotopy equivalence. □

LEMMA 6.3. (i) *If $2 \leq m \leq 2n$, $\mathrm{Map}_1^*(\mathbb{R}\mathrm{P}^m, \mathbb{C}\mathrm{P}^n)$ is $(2n-m)$-connected.*
(ii) *If $1 \leq m \leq 2n$, $V_{2n+1,m}$ is $(2n-m)$-connected.*
(iii) *If $1 \leq m \leq 2n$,* $\pi_{2n-m+1}(V_{2n+1,m}) = \begin{cases} \mathbb{Z} & \text{if } m \equiv 1 \pmod 2, \\ \mathbb{Z}/2 & \text{if } m \equiv 0 \pmod 2. \end{cases}$

PROOF. (i) Consider the restriction fibration sequence

$$(6.2) \qquad \Omega^m \mathbb{C}\mathrm{P}^n \xrightarrow{\hat{j}} \mathrm{Map}_1^*(\mathbb{R}\mathrm{P}^m, \mathbb{C}\mathrm{P}^n) \xrightarrow{r} \mathrm{Map}_1^*(\mathbb{R}\mathrm{P}^{m-1}, \mathbb{C}\mathrm{P}^n).$$

We prove the assertion (i) by induction on m. First, consider the case $m = 2$. Since $\pi_1(\Omega^2\mathbb{C}\mathrm{P}^n) = 0$ and $\mathrm{Map}_1^*(\mathbb{R}\mathrm{P}^1, \mathbb{C}\mathrm{P}^n) = \Omega\mathbb{C}\mathrm{P}^n$, taking $m = 2$ in (6.2) and using the induced exact sequence

$$0 \to \pi_1(\mathrm{Map}_1^*(\mathbb{R}\mathrm{P}^2, \mathbb{C}\mathrm{P}^n)) \xrightarrow{r_*} \pi_1(\Omega\mathbb{C}\mathrm{P}^n) = \mathbb{Z} \xrightarrow{\partial} \mathbb{Z} = \pi_0(\Omega^2\mathbb{C}\mathrm{P}^n) \to 0,$$

we see that $\pi_1(\mathrm{Map}_1^*(\mathbb{R}\mathrm{P}^2, \mathbb{C}\mathrm{P}^n)) = 0$. Since $\pi_k(\Omega^2\mathbb{C}\mathrm{P}^n) \cong \pi_{k+2}(S^{2n+1}) = 0$ and $\pi_k(\Omega\mathbb{C}\mathrm{P}^n) \cong \pi_{k+1}(S^{2n+1}) = 0$ for $2 \leq k \leq 2n-2$, applying (6.2) with $m = 2$ we see that $\pi_k(\mathrm{Map}_1^*(\mathbb{R}\mathrm{P}^2, \mathbb{C}\mathrm{P}^n)) = 0$ for $2 \leq k \leq 2n-2$. Hence, the case $m = 2$ has been proved. Next, assume that $\mathrm{Map}_1^*(\mathbb{R}\mathrm{P}^{m-1}, \mathbb{C}\mathrm{P}^n)$ is $(2n-m+1)$-connected for some $m \geq 3$. Since $3 \leq m < 2n$, $\Omega^m\mathbb{C}\mathrm{P}^n = \Omega^m S^{2n+1}$ is $(2n-m)$-connected. Hence, from (6.2) we easily deduce that $\mathrm{Map}_1^*(\mathbb{R}\mathrm{P}^m, \mathbb{C}\mathrm{P}^n)$ is $(2n-m)$-connected. We have proved (i).

(ii) The assertion (ii) can be easily proved by induction on m by making use of the following fibration sequence:

$$S^{2n-m+1} \to V_{2n+1,m} \to V_{2n+1,m-1}. \tag{6.3}$$

We omit the details.

(iii) First, let $m \equiv 0 \pmod 2$. It is known that $H^{2n-m+2}(V_{2n+1,m}, \mathbb{Z}/p) = 0$ for any odd prime $p \geq 3$,

$$H^*(V_{2n+1,m}, \mathbb{Z}/2) = E[x_j : 2n-m+1 \leq j \leq 2n] \quad (|x_j| = j)$$

and that $Sq^1(x_{2n-m+1}) = x_{2n-m+2}$. Hence, the $(2n-m+2)$-skeleton of $V_{2n+1,m}$ is $S^{2n-m+1} \cup_2 e^{2n-m+2}$ (up to homotopy equivalence), and we have the isomorphism $\pi_{2n-m+1}(V_{2n-m+1}) = \mathbb{Z}/2$. If $m \equiv 1 \pmod 2$, (6.3) induces the exact sequence

$$\pi_{N+1}(V_{2n+1,m-1}) = \mathbb{Z}/2 \xrightarrow{\partial} \pi_N(S^{2n-m+1}) = \mathbb{Z} \to \pi_N(V_{2n+1,m}) \to 0,$$

where $N = 2n-m+1$. Hence, $\pi_{2n-m+1}(V_{2n+1,m}) \cong \mathbb{Z}$. Thus (iii) has also been proved. $\square$

COROLLARY 6.4. *If $1 \leq m \leq 2n$, $\mathrm{Alg}_1^*(\mathbb{R}\mathrm{P}^m, \mathbb{C}\mathrm{P}^n)$ is $(2n-m)$-connected and*

$$\pi_{2n-m+1}(\mathrm{Alg}_1^*(\mathbb{R}\mathrm{P}^m, \mathbb{C}\mathrm{P}^n)) = \begin{cases} \mathbb{Z} & \text{if } m \equiv 1 \pmod 2, \\ \mathbb{Z}/2 & \text{if } m \equiv 0 \pmod 2. \end{cases}$$

PROOF. Since there is a homotopy equivalence $V_{2n+1,m} \simeq \mathrm{Alg}_1^*(\mathbb{R}\mathrm{P}^m, \mathbb{C}\mathrm{P}^n)$, the assertion follows from Lemma 6.3. $\square$

DEFINITION 6.5. For $1 \leq m \leq 2n$, we define the map $\tilde{i}_m : V_{2n+1,m} \to \mathrm{Map}_1^*(\mathbb{R}\mathrm{P}^m, \mathbb{C}\mathrm{P}^n)$ by $\tilde{i}_m = i_1^{\mathbb{C}} \circ \varphi_{m,n}$.

For $2 \leq m \leq 2n$, it is easy to verify that the following diagram is commutative

$$\begin{array}{ccccc} S^{2n-m+1} & \xrightarrow{\tilde{j}} & V_{2n+1,m} & \longrightarrow & V_{2n+1,m-1} \\ \hat{s}_m \downarrow & & \tilde{i}_m \downarrow & & \tilde{i}_{m-1} \downarrow \\ \Omega^m\mathbb{C}\mathrm{P}^n & \xrightarrow{\hat{j}} & \mathrm{Map}_1^*(\mathbb{R}\mathrm{P}^m, \mathbb{C}\mathrm{P}^n) & \xrightarrow{r} & \mathrm{Map}_1^*(\mathbb{R}\mathrm{P}^{m-1}, \mathbb{C}\mathrm{P}^n) \end{array} \tag{6.4}$$

where we identify $S^{2n-m+1} \cong \frac{O(2n+2-m)}{O(2n+1-m)}$ and the two rows are fibration sequences.

LEMMA 6.6. *If $2 \leq m \leq 2n$, the map $\hat{s}_m : S^{2n-m+1} \to \Omega^m\mathbb{C}\mathrm{P}^n$ is a homotopy equivalence up to dimension $D_{\mathbb{C}}(1; m, n) = 4n-2m+1$.*

PROOF. By means of a method similar to the one used in the proof of [**18**, Lemma 3.1] we can show that $\hat{s}_{m*} : \pi_{2n-m+1}(S^{2n-m+1}) \xrightarrow{\cong} \pi_{2n-m+1}(\Omega^m S^{2n+1}) \cong \pi_{2n-2m+1}(\Omega^m\mathbb{C}\mathrm{P}^n)$ is an isomorphism. Thus we can identify $\hat{s}_m$ with the m-fold suspension $E^m : S^{2n-m+1} \to \Omega^m S^{2n+1} \simeq \Omega^m\mathbb{C}\mathrm{P}^n$ (up to homotopy equivalence). Hence $\hat{s}_m$ is a homotopy equivalence up to dimension $4n-2m+1$. $\square$

LEMMA 6.7. *If $n \geq 2$, $\tilde{i}_{1*} : \pi_k(V_{2n+1,1}) \to \pi_k(\mathrm{Map}_1^*(\mathbb{R}\mathrm{P}^1, \mathbb{C}\mathrm{P}^n))$ is an isomorphism for any $2 \leq k < 4n-1$ and an epimorphism for $k = 4n-1$.*

PROOF. After identifications (up to homotopy equivalence) $V_{2n+1,1} = S^{2n}$ and $\mathrm{Map}_1^*(\mathbb{R}\mathrm{P}^1, \mathbb{C}\mathrm{P}^n) = \Omega\mathbb{C}\mathrm{P}^n = \Omega S^{2n+1} \times S^1$, the map $\tilde{i}_1$ can be viewed as a map $\tilde{i}_1 : S^{2n} \to \Omega\mathbb{C}\mathrm{P}^n$. Let $q_1 : \Omega\mathbb{C}\mathrm{P}^n = \Omega S^{2n+1} \times S^1 \to \Omega S^{2n+1}$ denote the projection onto the first factor. Note that the composite $q_1 \circ \tilde{i}_1$ can be identified with the map $S^{2n} \simeq Q^1_{(2n+1)}(\mathbb{R}) \to \Omega S^{2n+1}$ given in [**9**, Corollary 5]. Hence it follows from [**9**, Corollary 5] that $q_1 \circ \tilde{i}_1$ is a homotopy equivalence up to dimension $N(1, 2n+2) = 4n-1$. Recalling that $q_{1*} : \pi_l(\Omega\mathbb{C}\mathrm{P}^n) \xrightarrow{\cong} \pi_l(\Omega S^{2n+1})$ is an isomorphism for all $l \geq 2$, we see that $\pi_k(\tilde{i}_1)$ is an isomorphism for any $2 \leq k < 4n-1$ and an epimorphism for $k = 4n-1$. □

PROPOSITION 6.8. *If $2 \leq m \leq 2n$, the map $\tilde{i}_m : V_{2n+1,m} \to \mathrm{Map}_1^*(\mathbb{R}\mathrm{P}^m, \mathbb{C}\mathrm{P}^n)$ is a homotopy equivalence up to dimension $D_{\mathbb{C}}(1; m, n) = 4n - 2m + 1$*

PROOF. The proof proceeds by induction on m. First, consider the case $m = 2$. Since $\pi_1(V_{2n+1,2}) = \pi_1(\mathrm{Map}_1^*(\mathbb{R}\mathrm{P}^2, \mathbb{C}\mathrm{P}^n)) = 0$ by Lemma 6.3, the map $\tilde{i}_2$ induces an isomorphism on $\pi_1(\ )$. Hence it suffices to show that $\tilde{i}_{2*} : \pi_k(V_{2n+1,2}) \to \pi_k(\mathrm{Map}_1^*(\mathbb{R}\mathrm{P}^2, \mathbb{C}\mathrm{P}^n))$ is an isomorphism for any $2 \leq k < D_{\mathbb{C}}(1; 2, n) = 4n - 1$ and an epimorphism for $k = D_{\mathbb{C}}(1; 2, n)$. However, recalling the commutative diagram (6.4) for $m = 2$, and using the Five Lemma, Lemma 6.6 and Lemma 6.7, we see that $\pi_k(\tilde{i}_2)$ is an isomorphism for any $2 \leq k < 4n-1 = D_{\mathbb{C}}(1; 2, n)$ and an epimorphism for $k = 4n - 1 = D_{\mathbb{C}}(1; 2, n)$. Hence, the case $m = 2$ is proved.

Now assume that the map $\tilde{i}_{m-1}$ is a homotopy equivalence up to dimension $D_{\mathbb{C}}(1; m-1, n) = 4n - 2m + 3$ for some $m \geq 3$. Since, by Lemma 6.6, $\hat{s}_m$ is a homotopy equivalence up to dimension $D_{\mathbb{C}}(1; m, n) = 4n - 2m + 1$, the Five Lemma and the diagram (6.4) imply that the map $\tilde{i}_m$ is a homotopy equivalence up to dimension $D_{\mathbb{C}}(1; m, n)$. □

Proof of Theorem 1.8. (i) Since $i_1^{\mathbb{C}} = i_{1,\mathbb{C}} \circ \Psi_1^{\mathbb{C}}$ and the projection map $\Psi_1^{\mathbb{C}} : A_1^{\mathbb{C}}(m, n) \xrightarrow{\cong} \mathrm{Alg}_1^*(\mathbb{R}\mathrm{P}^m, \mathbb{C}\mathrm{P}^n)$ is a homeomorphism, it suffices to show that $i_1^{\mathbb{C}}$ is a homotopy equivalence up to dimension $D_{\mathbb{C}}(1; m, n)$. As $\tilde{i}_m = i_1^{\mathbb{C}} \circ \varphi_{m,n}$ and $\varphi_{m,n}$ is a homotopy equivalence, this follows from Proposition 6.8. Hence, (i) has been proved.

(ii) Since $\hat{s}_{2n*} : \pi_1(S^1) \to \pi_1(\Omega^{2n}\mathbb{C}\mathrm{P}^n)$ is an epimorphism by Lemma 6.6 and $\pi_1(S^1) \cong \mathbb{Z} \cong \pi_1(\Omega\mathbb{C}\mathrm{P}^n)$, $\hat{s}_{2n}$ induces an isomorphism on $\pi_1(\ )$. Consider the following commutative diagram of the exact sequences induced from (6.4) for $m = 2n$

$$\begin{array}{ccccccc}
\pi_2(V_{2n+1,2n-1}) & \xrightarrow{\partial} & \pi_1(S^1) & \longrightarrow & \pi_1(V_{2n+1,2n}) = \mathbb{Z}/2 & \longrightarrow & 0 \\
\tilde{i}_{2n-1*}\downarrow\cong & & \hat{s}_{2n*}\downarrow\cong & & \tilde{i}_{2n*}\downarrow & & \\
\pi_2(\mathrm{Map}_1^*) & \xrightarrow{\partial'} & \pi_1(\Omega\mathbb{C}\mathrm{P}^n) & \longrightarrow & \pi_1(\mathrm{Map}_1^*(\mathbb{R}\mathrm{P}^{2n}, \mathbb{C}\mathrm{P}^n)) & \longrightarrow & 0
\end{array}$$

where $\mathrm{Map}_1^* = \mathrm{Map}_1^*(\mathbb{R}\mathrm{P}^{2n-1}, \mathbb{C}\mathrm{P}^n)$. Since, by Proposition 6.8, the induced homomorphism $\tilde{i}_{2n-1*} : \pi_2(V_{2n+1,2n-1}) \xrightarrow{\cong} \pi_2(\mathrm{Map}_1^*(\mathbb{R}\mathrm{P}^{2n-1}, \mathbb{C}\mathrm{P}^n))$ is an isomorphism, $\tilde{i}_{2n}$ induces an isomorphism on $\pi_1(\ )$. □

PROOF OF COROLLARY 1.9. The assertion (i) follows from Lemma 4.3 and Theorem 1.5. The assertion (ii) also easily follows from Corollary 4.6, Lemma 6.2, Lemma 6.3, Proposition 6.8 and Theorem 1.8. □

References

1. M. Adamaszek, A. Kozlowski and K. Yamaguchi, Spaces of algebraic and continuous maps between real algebraic varieties, preprint (math.AT; arXiv: 0809.4893).
2. C. P. Boyer, J. C. Hurtubise, R. J. Milgram, Stability Theorems for spaces of rational curves, Int. J. Math. **12** (2001), 223-262
3. R. L. Cohen, J. D. S. Jones and G. B. Segal, Stability for holomorphic spheres and Morse Theory, Contemporary Math. **258** (2000), 87–106.
4. F. R. Cohen, T. J. Lada and J. P. May, The homology of iterated loop spaces, Lecture Notes in Math., **533** Springer-Verlag, 1976.
5. M. C. Crabb and W. A. Sutherland, Function spaces and Hurewicz-Radon numbers, Math. Scand. **55** (1984), 67–90.
6. M. Goresky and R. MacPherson Stratified Morse Theory, A Series of Modern Surveys in Math., Springer-Verlag, 1980.
7. M. Gromov, Oka's principle for holomorphic sections of elliptic bundles, J. Amer. Math. Soc. **2** (1989) 851-897
8. M. A. Guest, A. Kozlowski and K. Yamaguchi, Spaces of polynomials with roots of bounded multiplicity, Fund. Math. **116** (1999), 93–117.
9. A. Kozlowski and K. Yamaguchi, Topology of complements of discriminants and resultants, J. Math. Soc. Japan **52** (2000), 949–959.
10. A. Kozlowski and K. Yamaguchi, Spaces of holomorphic maps between complex projective spaces of degree one, Topology Appl. **132** (2003), 139–145.
11. J. Mostovoy, Spaces of rational loops on a real projective space, Trans. Amer. Math. Soc. **353** (2001), 1959–1970.
12. J. Mostovoy, Spaces of rational maps and the Stone-Weierstrass Theorem, Topology **45** (2006), 281–293.
13. S. Sasao, The homotopy of Map($\mathbb{C}P^m, \mathbb{C}P^n$), J. London Math. Soc. **8** (1974), 193–197.
14. G. B. Segal, The topology of spaces of rational functions, Acta Math. **143** (1979), 39–72.
15. V. P. Snaith, A stable decomposition of $\Omega^n S^n X$, J. London Math. Soc. **2** (1974), 577–583.
16. V. A. Vassiliev, Complements of Discriminants of Smooth Maps, Topology and Applications, Amer. Math. Soc., Translations of Math. Monographs **98**, 1992 (revised edition 1994).
17. K. Yamaguchi, Complements of resultants and homotopy types, J. Math. Kyoto Univ. **39** (1999), 675–684.
18. K. Yamaguchi, The topology of spaces of maps between real projective spaces, J. Math. Kyoto Univ. **43** (2003), 503–507.
19. K. Yamaguchi, The homotopy of spaces of maps between real projective spaces, J. Math. Soc. Japan **58** (2006), 1163–1184; *ibid.* **59** (2007), 1235–1237.

Tokyo Denki University, Inzai, Chiba 270-1382 Japan
E-mail address: akoz@mimuw.edu.pl

The University of Electro-Communications, Chofu Tokyo 182-8585 Japan
E-mail address: kohhei@im.uec.ac.jp

Contemporary Mathematics
Volume **519**, 2010

ON THE RATIONAL COHOMOLOGY OF THE TOTAL SPACE OF THE UNIVERSAL FIBRATION WITH AN ELLIPTIC FIBRE

KATSUHIKO KURIBAYASHI

ABSTRACT. Let $\mathcal{F}_M$ be the universal fibration having fibre M an elliptic space with vanishing odd rational cohomology. We consider the rational cohomology of the total space of $\mathcal{F}_M$ by using a function space model due to Haefliger, Brown and Szczarba when Halperin's conjecture is affirmatively solved for the fibre. The calculation enables one to deduce that the cohomology of the classifying space of the self-homotopy equivalences of a c-symplectic space M is generated only by the Kedra-McDuff μ-classes if the cohomology of M is generated by a single element.

1. Introduction

Let M be an elliptic space, namely a simply-connected space whose rational homotopy and cohomology are finite-dimensional. Suppose further that the Euler characteristic $\sum_i (-1)^i \dim H^*(M;\mathbb{Q})$ is positive. Following Lupton [19], we call such a space *positively elliptic*. Then Halperin's conjecture states that for any positively elliptic space M, each fibration $M \to E \xrightarrow{p} B$ with a simply-connected base is TNCZ; that is, the Leray-Serre spectral sequence with coefficients in the rational field for the fibration collapses at the E_2-term. In this case the cohomology $H^*(E;\mathbb{Q})$ is isomorphic to the tensor product $H^*(B;\mathbb{Q}) \otimes H^*(M;\mathbb{Q})$ as an $H^*(B;\mathbb{Q})$-module, where the module structure is defined by the induced map $p^* : H^*(B;\mathbb{Q}) \to H^*(E;\mathbb{Q})$. We observe that, as an algebra, $H^*(E;\mathbb{Q})$ is not isomorphic in general to the tensor product with the natural algebraic structure induced by the products on the factors.

We mention that the conjecture has been affirmatively solved in some cases; see [35], [27], [31], [20] and [22]. Our interest here lies in investigating the algebra structure of the rational cohomology of the total space of a fibration with a fibre for which Halperin's conjecture is affirmatively solved.

In the rest of this section, we describe our results. Let $\mathrm{aut}_1(M)$ be the component of the monoid of the self-homotopy equivalences of a space M containing the identity and $M \xrightarrow{\iota} M_{\mathrm{aut}_1(M)} \xrightarrow{\pi} B\mathrm{aut}_1(M)$ the universal M-fibration [24]. Suppose that M is a simply-connected space whose rational cohomology is isomorphic to an algebra of the form

$$\mathbb{Q}[x_1, ..., x_q]/(u_1, ..., u_q), \tag{1.1}$$

2000 Mathematics Subject Classification: 55P62, 57R19, 57R20, 57T35.
Key words and phrases. Self-homotopy equivalence, c-symplectic manifold, homogeneous space, Sullivan model, evaluation map, characteristic class, the Eilenberg-Moore spectral sequence.

This research was partially supported by a Grant-in-Aid for Scientific Research (C)20540070 from Japan Society for the Promotion of Science.

Department of Mathematical Sciences, Faculty of Science, Shinshu University, Matsumoto, Nagano 390-8621, Japan e-mail:kuri@math.shinshu-u.ac.jp

©2010 American Mathematical Society

where $\{u_1, ..., u_q\}$ is a regular sequence. Observe that the cohomology of every positively elliptic space is of the form (1.1) for some q and conversely, a simply connected space whose cohomology has the form (1.1) is positively elliptic; see [9] and [4].

In what follows, for an augmented algebra A, we denote by A^+ the augmentation ideal. Let $\{m_j\}_{j\in J}$ be an arbitrary basis of the cohomology $H^*(M;\mathbb{Q})$ with $J = J'\coprod\{0\}$ and $m_0 = 1$. By using a minimal model for M and the basis $\{m_j\}_{j\in J}$, we can construct a model for $\mathrm{aut}_1(M)$ due to Haefliger [8], Brown and Szczarba [1], which is called the HBS model. Let $ev : \mathrm{aut}_1(M) \times M \to M$ be the evaluation map defined by $ev(\gamma, x) = \gamma(x)$ for $\gamma \in \mathrm{aut}_1(M)$ and $x \in M$. Then an explicit model for the evaluation map is constructed with the HBS model for $\mathrm{aut}_1(M)$; see [2], [16] and [17]. Such the models enable one to determine $H^*(B\mathrm{aut}_1(M);\mathbb{Q})$-*linear parts* of the relations in the algebra $H^*(M_{\mathrm{aut}_1(M)};\mathbb{Q})$. More precisely we establish

Theorem 1.1. *Let M be a simply-connected space whose rational cohomology is of the form (1.1). Suppose that all derivations of negative degree of $H^*(M;\mathbb{Q})$ vanish. Then there exist a subset S of the set $\{1, ..., q\}\times J$ containing $S_0 := \{(i, 0)| 1 \le i \le q\}$ and indecomposable elements ν_s of $H^*(B\mathrm{aut}_1(M);\mathbb{Q})$ indexed by S such that*

$$H^*(B\mathrm{aut}_1(M);\mathbb{Q}) \cong \mathbb{Q}[\nu_s|\ s \in S]$$

as an algebra and

$$H^*(M_{\mathrm{aut}_1(M)};\mathbb{Q}) \cong \mathbb{Q}[\nu_s|\ s \in S] \otimes \mathbb{Q}[x_1, ..., x_q] \Bigg/ \Bigg(u_i - \sum_{s=(i,j)\in S} \nu_s m_j + D_i\Bigg)$$

as an $H^*(B\mathrm{aut}_1(M);\mathbb{Q})$-algebra, where each D_i is an appropriate decomposable element in the ideal generated by $\pi^*(H^*(B\mathrm{aut}_1(M);\mathbb{Q})^+ \cdot H^*(B\mathrm{aut}_1(M);\mathbb{Q})^+)$, $\iota^*(x_i) = x_i$ for $i = 1, ..., q$ and $\deg \nu_{(i,j)} = \deg u_i - \deg m_j$ for $(i, j) \in S$. Moreover one has

$$\begin{aligned}\sharp\{\nu_s| \deg \nu_s = 2k, s \in S\} &= \mathrm{rank} \oplus_{a-b=2k-1} \pi_a(M) \otimes H^b(M;\mathbb{Q}) \\ &\quad -\mathrm{rank} \oplus_{a-b=2k} \pi_a(M) \otimes H^b(M;\mathbb{Q}).\end{aligned}$$

For any $s \in S_0$, the element ν_s is decomposable in $H^*(M_{\mathrm{aut}_1(M)};\mathbb{Q})$. Thus we have

Corollary 1.2. *As an algebra,*

$$H^*(M_{\mathrm{aut}_1(M)};\mathbb{Q}) \cong \mathbb{Q}[\nu_s|\ s \in S - S_0] \otimes \mathbb{Q}[x_1, ..., x_q].$$

The derivations of $H^*(M;\mathbb{Q})$ are closely related with the Halperin conjecture. Indeed, the result [27, Theorem A] due to Meier asserts that the conjecture is true for a positively elliptic fibre M if and only if all derivations of negative degree of $H^*(M;\mathbb{Q})$ vanish. Thus Theorem 1.1 is applicable to spaces satisfying Halperin's conjecture, for example, the homogeneous space G/H for which G is a Lie group and H is a subgroup with rank $G =$ rank H; see [31].

We mention that Lemma 3.1 below describes a necessary and sufficient condition for the vector space of negative derivations of $H^*(M;\mathbb{Q})$ to be trivial in terms of the differential δ of the HBS model for $\mathrm{aut}_1(M)$. We also observe that the subset S in Theorem 1.1 is determined when choosing a basis for the image of the linear part of δ; see Section 3 for more details.

We can arrange the given basis $\{m_j\}_{j\in J}$ of $H^*(M;\mathbb{Q})$ so that $D_i = 0$ for any i. This follows from deformation theory as we see in [10, Section 3]. It is mentioned that the suitable basis of $H^*(M;\mathbb{Q})$ are chosen by considering the vector space of *infinitesimal deformations* $T^1_{\mathbb{Q}}(H^*(M;\mathbb{Q}))$ of $H^*(M;\mathbb{Q})$. By virtue of [10, Theorem 3.1], the cohomology ring $H^*(M_{\mathrm{aut}_1(M)};\mathbb{Q})$ is viewed as *an equivariant versal deformation* of $H^*(M;\mathbb{Q})$ along the ring $H^*(B\mathrm{aut}_1(M);\mathbb{Q})$. Moreover a formula of the rational homotopy group of $M_{\mathrm{aut}_1(M)}$ is described in [10, Theorem D] in terms of the infinitesimal deformations. The proof of [10, Theorem D] then yields Corollary 1.2.

The novelty here is that the generators ν_s of $H^*(B\mathrm{aut}_1(M);\mathbb{Q})$ are related to those of the HBS model for $\mathrm{aut}_1(M)$ via the Eilenberg-Moore spectral sequence converging to $H^*(M_{\mathrm{aut}_1(M)};\mathbb{Q})$; see Section 5. We also stress that our computation of $H^*(M_{\mathrm{aut}_1(M)};\mathbb{Q})$ is started with an arbitrary basis of $H^*(M;\mathbb{Q})$. In addition, we obtain the algebra structure of $H^*(M_{\mathrm{aut}_1(M)};\mathbb{Q})$ without depending on deformation theory.

We refer the reader to [5, Theorems 1, 2 and 3], [27, Proposition 1] and [33, Theorem 3.1] for the rational homotopy group of $\mathrm{aut}_1(M)$ for an elliptic space M and a more general two-stage space.

Let $M \to E \to B$ be a fibration over a simply-connected space B. Assume that the cohomology of M is of the form (1.1) and that all derivations of negative degree of $H^*(M;\mathbb{Q})$ vanish. As mentioned in [27, (2.12)], the Eilenberg-Moore spectral sequence argument enables us to conclude that, as an $H^*(B;\mathbb{Q})$-algebra,

$$\begin{aligned} H^*(E;\mathbb{Q}) &\cong H^*(B;\mathbb{Q}) \otimes_{H^*(B\mathrm{aut}_1(M);\mathbb{Q})} H^*(M_{\mathrm{aut}_1(M)};\mathbb{Q}) \\ &\cong H^*(B;\mathbb{Q}) \otimes \mathbb{Q}[x_1, ..., x_q] \Big/ \left(u_i - \sum_{s=(i,j)\in S} f^*(\nu_s) m_j + f^*(D_i) \right), \end{aligned}$$

where $f : B \to B\mathrm{aut}_1(M)$ is the classifying map of the given fibration. In particular, one can obtain the algebra structure of $H^*(E;\mathbb{Q})$ if B is a suspension space because $f^*(D_i) = 0$ for $i = 1, ..., q$.

In a particular case, the $H^*(B\mathrm{aut}_1(M);\mathbb{Q})$-algebra structure of $H^*(M_{\mathrm{aut}_1(M)};\mathbb{Q})$ is determined explicitly.

Proposition 1.3. *Let M be a simply-connected space such that $H^*(M;\mathbb{Q}) \cong \mathbb{Q}[a]/(a^{m+1})$ with* $\deg a = l$. *Then*

$$H^*(B\mathrm{aut}_1(M);\mathbb{Q}) \cong \mathbb{Q}[\nu_2, ..., \nu_{m+1}]$$

as an algebra and

$$H^*(M_{\mathrm{aut}_1(M)};\mathbb{Q}) \cong \mathbb{Q}[\nu_2, ..., \nu_{m+1}] \otimes \mathbb{Q}[a] \Big/ \left(a^{m+1} - \sum_{s=2}^{m+1} \nu_s a^{m-s+1} \right)$$

as an $H^*(B\mathrm{aut}_1(M);\mathbb{Q})$-algebra, where $\deg \nu_k = lk$.

This result also follows from the computation in [6, Section 3] due to Gatsinzi.

The following subject is characterization of the indecomposable elements of $H^*(B\mathrm{aut}_1(M);\mathbb{Q})$ mentioned in Theorem 1.1. Following Kedra and McDuff [13], we introduce characteristic classes of $B\mathrm{aut}_1(M)$ with fibre integration of the universal M-fibration. The calculation of $H^*(M_{\mathrm{aut}_1(M)};\mathbb{Q})$ in Theorem 1.1 allows us to conclude that the characteristic classes coincide with the indecomposable elements ν_s of $H^*(B\mathrm{aut}_1(M);\mathbb{Q})$ modulo decomposable elements. To see this more precisely, we recall that (M, a) is a cohomologically symplectic (c-symplectic) space with formal dimension $2m$ if M satisfies Poincaré duality over $\mathbb{Q}$ and a is a class in $H^2(M;\mathbb{Q})$ such that $a^m \neq 0$; see [21].

Assume further that M is a closed manifold and let $\mathcal{H}_a$ denote the group of diffeomorphisms of M that fix a. Kedra and McDuff defined in [13, Section 3] cohomology classes, which are called the μ-classes, of the classifying space $B\mathcal{H}_a$ provided $H^1(M;\mathbb{Q}) = 0$. These classes are generalization of the characteristic classes of the classifying space of the group of Hamiltonian symplectomorphisms due to Reznikov [30] and Januszkiewicz and Kedra [12]. By the same way, for a symplectic space (M, a) mentioned above, we can define characteristic classes μ_k of the classifying space $B\mathrm{aut}_1(M)$ with degree $2k$ for $2 \leq k \leq m+1$. The class μ_k is also called the kth μ-class; see Section 6 for more details.

We can characterize the generators of $H^*(B\mathrm{aut}_1(M);\mathbb{Q})$ in Proposition 1.3 by μ-classes if M is a c-symplectic space.

Proposition 1.4. *Let M be a simply-connected c-symplectic space whose rational cohomology M is of the form $\mathbb{Q}[a]/(a^{m+1})$. Then $H^*(B\mathrm{aut}_1(M);\mathbb{Q})$ is generated by Kedra-McDuff μ-classes:*

$$H^*(B\mathrm{aut}_1(M);\mathbb{Q}) \cong \mathbb{Q}[\mu_2, \mu_3, ..., \mu_{m+1}].$$

We give a computational example. Let G be a Lie group, K and H subgroups of G. We define a map

$$\lambda_{K,G/H} : K \to \mathrm{aut}_1(G/H)$$

by $\lambda_{K,G/H}(g)(m) = gm$ with the left translation of K on G/H, where $g \in K$ and $m \in G/H$. We observe that $\lambda_{K,G/H}$ is a monoid map and hence it induces the map $B\lambda_{K,G/H} : BK \to B\mathrm{aut}_1(G/H)$ between classifying spaces. Consider the real Grassmann manifold M of the form $SO(2m+1)/SO(2) \times SO(2m-1)$. Since $H^*(M) \cong \mathbb{Q}[\chi]/(\chi^{2m})$ as an algebra, it follows from Proposition 1.4 that

$$H^*(B\mathrm{aut}_1(M);\mathbb{Q}) \cong \mathbb{Q}[\mu_2, \mu_3, ..., \mu_{2m}].$$

Observe that $\chi \in H^2(M;\mathbb{Q})$ is the element which comes from the Euler class $\chi \in H^2(BSO(2);\mathbb{Q})$ via the map

$$j^* : H^*(B(SO(2) \times SO(2m-1);\mathbb{Q}) \cong \mathbb{Q}[\chi, p'_1, ..., p'_{m-1}] \to H^*(M;\mathbb{Q}),$$

where j is the fibre inclusion of the fibration

$$M \xrightarrow{j} B(SO(2) \times SO(2m-1)) \xrightarrow{Bi} BSO(2m+1).$$

Recall that the rational cohomology of $BSO(2m+1)$ is a polynomial algebra generated by Pontrjagin classes; that is, $H^*(BSO(2m+1);\mathbb{Q}) \cong \mathbb{Q}[p_1, ..., p_m]$, where $\deg p_i = 4i$. We relate the Pontrjagin classes to the μ-classes with the map induced by $\lambda_{SO(2m+1),M}$. More precisely, we have

Proposition 1.5. $(B\lambda_{SO(2m+1),M})^*(\mu_{2i}) \equiv p_i$ *modulo decomposable elements.*

Let $\mathrm{Diff}_1(M)$ be the identity component of the group of diffeomorphisms of a manifold M and $(\mathcal{H}_a)_1$ the subgroup of $\mathrm{Diff}_1(M)$ which fix the class a. Let $\mathrm{Homeo}_1(M)$ denote the identity component of the group of homeomorphisms of M. The naturality of the integration along the fibre implies that the kth Kedra-McDuff μ-classes of $B\mathrm{Diff}_1(M)$, $B(\mathcal{H}_a)_1$ and of $B\mathrm{Homeo}_1(M)$ are extendable to the class μ_k in $H^{2k}(B\mathrm{aut}_1(M);\mathbb{Q})$. Thus an algebraic property of μ-classes is deduced as a corollary to Proposition 1.5.

Corollary 1.6. *Let M be the Grassmann manifold of the form $SO(2m{+}1)/SO(2)\times SO(2m{-}1)$. Then the μ-classes μ_2, μ_4,.., μ_{2m} in the cohomology $H^*(B\mathrm{Diff}_1(M);\mathbb{Q})$, $H^*(B(\mathcal{H}_a)_1;\mathbb{Q})$ and $H^*(B\mathrm{Homeo}_1(M);\mathbb{Q})$ are algebraically independent.*

The proof of Proposition 1.5 also allows us to conclude that the image of the kth μ-class by the induced map

$$(B\lambda_{SU(m+1),\mathbb{C}P^m})^* : H^*(B\mathrm{aut}_1(\mathbb{C}P^m);\mathbb{Q}) \to H^*(BSU(m+1);\mathbb{Q})$$

coincides with the kth Chern class up to sign modulo decomposable elements; see the proof of [13, Proposition 1.7]. Observe that $\mathbb{C}P^m \cong U(m+1)/U(1)\times U(m)$. In order to prove Proposition 1.5, we will heavily rely on an explicit model for the left translation by a Lie group on a homogeneous space, which is investigated in [17, Sections 3 and 4]. This also illustrates usefulness of such an algebraic model.

The rest of this manuscript is set out as follows. In Section 2, we recall briefly the HBS model for $\mathrm{aut}_1(M)$ and a model for the evaluation map. A minimal model for the monoid is constructed in Section 3. Section 4 is devoted to constructing an approximation to the Eilenberg-Moore spectral sequence. In Section 5, we prove Theorem 1.1. After recalling μ-classes, we prove Propositions 1.4 and 1.5 in Section 6.

2. Models for the monoid $\mathrm{aut}_1(M)$ and for the evaluation map

In what follows, $H^*(-)$ and $H_*(-)$ denote the cohomology and homology with coefficients in the rational field, respectively.

Let M be a formal space with $\dim H^*(M) < \infty$. We recall a Sullivan model for $\mathrm{aut}_1(M)$ and a model for the evaluation map $ev : \mathrm{aut}_1(M)\times M \to M$; see [8], [1], [5], [2] and [17] for more general function space models.

Let $\alpha : (\wedge V, d) \xrightarrow{\simeq} A_{PL}(M)$ be a minimal model for M, where $A_{PL}(M)$ denotes the commutative differential graded algebra of differential polynomial forms on M. Since M is formal by assumption, there exists a quasi-isomorphism $\eta : (\wedge V, d) \xrightarrow{\simeq} H^*(M)$.

Consider a differential graded algebra of the form $(\wedge(\wedge V\otimes B_*), D = d\otimes 1)$, where $B_* = H_{-*}(M)$. Let Δ be the coproduct of $H_*(M)$. We denote by I the ideal of $\wedge(\wedge V\otimes H_*(M))$ generated by $1\otimes 1_* - 1$ and all elements of the form

$$a_1a_2\otimes\beta - \sum_i (-1)^{|a_2||\beta_i'|}(a_1\otimes\beta_i')(a_2\otimes\beta_i''),$$

where $a_1, a_2 \in \wedge V$, $\beta\in B_*$ and $\Delta(\beta) = \sum_i \beta_i'\otimes\beta_i''$. The result [1, Theorem 3.5] implies that the composite $\rho : \wedge(V\otimes B_*) \hookrightarrow \wedge(\wedge V\otimes B_*) \to \wedge(\wedge V\otimes B_*)/I$ is an isomorphism of graded algebras. Moreover, it follows from [1, Theorem 3.3] that $DI\subset I$. Thus we have a DGA (E,δ) of the form

$$E = \wedge(V\otimes B_*) \ \text{ and } \ \delta = \rho^{-1}D\rho.$$

Observe that, for elements $v\in V$ and $e\in B_*$, if $d(v) = v_1\cdots v_m$ with $v_i\in V$ and $\Delta^{(m-1)}(e) = \sum_j e_{j_1}\otimes\cdots\otimes e_{j_m}$, then

$$\delta(v\otimes e) \ = \ \textstyle\sum_j \pm(v_1\otimes e_{j_1})\cdots(v_m\otimes e_{j_m}). \tag{2.1}$$

Here the sign is determined by the Koszul rule; that is, $ab = (-1)^{\deg a\deg b}ba$ in a graded algebra.

We define a DGA map $u : \wedge(\wedge V \otimes B_*)/I \to \mathbb{Q}$ by

$$u(a \otimes b) = (-1)^{\tau(|a|)} b(\eta(a)),$$

where $\tau(n) = [(n+1)/2]$, $a \in \wedge V$ and $b \in B_*$. Let M_u be the ideal of E generated by the set

$$\{\omega \mid \deg \omega < 0\} \cup \{\delta\omega \mid \deg \omega = 0\} \cup \{\omega - u(\omega) \mid \deg \omega = 0\}.$$

Then the result [1, Theorem 6.1] asserts that $(E/M_u, \delta)$ is a model for $\mathrm{aut}_1(M)$. This means that there exists a quasi-isomorphism $\xi : (E/M_u, \delta) \to A_{PL}(\mathrm{aut}_1(M))$. We call the DGA $(E/M_u, \delta)$ the HBS model for $\mathrm{aut}_1(M)$.

The proof of [16, Theorem 4.5] and [11, Remark 3.4] enable us to construct a model for the evaluation map $ev : \mathrm{aut}_1(M) \times M \to M$; see also [2]. The explicit form of the model is described in Proposition 2.1 below.

Let $\{m_j\}_{j\in J}$ be a basis of $H^*(M)$ and $\{(m_j)_*\}_{j\in J}$ the dual basis to $\{m_j\}_{j\in J}$. Then there exists a set $\{m'_j\}_{j\in J}$ of linearly independent elements in $\wedge V$ such that $\eta(m'_j) = m_j$.

Proposition 2.1. [17, Proposition 2.3, Remark 2.5(ii)] *With the same notation as above, we define a map $m(ev) : (\wedge V, d) \to (E/M_u, \delta) \otimes \wedge V$ by*

$$m(ev)(x) = \sum_j (-1)^{\tau(|m_j|)} \pi(x \otimes (m_j)_*) \otimes m'_j,$$

for $x \in \wedge V$, where $\pi : E \to E/M_u$ denotes the natural projection. Then $m(ev)$ is a Sullivan representative for the evaluation map $ev : \mathrm{aut}_1(M) \times M \to M$; that is, there exists a homotopy commutative diagram

$$\begin{array}{ccc}
A_{PL}(M) & \xrightarrow{ev^*} & A_{PL}(\mathrm{aut}_1(M) \times M) \\
 & & \uparrow \simeq \\
\alpha \uparrow \simeq & & A_{PL}(\mathrm{aut}_1(M)) \otimes A_{PL}(M) \\
 & & \simeq \uparrow \xi\otimes\alpha \\
\wedge V & \xrightarrow[m(ev)]{} & (E/M_u, \delta) \otimes \wedge V,
\end{array}$$

in which $\xi : (E/M_u, \delta) \xrightarrow{\simeq} A_{PL}(\mathrm{aut}_1(M))$ is the Sullivan model for $\mathrm{aut}_1(M)$ *mentioned above.*

Example 2.2. Let M be a space whose rational cohomology is isomorphic to the truncated algebra $\mathbb{Q}[x]/(x^m)$, where $\deg x = l$. Recall the model $(E/M_u, \delta)$ for $\mathrm{aut}_1(M)$ mentioned in [11, Example 3.6]. Since the minimal model for M has the form $(\wedge(x, y), d)$ with $dy = x^m$, it follows that

$$E/M_u = \wedge(x \otimes 1_*, y \otimes (x^s)_*; 0 \le s \le m-1)$$

with $\delta(x \otimes 1_*) = 0$ and $\delta(y \otimes (x^s)_*) = (-1)^s \binom{m}{s} (x \otimes 1_*)^{m-s}$, where $\deg x \otimes 1_* = l$ and $\deg(y \otimes (x^s)_*) = lm - ls - 1$. Then the rational model $m(ev)$ for the evaluation map $ev : \mathrm{aut}_1(M) \times M \to M$ is given by $m(ev)(x) = (x \otimes 1_*) \otimes 1 + 1 \otimes x$ and

$$m(ev)(y) = \sum_{s=0}^{m-1} (-1)^s (y \otimes (x^s)_*) \otimes x^s + 1 \otimes y.$$

3. A minimal model for $\mathrm{aut}_1(M)$

Let M be a space whose rational cohomology is isomorphic to an algebra of the form (1.1). Observe that M is a formal space with $\dim H^*(M) < \infty$. Thus the construction in Section 2 of the HBS model for $\mathrm{aut}_1(M)$ is applicable. We here construct a relevant minimal model for $\mathrm{aut}_1(M)$ with the HBS model.

We first take a minimal model $(\wedge V, d)$ for M for which $\wedge V = \wedge(x_1, ..., x_q, \rho_1, ..., \rho_q)$ and $d(\rho_i) = u_i$ for $i = 1, ..., q$. Let $\{m_j\}_{j \in J}$ be a basis of $H^*(M)$ and T the set $\{1, ..., q\} \times J$. Consider the HBS model E/M_u for $\mathrm{aut}_1(M)$ mentioned in Section 2. It is readily seen that the model has the form

$$E/M_u = (\wedge(\rho_i \otimes (m_j)_*; (i,j) \in T_1, x_i \otimes (m_j)_*; (i,j) \in T_2), \delta),$$

where $T_1 = \{(i,j) \in T | \deg \rho_i \otimes (m_j)_* \geq 1\}$ and $T_2 = \{(i,j) \in T | \deg x_i \otimes (m_j)_* \geq 1\}$. We define the linear part δ_0 of the differential δ by $\delta(x) - \delta_0(x) \in (E/M_u)^+ \cdot (E/M_u)^+$. Then the part δ_0 defines a linear map

$$\delta_{0,\mathrm{res}} : \mathbb{Q}\{\rho_i \otimes (m_j)_*; (i,j) \in T_1\} \to \mathbb{Q}\{x_i \otimes (m_j)_*; (i,j) \in T_2\},$$

where $\mathbb{Q}\{U\}$ denotes the vector space generated by a set U.

As mentioned in Introduction, we are able to relate Halperin's conjecture with a property of the linear part $\delta_{0,\mathrm{res}}$. Let $\mathrm{Der}_{\mathbb{Q}}(H^*(M))_-$ be the vector space of all derivations of negative degree of $H^*(M)$.

Lemma 3.1. *The vector space* $\mathrm{Der}_{\mathbb{Q}}(H^*(M;\mathbb{Q}))_-$ *vanishes if and only if* $\delta_{0,res}$ *is surjective.*

Proof. It follows from [27, (2.6) Proposition (ii)] that the vector space of all derivations of negative degree of $H^*(M)$ is isomorphic to $\pi_{\mathrm{even}}(\mathrm{aut}_1(M_{\mathbb{Q}}))$. Moreover the result [4, Theorem 15.11] enables us to conclude that $\mathrm{Coker}\delta_{0,\mathrm{res}} \cong \pi_{\mathrm{even}}(\mathrm{aut}_1(M_{\mathbb{Q}}))$. This completes the proof. □

In order to construct a minimal model for E/M_u, we arrange bases $\rho_i \otimes (m_j)_*$ and $x_i \otimes (m_j)_*$ as follows. Fix total orders on the sets T_1 and T_2, respectively. Using the orders, we define column vectors $(\delta_{0,\mathrm{res}}(\rho_i \otimes (m_j)_*))$ and $(x_i \otimes (m_j)_*)$.

Suppose that $\mathrm{Der}_{\mathbb{Q}}(H^*(M;\mathbb{Q}))_- = 0$. Then it follows from Lemma 3.1 that there exists a (k,l)-matrix A with full rank such that

$$A(\delta_{0,\mathrm{res}}(\rho_i \otimes (m_j)_*)) = (x_i \otimes (m_j)_*),$$

where $k = \sharp T_2$ and $l = \sharp T_1$. Thus, changing the order on T_1 if necessary, we can find a regular (k,k)-matrix B such that $BA = (E, *)$, where E is the identity matrix. Put $B(x_i \otimes (m_j)_*) = (v_l)$. We choose the pair $(s,t) \in T_1$ which satisfies the condition that

$$\sharp\{(i,j) \in T_1 | (i,j) \leq (s,t)\} = k.$$

We have

$$\begin{pmatrix} 1 & & & 0 & \cdots & 0 \\ & 1 & & \vdots & & \vdots \\ & & \ddots & \vdots & & \vdots \\ & & 1 & 0 & \cdots & 0 \end{pmatrix} \begin{pmatrix} \vdots \\ (\delta_{0,\mathrm{res}}(\rho_{i'} \otimes (m_{j'})_*) + \delta_{0,\mathrm{res}} w_{i',j'} \\ \vdots \\ (\delta_{0,\mathrm{res}}(\rho_s \otimes (m_t)_*) + \delta_{0,\mathrm{res}} w_{s,t} \\ \vdots \\ \delta_{0,\mathrm{res}}(\rho_i \otimes (m_j)_*) \\ \vdots \end{pmatrix} = \begin{pmatrix} v_1 \\ \vdots \\ \vdots \\ \vdots \\ v_k \end{pmatrix}$$

for which each $w_{i',j'}$ is a linear combination of elements $\rho_i \otimes (m_j)_*$ with $(i,j) > (s,t)$. Without loss of generality, we can assume that $(i,0) > (s,t)$ for $i = 1, ..., q$ because $\delta_{0,\mathrm{res}}(\rho_i \otimes 1_*) = 0$ for any i. We write $(b_1, ..., b_k) = (\cdots, \rho_{i'} \otimes (m_{j'})_* + w_{i',j'}, \cdots, \rho_s \otimes (m_t)_* + w_{s,t})$. Observe that $b_1, ..., b_k, \rho_i \otimes (m_j)_*$; $(i,j) > (s,t)$ are linearly independent. Since $v_1, ..., v_k$ generate the image of $\delta_{0,\mathrm{res}}$, it follows that for any (i,j) greater than (s,t),

$$\delta_{0,\mathrm{res}}(\rho_i \otimes (m_j)_*) - \delta_{0,\mathrm{res}}(\sum_l \alpha_{il} b_l) = 0$$

for some $\alpha_{il} \in \mathbb{Q}$. Thus we have

$$E/M_u \cong \wedge(b_1, ..., b_k, ..., \rho_i \otimes (m_j)_* + \sum_l \alpha_{il} b_l, ..., v_1, ..., v_k), \delta)$$

for which $\delta b_i \equiv v_i$ and $\delta(\rho_i \otimes (m_j)_* + \sum_l \alpha_{il} b_l) \equiv 0$ modulo decomposable elements. Put $S := \{(i,j) \in T_1 | (i,j) > (s,t)\}$. It is immediate that $S_0 \subset S$. We then have a retraction

$$r : E/M_u \to \wedge Z := (\wedge(\rho_i \otimes (m_j)_*; (i,j) \in S), 0)$$

defined by $r((\rho_i \otimes (m_j)_*) = 0$ if $(i,j) \notin S$, $r(v_i) = 0$ and $r((\rho_i \otimes (m_j)_*) = \rho_i \otimes (m_j)_*$ if $(i,j) \in S$. Since each term of the decomposable elements $\delta b_i - v_i$ and $\delta(\rho_i \otimes (m_j)_* + \sum_l \alpha_{il} b_l)$ contain v_l for some l as a factor, we see that γ is a well-defined DGA map.

The fact that r is a quasi-isomorphism follows from the usual spectral sequence argument.

4. An approximation to the Eilenberg-Moore spectral sequence

In order to prove Theorem 1.1, we introduce a spectral sequence. Let $C_*(M)$ denote the normalized chain complex of a space M with coefficients in the rational field. By definition, the total space $M_{\mathrm{aut}_1(M)}$ of the universal M-fibration is regarded as the realization $|B_*(*, \mathrm{aut}_1(M), M)|$ of the geometric bar construction $B_*(*, \mathrm{aut}_1(M), M)$, which is a simplicial topological space with $B_i(*, \mathrm{aut}_1(M), M) = * \times \mathrm{aut}_1(M)^{\times i} \times M$; see [24, Proposition 7.9]. The result [24, Theorem 13.9] allows us to obtain natural quasi-isomorphisms which connect with $C_*(|B(*, \mathrm{aut}_1(M), M)|)$ and the algebraic bar construction of the form $B(C_*(*), C_*(\mathrm{aut}_1(M)), C_*(M))$ for which

$$B(C_*(*), C_*(\mathrm{aut}_1(M)), C_*(M))_k = \oplus_{i+j=k}(C_*(*) \otimes C_*(\mathrm{aut}_1(M))^{\otimes i} \otimes C_*(M))_j.$$

Moreover the Eilenberg-Zilber map gives rise to a quasi-isomorphism from the bar complex to the total complex $\mathrm{Total}C_*(B_*(*, \mathrm{aut}_1(M), M))$. Observe that

$$\mathrm{Total}C_*(B_*(*, \mathrm{aut}_1(M), M))_k = \oplus_{i+j=k} C_j B_i(*, \mathrm{aut}_1(M), M).$$

In consequence, by virtue of [4, Corollary 10.10], we have natural quasi-isomorphisms which connect $C^*(M_{\mathrm{aut}_1(M)}) = C^*(|B_*(*, \mathrm{aut}_1(M), M)|)$ with the total complex of a double complex $\mathcal{B} = \{\mathcal{B}^{i,j}, d_i, \delta_j\}$ of the form

$$\mathcal{B}^{i,j} = \left(A_{PL}(\mathrm{aut}_1(M)^{\times i} \times M)\right)^j.$$

In particular, $d_0 : \mathcal{B}^{0,*} \to \mathcal{B}^{1,*}$ is regarded as the map

$$(pr_2)^* - ev^* : A_{PL}(M) \to A_{PL}(\mathrm{aut}_1(M) \times M),$$

where the maps pr_2 and ev from $\mathrm{aut}_1(M) \times M \to M$ to M denote the second projection and the evaluation map, respectively.

We define a double complex $\mathcal{C} = \{\mathcal{C}^{i,j}, d_i, \delta_j\}$ by truncating the double complex $\{\mathcal{B}^{i,j}\}$ for $i \geq 2$; that is, $\mathcal{C}^{i,j} = \mathcal{B}^{i,j}$ for $i \leq 0, 1$ and $\mathcal{C}^{i,j} = 0$ for $i \geq 2$.

Let $\{E_r, d_r\}$ be the Eilenberg-Moore spectral sequence converging to the rational cohomology $H^*(M_{\mathrm{aut}_1(M)})$ with

$$E_2^{*,*} \cong \mathrm{Cotor}^{*,*}_{H^*(\mathrm{aut}_1(M))}(\mathbb{Q}, H^*(M))$$

as an algebra. Observe that this spectral sequence is constructed with the double complex $\mathcal{B}$. The double complex $\mathcal{C}$ gives rise to a spectral sequence $\{\widetilde{E}_r, \widetilde{d}_r\}$ converging to $H^*(\mathrm{Total}(\mathcal{C}))$. Moreover, we see that the projection $q : \mathcal{B} \to \mathcal{C}$ induces the morphism of the spectral sequences

$$\{q_r\} : \{E_r, d_r\} \to \{\widetilde{E}_r, \widetilde{d}_r\}$$

and the morphism $\widehat{q} : H^*(M_{\mathrm{aut}_1(M)}) \to H^*(\mathrm{Total}(\mathcal{C}))$ of algebras.

Lemma 4.1. *For any $\alpha \in H^*(M_{\mathrm{aut}_1(M)})$, $\widehat{q}(\alpha) = 0$ if and only if $\alpha \in F^2H^*$. Here $\{F^pH^*\}_{p\geq 0}$ denotes the filtration of $H^*(M_{\mathrm{aut}_1(M)})$ associated with the spectral sequence $\{E_r, d_r\}$.*

Proof. By construction, we see that $q_2 : E_2^{p,*} \to \widetilde{E}_2^{p,*}$ is bijective for $p = 0$ and injective for $p = 1$. Since the spectral sequence $\{\widetilde{E}_r, \widetilde{d}_r\}$ collapses at the E_1-term, it follows that the map $q_r^{p,*}$ for $3 \leq r \leq \infty$ and $p \leq 1$ is injective. This completes the proof. □

Let M be a space as in Theorem 1.1 and E/M_u the HBS model for the monoid $\mathrm{aut}_1(M)$ mentioned in Section 2. Since M is formal, we can take a quasi-isomorphism $\eta : (\wedge V, d) \xrightarrow{\simeq} H^*(M; \mathbb{Q})$. Moreover let $m(ev) : \wedge V \to E/M_u \otimes \wedge V$ denote the model for the evaluation map $ev : \mathrm{aut}_1(M) \times M \to M$ described in Proposition 2.1. Recall the retraction $r : E/M_u \to \wedge Z$ mentioned in Section 3. We then have a commutative diagram

$$\begin{array}{ccc}
A_{PL}(M) & \xrightarrow{d^0} & A_{PL}(\mathrm{aut}_1(M) \times M) \\
\uparrow & & \varepsilon_0 \uparrow \simeq \\
 & & A_{PL}(\mathrm{aut}_1(M) \times M) \otimes \wedge(t, dt) \\
\alpha \,\Big|\, \simeq & \nearrow^{(pr_2)^* \circ \alpha - H} & \varepsilon_1 \downarrow \simeq \\
 & & A_{PL}(\mathrm{aut}_1(M) \times M) \\
 & & \uparrow \simeq \\
\wedge V & \xrightarrow[s - m(ev)]{} & E/M_u \otimes \wedge V \\
 & \searrow_{\eta \circ s - (r \otimes \eta) \circ m(ev)} & r \otimes \eta \downarrow \simeq \\
 & & \wedge Z \otimes H^*(M; \mathbb{Q}),
\end{array}$$

where $H : \wedge V \to A_{PL}(\mathrm{aut}_1(M) \times M) \otimes \wedge(t, dt)$ denotes the homotopy between the model $m(ev)$ for the evaluation map ev and the induced map $A_{PL}(ev)$ up to quasi-isomorphisms and s stands for the inclusion into the second factor.

Let $\mathcal{D}$ be the double complex associated with the differential graded $\mathbb{Q}$-module map

$$\eta \circ s - (r \otimes \eta) \circ m(ev) : (\wedge V, d_{\wedge V}) \to \wedge Z \otimes H^*(M; \mathbb{Q}).$$

Observe that $\mathcal{D}^{i,j} = 0$ for $i \geq 2$. The usual spectral sequence argument allows us to conclude that $H^*(\mathrm{total}\mathcal{C}) \cong H^*(\mathrm{total}\mathcal{D})$ as a vector pace. By using this identification, we shall prove Theorem 1.1.

5. Proof of Theorem 1.1

We use the same notation as in Sections 3 and 4 throughout this section.

Proof of Theorem 1.1. Recall from Section 3 the minimal model $\wedge Z$ for the monoid $\mathrm{aut}_1(M)$; that is,

$$\wedge Z = (\wedge(\rho_i \otimes (m_j)_*, (i, j) \in S), 0),$$

where $\{(m_j)_*\}_{j \in J}$ denotes the dual basis of $H^*(M; \mathbb{Q})$ and $\deg \rho_i \otimes (m_j)_* = \deg u_i - 1 - \deg m_j$. We observe that $\mathrm{Der}_{\mathbb{Q}}(H^*(M; \mathbb{Q}))_- = 0$ by assumption. Thus we have

$$H^*(\mathrm{aut}_1(M)) \cong \wedge(\rho_i \otimes (m_j)_*, (i, j) \in S),$$

This yields that, in the Leray-Serre spectral sequence for the universal fibration $\mathrm{aut}_1(M) \to E\mathrm{aut}_1(M) \overset{\pi'}{\to} B\mathrm{aut}_1(M)$, the element $\rho_i \otimes (m_j)_*$ is transgressive for any $(i, j) \in S$. In fact we have a commutative diagram

$$\begin{array}{ccccc}
\pi_*(\mathrm{aut}_1(M))^\sharp & \underset{\cong}{\xrightarrow{\delta^\sharp}} & \pi_{*+1}(E_{\mathrm{aut}_1(M)}, \mathrm{aut}_1(M))^\sharp & \underset{\cong}{\xleftarrow{\pi'^{\sharp}_*}} & \pi_{*+1}(B\mathrm{aut}_1(M))^\sharp \\
\uparrow h_G^\sharp & & \uparrow (-1)^{*+1} h^\sharp & & \uparrow (-1)^{*+1} h_{BG}^\sharp \\
H^*(\mathrm{aut}_1(M)) & \underset{\cong}{\xrightarrow{\delta}} & H^{*+1}(E_{\mathrm{aut}_1(M)}, \mathrm{aut}_1(M)) & \xleftarrow{\pi'^*} & H^{*+1}(B\mathrm{aut}_1(M)),
\end{array}$$

where h_G, h and h_{BG} denote the duals to Hurewicz maps. Since $H^*(\mathrm{aut}_1(M))$ and $H^*(B\mathrm{aut}_1(M))$ are exterior algebra and a polynomial algebra, respectively, it follows that h_G and h_{BG} are isomorphisms on vector subspaces of indecomposable elements. Thus we see that the element $\rho_i \otimes (m_j)_*$ is transgressive because the map $\delta^{-1}\pi'^*$ is the transgression by definition. Thus the result [29, Section 7 (2.27)] implies that the element $\rho_i \otimes (m_j)_*$is primitive for $(i, j) \in S$.

Let $\{\widehat{E}_r, \widehat{d}_r\}$ be the Eilenberg-Moore spectral sequence converging to the cohomology $H^*(B\mathrm{aut}_1(M))$ with $\widehat{E}_2^{*,*} \cong \mathrm{Cotor}^{*,*}_{H^*(\mathrm{aut}_1(M))}(\mathbb{Q},\mathbb{Q})$. Since the element $\rho_i \otimes (m_j)_*$ is primitive for $1 \le i \le m$, it follows that

$$\widehat{E}_2^{*,*} = \mathbb{Q}\big[[\rho_i \otimes (m_j)_*], (i,j) \in S\big],$$

where bideg $[\rho_i \otimes (m_j)_*] = (1, \deg \rho_i - \deg m_j)$. This allows us to conclude that, as algebras,

$$H^*(B\mathrm{aut}_1(M)) \cong \mathrm{Total}(\widehat{E}_2^{*,*}) \cong \mathbb{Q}\big[[\rho_i \otimes (m_j)_*], (i,j) \in S\big].$$

Recall the Eilenberg-Moore spectral sequence $\{E_r, d_r\}$ converging to the cohomology ring $H^*(M_{\mathrm{aut}_1(M)})$. We see that

$$\begin{aligned} E_2^{*,*} &\cong \mathrm{Cotor}^{*,*}_{H^*(\mathrm{aut}_1(M))}(\mathbb{Q}, H^*(M)) \\ &\cong \mathbb{Q}\big[[\rho_i \otimes (m_j)_*], (i,j) \in S\big] \otimes H^*(M) \end{aligned}$$

as bigraded algebras. For dimensional reasons, we see that the spectral sequence $\{E_r, d_r\}$ collapses at the E_2-term. Let $M \overset{\iota}{\to} M_{\mathrm{aut}_1(M)} \overset{\pi}{\to} B\mathrm{aut}_1(M)$ be the universal M-fibration. The naturality of the spectral sequence enables us to deduce that $\pi^*([\rho_i \otimes (m_j)_*]) = [\rho_i \otimes (m_j)_*]$ in $H^*(M_{\mathrm{aut}_1(M)})$. In the total complex of the double complex $\mathcal{D}$, we have

$$\begin{aligned} (d_{\wedge V} \pm (\eta s - (r \otimes \eta) \circ m(ev)))\rho_i &= d_{\wedge V}(\rho_i) + (-1)^{\deg \rho_i}(\eta s - (r \otimes \eta) \circ m(ev))(\rho_i) \\ &= ([\ ]u_i) + \sum_{(i,j) \in S} (-1)^{\deg m_j/2}[\rho_i \otimes (m_j)_*] \otimes m_j. \end{aligned}$$

This implies that $\widehat{q}\big(([\ ]u_i) + \sum_{(i,j)\in S}(-1)^{\deg m_j/2}[\rho_i \otimes (m_j)_*] \otimes ([\ ]m_j)\big) = 0$ in $H^*(\mathrm{Total}(\mathcal{C}))$. Therefore it follows from Lemma 4.1 that

$$([\ ]u_i) \equiv \sum_{(i,j)\in S} \nu_{(i,j)} \otimes m_j$$

modulo the ideal generated by $\pi^*(H^*(B\mathrm{aut}_1(M))^+ \cdot H^*(B\mathrm{aut}_1(M))^+)$ in $H^*(M_{\mathrm{aut}_1(M)})$, where $\nu_{(i,j)} = (-1)^{(\deg m_j/2)+1}[\rho_i \otimes (m_j)_*]$.

The latter half of Theorem 1.1 follows from the equalities

$$\begin{aligned} \sharp\{\nu_s |\, \deg \nu_s = 2k, s \in S\} &= \dim \pi_{2k}(B\mathrm{aut}_1(M_{\mathbb{Q}})) \\ &= \dim \pi_{2k-1}(\mathrm{aut}_1(M_{\mathbb{Q}})) \\ &= \sharp\{\rho_i \otimes (m_j)_* |\, \deg \rho_i \otimes (m_j)_* = 2k-1, (i,j) \in S\} \\ &= \mathrm{rank} \oplus_{a-b=2k-1} \pi_a(M) \otimes H^b(M;\mathbb{Q}) \\ &\quad -\mathrm{rank} \oplus_{a-b=2k} \pi_a(M) \otimes H^b(M;\mathbb{Q}). \end{aligned}$$

This completes the proof. □

Proof of Proposition 1.3. As is well-known, Halperin's conjecture is affirmatively solved for the given space M. Then Theorem 1.1 is applicable. We choose a basis of the form $\{a^i\}_{0\le i \le m}$ of $H^*(M;\mathbb{Q})$. There is no element w in $\pi^*(H^*(B\mathrm{aut}_1(M);\mathbb{Q})^+ \cdot H^*(B\mathrm{aut}_1(M);\mathbb{Q})^+)$ such that $\deg a^{m+1} = \deg wa^m$. This yields that each term of D_1 does not have a^m as a factor. Replace the indecomposable element $\nu_{(1,j)}$ in Theorem 1.1 with $\mu_j = \nu_{(1,j)} + \xi_j$ for any $j \ge 2$, where ξ_j is the coefficient of a^{m-j+1} in D_1. We have the result. □

Remark 5.1. Let M be a space as in Theorem 1.1. Recall the subset S of $\{1, ..., q\} \times J$ mentioned in the construction of the minimal model for $\mathrm{aut}_1(M)$ in Section 3. Let $\rho : \{1, ..., q\} \times J \to J$ be the projection. Suppose that $\deg u_i \neq \deg wm_j$ for any $(i, j) \in \{1, ..., q\} \times (J \backslash \rho(S))$ and $w \in \pi^*(H^*(B\mathrm{aut}_1(M); \mathbb{Q})^+ \cdot H^*(B\mathrm{aut}_1(M); \mathbb{Q})^+)$. As in the proof of Proposition 1.3, we can arrange the indecomposable elements ν_s so that each D_i in Theorem 1.1 is trivial.

6. μ-CLASSES

In order to define μ-classes due to Kedra and McDuff, we first recall the coupling class. Let (M, a) be a c-symplectic space with formal dimension $k = 2m$. Consider the Leray-Serre spectral sequence $\{E_r, d_r\}$ for a fibration $M \overset{i}{\to} E \overset{\pi}{\to} B$ for which $\pi_1(B)$ act trivially on $H^k(M) = \mathbb{Q}$. Let $\{F^pH^*\}_{p\geq 0}$ denote the filtration of $\{E_r, d_r\}$. Then the integration along the fibre (the cohomology push forward) $\pi! : H^{p+k}(E) \to H^p(B)$ is defined by the composite

$$H^{p+k}(E) = F^0H^{p+k} = F^pH^{p+k} \twoheadrightarrow E_\infty^{p,q} \rightarrowtail \cdots \rightarrowtail E_2^{p,q} \cong H^p(B; H^k(M)) \cong H^p(B).$$

Let $\mathcal{G}$ denote the group $\mathcal{H}_a$ or the monoid $\mathrm{aut}_1(M)$. Let $M \overset{\iota}{\to} M_{\mathcal{G}} \overset{\pi}{\to} B\mathcal{G}$ be the universal M-fibration; see [24, Proposition 7.9]. Proposition 6.1 below follows from the proofs of [12, Proposition 2.4.2] and [13, Proposition 3.1].

Proposition 6.1. *Suppose that $H^1(M) = 0$, then the element $a \in H^2(M)$ is extendable to an element $\overline{a} \in H^2(M_{\mathcal{G}})$. Moreover, there exists a unique element $\widetilde{a} \in H^2(M_{\mathcal{G}})$ that restricts to $a \in H^2(M)$ and such that $\pi!(\widetilde{a}^{m+1}) = 0$. In fact the element $\widetilde{a}$ has the form*

$$\widetilde{a} = \overline{a} - \frac{1}{n+1}\pi^*\pi!(\overline{a}^{m+1}).$$

The class $\widetilde{\omega}$ in Proposition 6.1 is called *the coupling class.*

Definition 6.2. [13, Section 3.1] [12, Section 2.4] [30] We define $\mu_k \in H^{2k}(B\mathcal{G})$, which is called kth μ-class, by

$$\mu_k := \pi!(\widetilde{a}^{m+k}),$$

where $\widetilde{a}$ is the coupling class.

Proof of Proposition 1.4. We use the same notation as in the previous sections. It follows from Lemma 4.1 that

$$([\ ]a)^{m+1} \equiv \sum_{i=1}^{m}(-1)^{m-i+1}[y \otimes (a^{m-i})_*] \otimes ([\ ]a)^{m-i}$$

modulo the ideal generated by $\pi^*(H^*(B\mathrm{aut}_1(M))^+ \cdot H^*(B\mathrm{aut}_1(M))^+)$ in $H^*(M_{\mathrm{aut}_1(M)})$. Since $H^2(B\mathrm{aut}_1(M)) = 0$, the definition of the integration $\pi!$ enables us to deduce that $\pi!(([\ ]a)^{m+1}) = 0$. We can choose the element $([\ ]a)$ as the coupling class $\widetilde{a}$ mentioned in Proposition 6.1. By definition, for $2 \leq k \leq m+1$, we see that

$$\begin{aligned}
\mu_k &= \pi!(\widetilde{a}^{m+k}) = \pi!(\widetilde{a}^{m+1} \cdot \widetilde{a}^{k-1}) \\
&= \pi!\Big((\sum_{i=1}^{m}(-1)^{m-i+1}[y \otimes (a^{m-i})_*]\ \widetilde{a}^{m-i}) \cdot \widetilde{a}^{k-1}\Big) \\
&= \pi!(\cdots + (-1)^{m-k}[y \otimes (a^{m-k+1})_*]\ \widetilde{a}^m + \cdots) \\
&= (-1)^{m-k}[y \otimes (a^{m-k+1})_*]
\end{aligned}$$

modulo decomposable elements. We have the result. □

Remark 6.3. Let M be a c-symplectic space of the form $(\mathbb{C}P^m \times \mathbb{C}P^n, a_1+a_2)$. Then it follows that μ-classes do not generate the whole algebra $H^*(B\mathrm{aut}_1(M))$. To see this, we choose minimal models $(\wedge(y_1, a_1), dy_1 = a_1^{m+1})$ and $(\wedge(y_2, a_2), dy_2 = a_1^{n+1})$ for the projective spaces $\mathbb{C}P^m$ and $\mathbb{C}P^n$, respectively. Suppose that $m \geq n$. Then the same argument as in [11, Example 3.6] allows us to conclude that $\mathrm{aut}_1(M)$ admits a minimal model of the form

$$\begin{gathered}\wedge(y_1 \otimes 1_*, y_1 \otimes (a_1)_*, ..., y_1 \otimes (a_1^{m-1})_*, y_2 \otimes 1_*, y_2 \otimes (a_2)_*, ..., y_2 \otimes (a_2^{n-1})_*, \\ y_1 \otimes (a_1^{k_1}a_2^{l_1})_*, y_2 \otimes (a_1^{k_2}a_2^{l_2})_*; 1 \leq k_1 + l_1 \leq m, 0 < l_1 \leq n, 1 \leq k_2 + l_2 \leq n, k_2 \neq 0)\end{gathered}$$

with the trivial differential. This yields that $H^2(B\mathrm{aut}_1(M))$ is isomorphic to

$$\mathbb{Q}\{[y_1 \otimes (a_1^{k_1}a_2^{l_1})_*], [y_2 \otimes (a_1^{k_2}a_2^{l_2})_*]; k_1 + l_1 = m, 0 < l_1 \leq n, k_2 + l_2 = n, k_2 \neq 0\}$$

as a vector space. Thus any of μ-classes does not detect an element in $H^2(B\mathrm{aut}_1(M))$ since the degrees of the μ-classes are greater than 4.

We also determine the $H^*(B\mathrm{aut}_1(M); \mathbb{Q})$-algebra structure of $H^*(M_{\mathrm{aut}_1(M)}; \mathbb{Q})$. It follows from Remark 5.1 that

$$H^*(B\mathrm{aut}_1(M); \mathbb{Q}) \cong \mathbb{Q}\left[\begin{array}{lll} \nu(k_1', 0),\ \nu(k_1, l_1) & ; & 0 \leq k_1' \leq m-1, 1 \leq k_1 + l_1 \leq m, 1 \leq l_1 \leq n \\ \eta(0, l_1'),\ \eta(k_2, l_2) & ; & 0 \leq l_1' \leq n-1, 1 \leq k_2 + l_2 \leq n, k_2 \neq 0 \end{array}\right]$$

and, as an $H^*(B\mathrm{aut}_1(M); \mathbb{Q})$-algebra,

$$H^*(M_{\mathrm{aut}_1(M)}; \mathbb{Q}) \cong H^*(B\mathrm{aut}_1(M); \mathbb{Q}) \otimes \mathbb{Q}[a_1, a_2]\,/(\rho_1, \rho_2),$$

where

$$\begin{aligned} \rho_1 &= a_1^{m+1} - \sum_{k_1'=0}^{m-1} \nu(k_1', 0)a_1^{k_1'} - \sum_{1 \leq k_1+l_1 \leq m, 1 \leq l_1 \leq n} \nu(k_1, l_1)a_1^{k_1}a_2^{l_1} \\ \rho_1 &= a_2^{n+1} - \sum_{l_1'=0}^{n-1} \eta(0, l_1')a_2^{l_1'} - \sum_{1 \leq k_2+l_2 \leq n, k_2 \neq 0} \eta(k_2, l_2)a_1^{k_2}a_2^{l_2}, \end{aligned}$$

$\deg \nu(k_1', 0) = 2(m+1) - 2k_1'$, $\deg \nu(k_1, l_1) = 2(m+1) - 2(k_1 + l_1)$, $\deg \eta(0, l_1') = 2(n+1) - 2l_1'$ and $\deg \eta(k_2, l_2) = 2(n+1) - 2(k_2 + l_2)$.

Proof of Proposition 1.5. We can take a Sullivan model $(\wedge V, d)$ for M such that $\wedge V = \wedge(\chi, p_1', ..., p_{m-1}', \tau_2, \tau_4, ..., \tau_{2m})$ and $d(\tau_{2i}) = (-1)^i(\chi^2 p_{i-1}' + p_i')$ for $1 \leq i \leq m$. In view of the rational model for $\lambda_{G,M} : SO(2m+1) \to \mathrm{aut}_1(M)$ mentioned in [17, Theorem 3.3], we have $(\lambda_{SO(2m+1),M})^*(\tau_{2m} \otimes (\chi^{2l})_*) = \tau_{2(m-l)}$ in the cohomology; see also [17, Section 8 (1)]. Thus the naturality of the Eilenberg-Moore spectral sequence allows us to deduce that $(B\lambda_{SO(2m+1),M})^*([\tau_{2m} \otimes (\chi^{2l})_*]) = [\tau_{2(m-l)}]$. The description of the μ-classes in the proof of Theorem 1.4 yields that

$$(B\lambda_{SO(2m+1),M})^*(\mu_{2(m-l)}) \equiv (B\lambda_{SO(2m+1),M})^*([\tau_{2m} \otimes (\chi^{2l})_*])$$

modulo decomposable elements. Let $\sigma^* : H^*(BSO(2m+1)) \to H^{*-1}(SO(2m+1))$ be the cohomology suspension. Without loss of generality, we can assume that $\sigma^*(p_{m-l}) = \tau_{2(m-l)}$; see [4, Proposition 15.13]. By virtue of [7, Corollary 3.12], we have $\sigma^*([\tau_{2(m-l)}]) = \tau_{2(m-l)}$. In our case, the cohomology suspension σ^* is injective on the vector subspace of indecomposable elements. This implies that $[\tau_{2(m-l)}] = p_{m-l}$. We have the result. □

Acknowledgments. The author thanks Toshihiro Yamaguchi for several helpful conversations during the early stages of this work. He is grateful to an anonymous referee for many comments on the proof of Theorem 1.1, which tell him that the holonomy operation as well as the HBS model for a function space is a useful tool for the study of fibrations.

References

[1] E. H. Brown Jr and R. H. Szczarba, On the rational homotopy type of function spaces, Trans. Amer. Math. Soc. **349**(1997), 4931-4951.

[2] U. Buijs and A. Murillo, Basic constructions in rational homotopy theory of function spaces, Ann. Inst. Fourier (Grenoble) **56**(2006), 815-838.

[3] U. Buijs and A. Murillo, The rational homotopy Lie algebra of function spaces, Comment. Math. Helv. **83**(2008), 723-739.

[4] Y. Félix, S. Halperin and J. -C. Thomas, Rational Homotopy Theory, Graduate Texts in Mathematics **205**, Springer-Verlag.

[5] Y. Félix and J. -C. Thomas, The monoid of self-homotopy equivalences of some homogeneous spaces, Exposiotiones Math. **12**(1994), 305-322.

[6] J. -B. Gatsinzi, On the genus of elliptic fibrations, Proc. Amer. Math. Soc. **132**(2003), 597-606.

[7] V. K. A. M. Gugenheim and J. P. May, On the theory and applications of differential torsion products, Mem. Amer. Math. Soc. 142, 1974.

[8] A. Haefliger, Rational homotopy of space of sections of a nilpotent bundle, Trans. Amer. Math. Soc. **273**(1982), 609-620.

[9] S. Halperin, Finiteness in the minimal models of Sullivan, Trans. Amer. Math. Soc. **230**(1977), 173-199.

[10] V. Hauschild, Deformations and the rational homotopy of the monoid of fibre homotopy equivalences, Illinois Journal of Math. **37**(1993), 537-560.

[11] Y. Hirato, K. Kuribayashi and N. Oda, A function space model approach to the rational evaluation subgroups, Math. Z. **258**(2008), 521-555.

[12] T. Januszkiewicz and J. Kedra, Characteristic classes of smooth fibrations, preprint (2002) `arXiv:math/0209288v1`.

[13] J. Kedra and D. McDuff, Homotopy properties of Hamiltonian group actions, Geometry & Topology, **9**(2005), 121-162.

[14] K. Kuribayashi, On the mod p cohomology of the spaces of free loops on the Grassmann and Stiefel manifolds, J. Math. Soc. Japan 43(1991), 331-346.

[15] K. Kuribayashi, Module derivations and the adjoint acton of finite loop space, J. Math. Kyoto Univ. **39** (1999) 67–85.

[16] K. Kuribayashi, A rational model for the evaluation map, Georgian Mathematical Journal **13**(2006), 127-141.

[17] K. Kuribayashi, Rational visibility of a Lie group in the monoid of self-homotopy equivalences of a homogeneous space, preprint (2008), submitted, `http://marine.shinshu-u.ac.jp/kuri/dvi/visibility-neo3.pdf`.

[18] F. Lalonde and D. McDuff, Symplectic structures on fibre bundles, Topology, **42**(2003), 309-347.

[19] G. Lupton, Variations on a conjecture of Halperin, Homotopy and geometry (Warsaw, 1997), 115-135, Banach Center Publ., 45, Polish Acad. Sci., Warsaw, 1998.

[20] G. Lupton, Note on a conjecture of Stephen Halperin, Lecture Notes in Math., vol. 1440, Springer-Verlag (1990), 148-163.

[21] G. Lupton and J. Oprea, Cohomologically symplectic space: Toral actions and the Gottlieb group, Trans. Amer. Math. Soc. **347**(1995), 261-288.

[22] M. Markl, Towards one conjecture on collapsing of the Serre spectral sequence, Rend. Circ. Mat. Palermo (2) Suppl. **22**(1990), 151-159.

[23] H. Matsumura, Commutative ring theory, Cambridge Studies in Advanced Mathematics 8, Cambridge University Press, Cambridge, 1986.

[24] J.P.May, Classifying spaces and fibrations, Mem. Amer. Math. Soc. 155, 1975.
[25] J.P.May, Fiberwise localization and completion, Trans. Amer. Math. Soc. **258**(1980), 127-146.
[26] D. McDuff and D. Salamon, Introduction to Symplectic Topology, Oxford Mathematical Monographs, Clarendon Press, Oxford, 1995.
[27] W. Meier, Rational universal fibrations and flag manifolds. Math. Ann. **258**(1982), 329-340.
[28] W. Meier, Some topological properties of Kähler manifolds and homogeneous spaces, Math. Z. **183**(1983), 473-481.
[29] M. Mimura and H. Toda, Topology of Lie groups. II. Translations of Mathematical Monographs, 91. American Mathematical Society, Providence, RI, 1991.
[30] A. G. Reznikov, Characteristic classes in symplectic topology, Selecta Math. **3**(1997), 601–642.
[31] H. Shiga ans M. Tezuka, Rational fibrations, homogeneous spaces with positive Euler characteristics and Jacobians, Ann. Inst. Fourier (Grenoble) **37**(1987), 81-106.
[32] L. Smith, On the characteristic zero cohomology of the free loop space, Amer. J. Math. **103**(1981), 887-910.
[33] S. B. Smith, The rational homotopy Lie algebra of classifying spaces for formal two-stage spaces, J. Pure Appl. Algebra **160**(2001), 333–343.
[34] D. Tanré, Homotopie rationnelle: Modèles de Chen, Quillen, Sullivan, Lecture Notes in Math., no. 1025, Springer-Verlag, Berlin, 1983.
[35] J. -C. Thomas, Rational homotopy of Serre fibrations, Ann. Inst. Fourier (Grenoble) **31**(1981), 71-90.

Contemporary Mathematics
Volume **519**, 2010

On the Realizability of Gottlieb Groups

John Oprea and Jeff Strom

ABSTRACT. In this brief note, we examine various issues related to the realizability of Gottlieb groups by manifolds or finite complexes. We prove that any finitely generated abelian group is always realized by $G_1(M)$ for some closed manifold M. We also prove some first results about realizing higher Gottlieb groups.

1. Introduction

An element $\alpha \in \pi_n(X)$ is a **Gottlieb element** if there exists an extension A (called *an associated map*) in the diagram

$$\begin{array}{ccc} S^n \vee X & \xrightarrow{(\alpha, \mathrm{id}_X)} & X \\ \downarrow & \nearrow_{A} & \\ S^n \times X & & \end{array}$$

The set of all Gottlieb elements in $\pi_n(X)$ is a subgroup of $\pi_n(X)$ denoted $G_n(X)$. These groups were discovered and studied by Gottlieb in the early 1960's (see [**Go1, Go2, Op2**]) and have led to many interesting results in homotopy theory and fixed point theory (see [**FH, St, Ji**]). Since the original discovery of Gottlieb groups, the definition has been extended from $\pi_n(X)$ to general homotopy sets $[Z, X]$ and so-called *cyclic maps* $\alpha\colon Z \to X$ (see [**Var, Lim**]).

With this long and extensive history, it is surprising that the question of realizability of Gottlieb groups has never been addressed. In this paper we ask: which groups G can be realized as the Gottlieb group $G_n(X)$ for some space X? If X is an H-space, then the H-space multiplication immediately shows that every map $\alpha\colon Z \to X$ is cyclic. Since Eilenberg-MacLane spaces are H-spaces, we have

$$G_n(K(\pi, n)) = \pi_n(K(\pi, n)) \cong \pi,$$

so our question is trivial if we do not place any restrictions on the space X. For a finitely generated abelian group G, we ask

(1) Is there a finite CW complex X with $G_n(X) \cong G$?
(2) Is there a compact manifold M with $G_n(M) \cong G$?

©2010 American Mathematical Society

When the required spaces exist, we also ask how small the dimension can be. We show in Theorem 3.1 that for any finitely generated abelian group G, there is a compact manifold M with $G_1(M) \cong G$. For $n > 1$, our results are not as strong: for a given finite abelian group G, there is an N so that for each $n \geq N$, there is a compact manifold M with G isomorphic to a subgroup of $G_{2n}(M)$. We have not resolved the case of n odd, but our attempts have led us to the following question about stable homotopy groups.

PROBLEM 1.1. *Do the even dimensional stable homotopy groups $\pi_{\text{even}}(S^0)$ have an exponent at some prime p?*

We supposed the answer to be no, but were unable to find any evidence one way or the other. We show in Proposition 3.5 that if the answer is no, then every finitely generated abelian group is isomorphic to a subgroup of an odd-dimensional Gottlieb group $G_{2n+1}(X)$.

The proofs of the G_1 result and the case of n even are constructive, and so give upper bounds on the dimension of the space X required. We define a numerical invariant $d(G)$ of finitely generated abelian groups as the minimum dimension of all spaces X with $G_1(X) \cong G$. We find upper and lower estimates for $d(G)$ in terms of the structure of G and relate $d(G)$ to $\text{cat}(X)$, the Lusternik-Schnirelmann category of X.

The results of this paper are just the start of understanding what groups appear as Gottlieb groups of specific types of manifolds. Such an understanding will have important implications for questions such as what circle actions certain manifolds (e.g. symplectically aspherical manifolds) support. Furthermore, as with many algebraic invariants, realizability results for finite complexes or manifolds shed light on the powerful constraints imposed by geometry.

Acknowledgement. We thank Bill Dwyer for an important clarification about an earlier draft of this paper, and Bob Bruner, Fred Cohen and Doulas Ravenel for answering our questions about stable homotopy groups. We also thank the Midwest Topology Network for funding that allowed the second author to visit the first in September 2009 when this paper was completed.

2. Preliminaries

Let's begin by recalling some basic facts concerning Gottlieb groups which we will use in the sequel and which are particularly relevant for the realizability question.

PROPERTIES 2.1. *Let X and Y be two topological spaces.*

(1) *If $\alpha \in G_n(X)$, then $[\alpha, \beta] = 0$ for all $\beta \in \pi_\ell(X)$ for all ℓ where $[\alpha, \beta]$ is the Whitehead product. This follows because the map associated to α, $A\colon S^n \times X \to X$ gives a composition*

$$S^n \times S^\ell \xrightarrow{\text{id}\times\beta} S^n \times X \xrightarrow{A} X,$$

which is of type (α, β). Therefore, $[\alpha, \beta] = 0$ for any $\beta \in \pi_\ell(X)$ (see **[Whi]***).*

(2) *By (1), since the Whitehead product on the fundamental group is simply conjugation, The Gottlieb group $G_1(X)$ is contained in the center of the*

fundamental group (and hence is abelian). This means that the realizability question is reduced to the world of abelian groups.

(3) *If $X = K(\pi, 1)$, then $G_1(X) = \mathcal{Z}\pi$, the center of the fundamental group. Moreover, if X is a finite complex with $\chi(X) \neq 0$, then $\mathcal{Z}\pi = \{1\}$. This is known as* Gottlieb's theorem *and it provides insight into the structure of groups that can be the fundamental groups of finite $K(\pi, 1)$'s. See* [**Go1**].

(4) *The exponential correspondence applied to the diagram defining a Gottlieb element shows that $G_n(X)$ is equal to*

$$\text{Image}\left(\text{ev}_*\colon \pi_n(X^X, 1_X) \to \pi_n(X)\right),$$

where $\text{ev}\colon X^X \to X$ *is evaluation* $\text{ev}(f) = f(x_0)$ *at a specified basepoint $x_0 \in X$ (see* [**Go2, Op2**]*). This shows why the realizability question for Gottlieb groups has relevance for understanding the homotopy theory of mapping spaces. For instance, if X realizes $\mathbb{Z}/p \subseteq G_n(X)$, then either $\pi_n(X^X, 1_X)$ contains elements of order p or an infinite order element evaluates to an element of order p.*

(5) *$G_n(X \times Y) \cong G_n(X) \times G_n(Y)$ (see* [**Go2**]*). This shows that the realization problem reduces to understanding summands arising from the fundamental structure theorem for abelian groups.*

(6) *If X is an H-space (e.g. a Lie group), then the H-multiplication provides a composition $S^n \times X \to X \times X \to X$ giving $G_n(X) = \pi_n(X)$ for all n. In particular,*

$$G_n(K(G, n)) = \pi_n(K(G, n)) \cong G.$$

This says that, for the realizability question, we need to restrict the types of spaces we consider in order to avoid triviality.

(7) *Let $p\colon \widetilde{X} \to X$ be a covering map. The unique homotopy lifting property of covering maps shows that, for $\alpha \in \pi_1(\widetilde{X})$, $p_*(\alpha) \in G_1(X)$ implies that $\alpha \in G_1(\widetilde{X})$ (see* [**Go2**, Theorem 6.1]*). For higher degrees $n > 1$, using the fact that $p_*\colon \pi_n(\widetilde{X}) \to \pi_n(X)$ is an isomorphism, we have $G_n(X) \subseteq G_n(\widetilde{X})$ (see* [**Go2**, Theorem 6.2]*). These results show that the realizability question* up to subgroups *can be reduced to considering covering spaces.*

3. Realizing Groups as Gottlieb Groups

We show that for $n = 1$, every finitely generated abelian group is indeed the Gottlieb group of a closed manifold. Our result for $n > 1$ is considerably weaker. We show that if G is a finite abelian group, then for sufficiently large *even* integer n, G is a *subgroup* of some Gottlieb group $G_n(X)$.

3.1. The Case $n = 1$**.** The group $G_1(X)$ has proven to be not only the most useful Gottlieb group, but the most computable as well. In this sense, the following result is not surprising.[1]

THEOREM 3.1. *If G is a finitely generated abelian group, then there is a closed manifold X with $G_1(X) = \pi_1(X) \cong G$.*

[1] Marek Golasinski has informed us that he has independently proved the same result.

PROOF. Let $G \cong \mathbb{Z}^k \oplus \bigoplus_{j=1}^n \mathbb{Z}/m_j$. Because the torus T^k is a group, we have $G_1(T^k) = \pi_1(T^k) \cong \mathbb{Z}^k$. According to [**Op1**], if π is a finite group acting freely on an odd-dimensional sphere S^m, then $G_1(S^m/\pi) \cong Z\pi$, the center of π. Since any cyclic group is contained in S^1 and S^1 acts freely on S^3, there is a free action of $\mathbb{Z}/m_j$ on S^3 and $G_1(S^3/(\mathbb{Z}/m_j)) = \pi_1(S^3/(\mathbb{Z}/m_j)) \cong \mathbb{Z}/m_j$. We then set

$$X = T^k \times \prod_{j=1}^{n} \left(S^3/(\mathbb{Z}/m_j)\right).$$

Since the Gottlieb group of a product is the product of the Gottlieb groups, we see that $G_1(X) = \pi_1(X) \cong G$. □

3.2. Higher Dimensional Gottlieb Groups. We show that a finitely generated abelian group G is isomorphic to a subgroup of a Gottlieb group $G_{2n}(X)$ for sufficiently large n. We begin with some known facts.

PROPOSITION 3.2.

(1) *In any odd degree n and for each $m \in \mathbb{Z}^+$, $\mathbb{Z}^m$ can be realized as $G_n(X) \subseteq \pi_n(X)$ for a manifold X.*
(2) *In any even degree n and for each $m \in \mathbb{Z}^+$, $\mathbb{Z}^m$ can never be realized as $G_n(X) \subseteq \pi_n(X)$ for a finite complex X.*

PROOF. It is known (see [**Go2**]) that

$$G_{2k+1}(S^{2k+1}) = \begin{cases} 2\mathbb{Z} \subset \mathbb{Z} = \pi_{2k+1}(S^{2k+1}) & \text{for} \quad k \neq 0,1,3 \\ \mathbb{Z} = \pi_{2k+1}(S^{2k+1}) & \text{for} \quad k = 0,1,3. \end{cases}$$

(via the usual group and H-space structures). Hence, we see that $G_{2k+1}(X) = \mathbb{Z}$ is realized in all odd degrees by $G_{2k+1}(S^{2k+1})$. Let n be odd , and set $X = \prod_1^m S^n$, so that

$$G_n(X) \cong \prod_1^m G_n(S^n) \cong \mathbb{Z}^m.$$

In even degrees, we know by rational homotopy theory (see, [**FH**] or, for instance, [**Op2**] for an exposition) that $G_{2k}(X)$ must be a finite group when X is a finite complex. □

The realization question is thus reduced to realizing finite abelian groups. And since a finite abelian group is a product of cyclic groups, we are really reduced to finding – for each $m \in \mathbb{N}$ – spaces X for which $G_n(X) \cong \mathbb{Z}/m$. This in turn boils down to finding spaces with $G_n(X) \cong \mathbb{Z}/p^r$ for each prime p and each $r \in \mathbb{N}$. Now, certain known facts can help us with this problem. Namely, since the unitary group $U(k)$ is a topological group, we know that $G_s(U(k)) = \pi_s(U(k))$. Furthermore, by [**MT**, Corollary 6.14], $\pi_{2k}(U(k)) = \mathbb{Z}/k!$, so we obtain the following result.

THEOREM 3.3. *Let G be a finite abelian group. Then there is an integer $N > 0$ so that for each $n \geq N$ there is a space X_n with $G \subseteq G_{2n}(X_n)$.*

PROOF. Write $G \cong \bigoplus_{j=1}^m \mathbb{Z}/p_j^{r_j}$, let n_j be the smallest integer such that $(p_j^{r_j})|n_j!$, and let $N = \max\{n_j\}$. Since $\mathbb{Z}/k! \subseteq \pi_{2k}(U(k))$, we have $\mathbb{Z}/p_j^{r_j} \subseteq$

$\pi_{2n}(U(n))$ for $j = 1, \ldots, m$ and any $n \geq N$. Also, $\pi_k(U(n)) = G_k(U(n))$ for all k because $U(n)$ is an H-space. Consequently, if we set

$$X = \overbrace{U(n) \times U(n) \times \cdots \times U(n)}^{m}$$

then we have

$$G \cong \bigoplus_{j=1}^{m} \mathbb{Z}/p_j^{r_j} \subseteq \pi_{2n}(X) = G_{2n}(X).$$

Thus, G can be realized as a subgroup of a Gottlieb group $G_{2n}(X)$ if $n \geq N$. □

We have been unable to resolve the corresponding question for odd n, and so pose the following problem.

PROBLEM 3.4. *Is every finitely generated abelian group isomorphic to a subgroup of some $G_{2n+1}(X)$ for all sufficiently large n?*

Now, from [**GM**], we know that, if n is odd then $2\pi_k(S^n) \subseteq G_k(S^n)$ (and if $n = 1, 3, 7$, then $G_k(S^n) = \pi_k(S^n)$). In particular, we have an identification $G_k(S^n; p) = \pi_k(S^n; p)$ of p-primary components for any odd prime p and $k \geq 1$. As a consequence, a greater understanding of the homotopy groups of spheres will contain within it a concomitant understanding of the realizability of Gottlieb groups in the odd degree case.

Indeed, a positive solution to Problem 3.4 would follow from the following statement about the even stable stems $\pi_{2n}(S^0)$:

($\star$) *for each prime p and each $r \in \mathbb{N}$ there is an even number $k(p,r) \in 2\mathbb{Z}$ so that the stable homotopy group $\pi^S_{k(p,r)}(S^0)$ contains a copy of $\mathbb{Z}/p^r$.*

PROPOSITION 3.5. *If ($\star$) is true then the answer to Problem 3.4 is yes.*

PROOF. Assuming ($\star$) and given $G \cong \bigoplus_{j=1}^{m} \mathbb{Z}/p_j^{r_j}$, let

$$k_j = \begin{cases} k(p_j, r_j) & \text{if } p_j \text{ is an odd prime} \\ k(2, r_j + 1) & \text{if } p_j = 2. \end{cases}$$

Then set $N = \max(k_j) + 1$ so that

$$\mathbb{Z}/p_j^{r_j} \subseteq 2\pi_n(S^{n-k_j}) \subseteq G_n(S^{n-k_j}) \subseteq \pi_n(S^{n-k_j})$$

for any odd $n \geq N$. Setting $X = S^{n-k_1} \times S^{n-k_2} \times \cdots \times S^{n-k_m}$, we have

$$G \cong \bigoplus_{j=1}^{m} \mathbb{Z}/p_j^{r_j} \subseteq \bigoplus_{j=1}^{m} G_n(S^{n-k_j}) \cong G_n(X).$$

□

This is not to say that realizability necessarily depends on the solution to this old problem; there may well be new calculations for the homotopy groups of H-spaces or new fibrations with bases the unitary groups $U(n)$ that would allow the calculation $\pi_{2n}(U(n)) = \mathbb{Z}/n!$ to be transferred, via the connecting homomorphism, to the $(2n-1)^{\text{st}}$ homotopy group of the fibre. At this time, however, Problem 3.4 remains very much open.

4. A Numerical Invariant of Finitely Generated Abelian Groups

Theorem 3.1 suggests a numerical invariant of finitely generated abelian groups. Let us define, for such a group G,

$$d(G) = \min\{\dim(X) \mid G_1(X) \cong G\}.$$

The construction used in the proof of Theorem 3.1 gives an upper bound for $d(G)$. We can obtain a lower bound as well by recalling the following.

If X is a space with $H_1(X;\mathbb{Z})$ finitely generated, the *Hurewicz rank* of X is the number of $\mathbb{Z}$-summands of $H_1(X;\mathbb{Z})$ which are contained in $h(G_1(X))$, where $h\colon \pi_1(X) \to H_1(X;\mathbb{Z})$ is the Hurewicz map. (Note that Hurewicz rank is not simply $\mathrm{rank}(h(G_1(X))$ since $h(G_1(X))$ could contain multiples of generators of $\mathbb{Z}$ summands of $H_1(X;\mathbb{Z})$ which increase $\mathrm{rank}(h(G_1(X))$, but which do not increase Hurewicz rank.) According to [**Go5, Op3**], if the Hurewicz rank of X is s, then $X \simeq T^s \times Y$, where T^s is an s-torus.

Now recall that the *Lusternik-Schnirelmann category* of a space X is defined by setting $\mathrm{cat}(X) \leq n$ if there is an open covering of X, $U_1, \ldots, U_{n+1}$, with each U_i contractible to a point *in the space* X; and $\mathrm{cat}(X) = \infty$ if no such cover exists (see [**CLOT**]). The basic estimates for category are in terms of cuplength (for any coefficients) and dimension:

$$\mathrm{cup}(X) \leq \mathrm{cat}(X) \leq \dim(X).$$

Also, a simple argument based on the homotopy lifting property gives $\mathrm{cat}(\overline{X}) \leq \mathrm{cat}(X)$ for any covering $\overline{X} \to X$. We begin by relating the rank of the Gottlieb group $G_1(X)$ to $\mathrm{cat}(X)$.

THEOREM 4.1. *If $A \subseteq G_1(X)$ is a finitely generated free abelian group, then*

$$\mathrm{rank}(A) \leq \mathrm{cat}(X) \leq \dim(X).$$

If $G_1(X)$ is itself finitely generated, then $\mathrm{rank}(G_1(X)) \leq \mathrm{cat}(X) \leq \dim(X)$.

PROOF. First, let $A \subseteq G_1(X)$ be a finitely generated free abelian group with rank k. Let $p\colon \overline{X} \to X$ be the covering of X corresponding to $A \subseteq \pi_1(X)$. Then $G_1(\overline{X}) = \pi_1(\overline{X}) \cong A$ since $p_*(A) \subseteq G_1(X)$ by Property 2.1 (7). It follows that the Hurewicz rank of $\overline{X}$ is k and so $\overline{X} \simeq T^k \times Y$. Then properties of category give

$$k \leq k + \mathrm{cup}(Y) = \mathrm{cup}(T^k \times Y) \leq \mathrm{cat}(T^k \times Y) = \mathrm{cat}(\overline{X}) \leq \mathrm{cat}(X),$$

and verifies the result for $\mathrm{rank}(A) < \infty$; in particular when $\mathrm{rank}(G_1(X)) < \infty$. If $\mathrm{rank}(G_1(X)) = \infty$, then for each $n \in \mathbb{N}$ we can find a free abelian subgroup $A_n \subseteq G_1(X)$ with $\mathrm{rank}(A_n) = n$. Then Theorem 4.1 gives $n \leq \mathrm{cat}(X) \leq \dim(X)$. Since this is the case for *every* n, we conclude $\mathrm{cat}(X) = \dim(X) = \infty$. □

Now we come to our estimates for $d(G)$.

THEOREM 4.2. *Let G be a finitely generated abelian group. Suppose the free part of G is $\mathbb{Z}^k$ and that each Sylow subgroup of G has at most n cyclic summands. Then*

$$k \leq d(G) \leq k + 3n.$$

PROOF. The hypotheses on G allow us to write $G \cong \mathbb{Z}^k \oplus \bigoplus_{j=1}^n \mathbb{Z}/m_j$ with at most n finite cyclic summands. Then the construction used in the proof of Theorem 3.1 gives a $(k+3n)$-dimensional space X with $G_1(X) \cong G$. On the other hand, if X is a space with $G_1(X) \cong G$. then $\dim(X) \geq \mathrm{rank}(G) = k$ by Theorem 4.1. □

Theorem 4.2 implies the following computation for free abelian groups.

COROLLARY 4.3. *If G is a free abelian group, then* $d(G) = \mathrm{rank}(G)$.

We close this section with some observations.

REMARK 4.4. *Let* $G \cong \mathbb{Z}^k \oplus \bigoplus_{j=1}^n \mathbb{Z}/m_j$.

(1) *Suppose X is a space with $G_1(X) \cong G$. Then the cover corresponding to $G_1(X) \subseteq \pi_1(X)$ is $\tilde{X} \simeq T^k \times Y$ where $G_1(Y) \cong \bigoplus_{j=1}^n \mathbb{Z}/m_j$. While $\tilde{X}$ may not be a finite complex, it is finite dimensional and, since* $\dim(X) = \dim(\tilde{X}) = k + \dim(Y)$, *our efforts to find $d(G)$, should be concentrated on the torsion part of G.*

(2) *For a closed manifold Y, it is known* [**Op4**] *that* $\mathrm{cat}(Y) = \mathrm{rank}(G_1(Y)) = k$ *only when $Y = T^k$.*

(3) *It is a curious fact that the manifold* $X = T^k \times \prod_{j=1}^n \left(S^3/(\mathbb{Z}/m_j)\right)$ *used to realize the arbitrary G of Theorem 3.1 has* $\dim(X) = \mathrm{cat}(X)$, *a very special situation indeed.*

(4) *Suppose $G_1(X) \cong G$, with G as above. Then Theorem 4.1 gives* $k \leq \mathrm{cat}(X) \leq k + 3n$ *and hence* $0 \leq \mathrm{cat}(X) - \mathrm{rank}(G_1(X)) \leq 3n$. *That is, the number of finite cyclic factors, n, gives a bound on the difference between the category and the rank of the Gottlieb group.*

PROBLEM 4.5. *Determine $d(G)$ precisely for G any finitely presented abelian group.*

References

[CLOT] O. Cornea and G. Lupton and J. Oprea and D. Tanré, *Lusternik-Schnirelmann Category*, Amer. Math. Soc. Surveys and Mono. 103 (2003).

[FH] Y. Félix and S. Halperin, *Rational LS category and its applications* Trans. Amer. Math. Soc. 273 (1982), no. 1, 1–38.

[Go1] D. Gottlieb, *A certain subgroup of the fundamental group*, Amer. J. Math. 87 (1965) 840-856.

[Go2] D. Gottlieb, *Evaluation subgroups of homotopy groups*, Amer. J. of Math. 91 (1969) 729-756.

[Go3] D. Gottlieb, *On the construction of G-spaces and applications to homogeneous spaces*, Proc. Camb. Phil. Soc. 68 (1970) 321–327.

[Go4] D. Gottlieb, *The evaluation map and homology*, Michigan Math. J. 19 (1972), 289–297.

[Go5] D. Gottlieb, *Splitting off tori and the evaluation subgroup*, Israel J. Math. 66 (1989) 216-222.

[GM] M. Golasinski and J. Mukai, *Gottlieb groups of spheres*, Topology 47, no. 6, (2008) 399–430.

[Ji] B. J. Jiang, *Lectures on Nielsen fixed point theory*, Contemporary Mathematics, 14. American Mathematical Society, Providence, R.I., (1983).

[Lim] K. L. Lim, *On cyclic maps*, J. Austral. Math. Soc. (Series A) 32 (1982) 349–357.

[MT] M. Mimura and H. Toda, *Topology of Lie groups. I, II*, Translations of Mathematical Monographs, 91. American Mathematical Society, Providence, RI, (1991).

[Op1] J. Oprea, *Finite group actions on spheres and the Gottlieb group*, J. Korean Math. Soc. 28 (1991), no. 1, 65–78.

[Op2] J. Oprea, *Gottlieb groups, group actions, fixed points and rational homotopy*, Lecture Notes Series, 29. Seoul National University, Research Institute of Mathematics, Global Analysis Research Center, Seoul, 1995.

[Op3] J. Oprea, *A homotopical Conner-Raymond theorem and a question of Gottlieb*, Can. Math. Bull. 33 (1990) 219-229.

[Op4] J. Oprea, *Bochner-type theorems for the Gottlieb group and injective toral actions*, Lusternik-Schnirelmann category and related topics (South Hadley, MA, 2001), 175–180, Contemp. Math., 316, Amer. Math. Soc., Providence, RI, 2002.

[St] J. Stallings, *Centerless groups—an algebraic formulation of Gottlieb's theorem*, Topology 4 (1965) 129–134.

[Var] K. Varadarajan, *Generalized Gottlieb groups*, J. Indian Math. Soc. 33 (1969) 141–164.

[Whi] G. W. Whitehead, *Elements of Homotopy Theory*, Graduate Texts in Mathematics **61**, Springer-Verlag, 1978.

Department of Mathematics, Cleveland State University, Cleveland OH, 44115 USA

E-mail address: `j.oprea@csuohio.edu`

Department of Mathematics, Western Michigan University, Kalamazoo, MI, 49008-5200 USA

E-mail address: `Jeff.Strom@wmich.edu`

Contemporary Mathematics
Volume **519**, 2010

Localization of grouplike function and section spaces with compact domain

Claude L. Schochet and Samuel B. Smith

ABSTRACT. We show that standard results on the localization of function and section spaces due to Hilton-Mislin-Roitberg and Möller extend outside the CW category to the case of compact metric domain in the presence of a grouplike structure. We study applications in two cases directly generalizing the gauge group of a principal bundle. We prove an identity for the monoid $\mathrm{aut}(\xi)$ of fibre-homotopy self-equivalences of a Hurewicz fibration ξ — due to Gottlieb and Booth-Heath-Morgan-Piccinini in the CW category — in the compact case. This leads to an extended localization result for $\mathrm{aut}(\xi)$. We also obtain an extended localization theory for groups of sections $\Gamma(\zeta)$ of a fibrewise group ζ. We give two applications in rational homotopy theory.

1. Introduction

Recently, with various coauthors [**23, 19**], we studied the rational homotopy type of the topological group of unitaries of certain Banach algebras. In the functional analysis setting, it is natural to assume the domain of a function or section space is a compact metric space not necessarily of CW type. The results of these papers thus entailed extensions of the standard results on the localization of function and section spaces in the CW category to certain cases in which the domain is only a compact metric space. In this paper, we expand on these results to give an extended localization theory for certain grouplike function and section spaces arising as natural generalizations of the gauge group of a principal bundle

Let X be a space, G a connected CW topological group and $h\colon X \to BG$ a map. The *gauge group* $\mathcal{G}(P)$ corresponding to this data is the topological group of G-equivariant self-maps of $P\colon E \to X$ where P is the principal G-bundle induced by h. Fixing G and X, the gauge group classification problem is that of determining the number of distinct homotopy types or, alternately, the number of H-homotopy types corresponding to maps in $[X, BG]$. Complete results in special cases are given by Kono [**20**], Crabb-Sutherland [**6**] and Kono-Tsukuda [**21**] among others.

After rationalization, the gauge group classification problem admits a complete solution with considerable generality. By [**19**, Theorem D], when X is a compact

2000 *Mathematics Subject Classification.* 55P60,55P62, 55R70,55R10, 55Q52.
Key words and phrases. gauge group, fibrewise self-equivalences, fibrewise groups, fibrewise localization, rational homotopy theory .

©2010 American Mathematical Society

metric space and G is a homotopy finite, connected, CW group, the rationalization of the connected component of the identity of $\mathcal{G}(P)$ is H-commutative and independent of the classifying map $h\colon X \to BG$. When X is actually a finite CW complex, this result may be deduced from an identity for the gauge group due to Gottlieb, (1), below, combined with a localization result of Hilton-Mislin-Roitberg [**17**] for function spaces (see [**19**, Theorem 5.7]). Alternately, the result for X finite CW follows from a localization result for section spaces due to Möller [**24**] and a result of Crabb-Sutherland [**6**, Proposition 2.2]. The proof for X compact metric in [**19**] is based on extensions of these localization results. These extensions are obtained using a result of Eilenberg-Steenrod [**9**] which expresses $X = \varprojlim_j X_j$ as an inverse limit of finite complexes X_j. We apply this analysis here to give extended localization results for two natural generalizations of the gauge group.

The two generalizations we consider are related to two basic identities for the gauge group. First, let $\mathrm{map}(X,Y)$ denote the space of all continuous maps with the compact-open topology. Let $\mathrm{map}(X,Y;f)$ denote the path-component of a given map $f\colon X \to Y$. Then, for X a finite CW complex, there is an H-equivalence

$$\mathcal{G}(P) \simeq \Omega\mathrm{map}(X, BG; h) \tag{1}$$

(see [**12**] and [**1**, Proposition 2.4]). Second, let $\mathrm{Ad}(P)\colon E \times_G G^{\mathrm{ad}} \to X$ denote the adjoint bundle. Here the total space is the quotient of the product by the diagonal action where $G^{\mathrm{ad}} = G$ is a left-G space via the adjoint action and the projection is induced by the projection $E \to X$. This is an non-principal G-bundle. We have an isomorphism

$$\mathcal{G}(P) \cong \Gamma(\mathrm{Ad}(P)), \tag{2}$$

where the latter space is the group of sections with fibrewise multiplication (see [**1**, p.539]).

Now suppose given a Hurewicz fibration $\xi\colon E \to X$. We assume here and throughout that X has a distinguished, nondegenerate basepoint. By the fibre of ξ, we will mean the fibre over this basepoint. We consider the monoid $\mathrm{aut}(\xi)$ consisting of all fibre-homotopy self-equivalences of ξ covering the identity of X topologized as a subspace of $\mathrm{map}(E,E)$. The monoid $\mathrm{aut}(\xi)$ is a natural generalization of the gauge group to the fibre-homotopy setting. Unlike the gauge group, the rational H-homotopy type of this monoid is generally a nontrivial invariant of the fibre-homotopy theory of ξ (see [**11**]).

By [**2**, Theorem 3.3], the identity (1) extends to a corresponding identity for $\mathrm{aut}(\xi)$ when both X and the fibre F are finite CW complexes. Our first main result extends this identity, in turn, to the case X is compact metric. Recall the universal F-fibration, for F finite CW, may be identified, up to homotopy type, as a sequence $F \to B\mathrm{aut}_*(F) \to B\mathrm{aut}(F)$ where $\mathrm{aut}(F)$ and $\mathrm{aut}_*(F)$ are the monoids of free and based homotopy self-equivalences of F. The spaces $B\mathrm{aut}(F)$ and $B\mathrm{aut}_*(F)$ are the Dold-Lashof classifying spaces for these monoids. (See [**32, 25**] for the classification theory and [**13**] for the identification with Dold-Lashof [**8**].)

THEOREM 1. *Let X be a compact metric space, F a finite CW complex and $h\colon X \to B\mathrm{aut}(F)$ a map. Let $\xi\colon E \to X$ be the corresponding F-fibration. Then there is an H-equivalence*

$$\mathrm{aut}(\xi) \simeq \Omega\mathrm{map}(X, B\mathrm{aut}(F); h).$$

As a consequence, we obtain an extended localization theory for $\mathrm{aut}(\xi)$. Let $\mathbb{P}$ be a collection of primes. We say a space Y is *nilpotent* if Y is connected, has the homotopy type of a CW complex and has a nilpotent homotopy system (see [**17**, Definition II.2.1]). In this case, Y admits a $\mathbb{P}$-localization $\ell_Y\colon Y \to Y_{\mathbb{P}}$ [**17**, Theorem II.3A]. Given a map $f\colon X \to Y$ we write $f_{\mathbb{P}} = \ell_Y \circ f\colon X \to Y_{\mathbb{P}}$. Given a monoid G we write $G_\circ$ for the path component of the identity. Given a space Z with distinguished basepoint we write $\Omega_\circ Z$ for the space of loops based at the basepoint. For the function space $\mathrm{map}(X, Y; f)$ we assume f is the basepoint.

THEOREM 2. *Let X be a simply connected compact metric space, F a finite CW complex and $h\colon X \to B\mathrm{aut}(F)$ a map. Let $\xi\colon E \to X$ be the corresponding F-fibration. Then $\mathrm{aut}(\xi)_\circ$ is a nilpotent space and we have an H-equivalence*

$$(\mathrm{aut}(\xi)_\circ)_{\mathbb{P}} \simeq \Omega_\circ \mathrm{map}(X, (B\mathrm{aut}(F)_\circ)_{\mathbb{P}}; (\tilde{h})_{\mathbb{P}}).$$

Here $B\mathrm{aut}(F)_\circ$ — the Dold-Lashof classifying space of $\mathrm{aut}(F)_\circ$ — is the universal cover of $B\mathrm{aut}(F)$ and $\tilde{h}\colon X \to B\mathrm{aut}(F)_\circ$ is the unique lifting of h.

The second generalization we consider is based on (2) in which the gauge group corresponds to a group of sections. Recall $\zeta\colon E \to X$ is a *fibrewise group* if there is a fibrewise map $m\colon E \times_X E \to E$ over X, a section $e\colon X \to E$ and a map $i\colon E \to E$ over X satisfying: (i) m is associative, (ii) e is a two-sided unit and (iii) i is an inverse with respect to the maps m and e. The space of sections $\Gamma(\zeta)$ is then a topological group with the multiplication of sections induced by m. More generally, relaxing the group axioms to require identities only up to homotopy, ζ is a *fibrewise grouplike space* ([**5**, p.62]) and $\Gamma(\zeta)$ is a grouplike space. Our motivating example is the adjoint bundle $\mathrm{Ad}(P)$ of a principal G-bundle P, as above. Observe that, if $P\colon E \to X$ is classified by a map $h\colon X \to BG$, then $\mathrm{Ad}(P)$ is the pullback by h of the universal G-adjoint bundle $EG \times_G G^{\mathrm{ad}} \to BG$ which is, in particular, a CW fibration.

Suppose generally that $\zeta\colon E \to X$ is a fibrewise grouplike space with connected grouplike fibre G and, further, that ζ is the pullback of a CW fibration. In this case, we may still identify a fibrewise $\mathbb{P}$-localization $\zeta \to \zeta_{(\mathbb{P})}$ of ζ (see the remarks preceding Theorem 2.4, below). Our third main result extends [**24**, Theorem 5.3] from the case the base space is finite CW to the case of compact metric base in this context.

THEOREM 3. *Let $\zeta\colon E \to X$ be a fibrewise grouplike space with connected, CW grouplike fibre G and base X a compact metric space. Suppose ζ is the pullback of a CW fibration. Then $\Gamma(\zeta)_\circ$ is a nilpotent space and the map*

$$\Gamma(\zeta)_\circ \to \Gamma(\zeta_{(\mathbb{P})})_\circ$$

induced by a fibrewise $\mathbb{P}$-localization $\zeta \to \zeta_{(\mathbb{P})}$ is a $\mathbb{P}$-localization map.

REMARK 1.1. In many circumstances, the fibrewise $\mathbb{P}$-localization $\zeta_{(\mathbb{P})}$ of a fibrewise grouplike space $\zeta\colon E \to X$ will itself be a fibrewise grouplike space over X and the fibre map $\zeta \to \zeta_{(\mathbb{P})}$ will be equivariant. In this case, the equivalence of Theorem 3 is actually an H-equivalence. For example, this is the case for the adjoint bundle and, generally, when ζ is a CW fibration. For general compact metric X, the fibrewise grouplike structure for $\zeta_{(\mathbb{P})}$ is not assured due to the lack of uniqueness of fibrewise $\mathbb{P}$-localization outside the CW category.

The paper is organized as follows. We prove Theorems 1-3 in Section 2. In Section 3, we apply our results to obtain two consequences. We prove that the rationalization of $\mathrm{aut}(\xi)_\circ$ is H-commutative and independent of the classifying map for fibrations ξ with fibre satisfying a famous conjecture in rational homotopy theory (Theorem 3.3). This result extends [**10**, Theorem 4]. As an application of Theorem 3, we extend [**19**, Theorem F] on the classification of projective gauge groups (Theorem 3.6). In Section 4, we deduce based versions of our main results.

Acknowledgements. We are grateful to our coauthors in the papers [**11, 19, 23**] for their part in discovering these ideas. We thank Peter Booth for very helpful discussions of his work.

2. Localization of Function and Section Spaces

The standard results on $\mathbb{P}$-localization of function spaces are as follows. By Milnor [**27**], the path components of $\mathrm{map}(X, Y)$ are of CW homotopy type when X is a compact metric space and Y is a CW complex. By Hilton-Mislin-Roitberg, when X is finite CW and Y is a nilpotent space then $\mathrm{map}(X, Y; f)$ is itself a nilpotent space and

$$\mathrm{map}(X, Y; f)_{\mathbb{P}} \simeq \mathrm{map}(X, Y_{\mathbb{P}}; f_{\mathbb{P}})$$

[**17**, Theorem II.3.11]. Recall we write $f_{\mathbb{P}} = \ell_Y \circ f \colon X \to Y_{\mathbb{P}}$.

These results all hold for spaces of basepoint-preserving functions, as well. In what immediately follows, we will focus on the basepoint free case. We discuss the based case in Section 4. The preceding results also hold with alternate hypotheses: namely, X may be any CW complex when Y is a finite Postnikov piece (see [**18**]). We will not consider these alternate hypotheses here as they are not amenable to our methods.

The results of Hilton-Mislin-Roitberg were extended to section spaces by Möller [**24**] (see also, Scheerer [**28**]). Let $\xi \colon E \to X$ be a fibration of connected CW complexes with X finite. Let F be the fibre over a basepoint of X. If F is a nilpotent space then ξ admits a fibrewise $\mathbb{P}$-localization which is a fibre map $\xi \to \xi_{(\mathbb{P})}$ over X inducing $\mathbb{P}$-localization on the fibres [**26**]. By [**24**, Theorem 5.3], under these hypotheses, the component of $\Gamma(\xi; s)$ corresponding to a given section $s \colon X \to E$ is a nilpotent space and

$$\Gamma(\xi; s)_{\mathbb{P}} \simeq \Gamma(\xi_{(\mathbb{P})}; s_0)$$

where s_0 is the section of $\xi_{(\mathbb{P})}$ induced by s.

Our extensions of these localization results depend on a classical result of Eilenberg-Steenrod in [**9**]. Let X be a compact metric space. By [**9**, Theorem X.10.1], there is an inverse system of finite complexes X_j with structure maps $g_{ij} \colon X_j \to X_i$ for $i \leq j$ and compatible maps $g_j \colon X \to X_j$ such that the induced map

$$g \colon X \to \varprojlim_j X_j$$

is a homeomorphism. Further, given any map $f \colon X \to Y$ for Y a CW complex, there is an index m and a map $f_m \colon X_m \to Y$ such that f is homotopic to $f_m \circ g_m$. [**9**, Theorem X.11.9].

We apply this result to study the function space $\mathrm{map}(X, Y; f)$, as follows. Choose m for $f \colon X \to Y$ as above. Given $j \geq m$ define $f_j \colon X_j \to Y$ by setting $f_j =$

$f_m \circ g_{mj}$. Restricting to indices $j \geq m$, we obtain a direct system $\mathrm{map}(X_j;Y;f_j)$ with structure maps $(g_{ij})^*\colon \mathrm{map}(X_i,Y;f_i) \to \mathrm{map}(X_j,Y;f_j)$ and compatible maps

$$(g_j)^*\colon \mathrm{map}(X_j,Y;f) \to \mathrm{map}(X,Y;f_j)$$

both induced by precomposition. We have:

THEOREM 2.1. ([**19**, Theorem 6.4]) *Let X be a compact metric space and Y a CW complex. Then, for all $n \geq 1$, the maps $(g_j)^*$ above induce an isomorphism*

$$\varinjlim_j \pi_n(\mathrm{map}(X_j,Y;f_j)) \cong \pi_n(\mathrm{map}(X,Y;f)).$$

□

This result leads to an extension of the Hilton-Mislin-Roitberg result, mentioned above, provided the space $\mathrm{map}(X,Y;f)$ is known *a priori* to be a nilpotent space [**19**, Theorem 7.1]. Essentially the same proof yields the following:

THEOREM 2.2. *Let X be a compact metric space. Let Y be a nilpotent space with $\mathbb{P}$-localization $\ell_Y\colon Y \to Y_{\mathbb{P}}$. Let $f\colon X \to Y$ be a given map. Then ℓ_Y induces an H-equivalence*

$$(\Omega_\circ \mathrm{map}(X,Y;f))_{\mathbb{P}} \simeq \Omega_\circ \mathrm{map}(X,Y_{\mathbb{P}};f_{\mathbb{P}}).$$

PROOF. Consider the commutative square

$$\begin{array}{ccc}
\varinjlim_j \pi_n(\mathrm{map}(X_j,Y;f_j)) & \xrightarrow{\cong} & \pi_n(\mathrm{map}(X,Y;f)) \\
\downarrow & & \downarrow \\
\varinjlim_j \pi_n(\mathrm{map}(X_j,Y_{\mathbb{P}};(f_j)_{\mathbb{P}})) & \xrightarrow{\cong} & \pi_n(\mathrm{map}(X,Y_{\mathbb{P}};f_{\mathbb{P}}))
\end{array}$$

with vertical maps induced by ℓ_Y. That the left vertical map is a $\mathbb{P}$-localization map follows from [**17**, Theorem II.3.11]. Thus the right vertical map is a $\mathbb{P}$-localization map, as well. Looping, we see that ℓ_Y induces a weak H-equivalence $(\Omega_\circ \mathrm{map}(X,Y;f))_{\mathbb{P}} \simeq \Omega_\circ \mathrm{map}(X,Y_{\mathbb{P}};f_{\mathbb{P}})$ and so an honest H-equivalence since the spaces are CW, again by [**27**] . □

Next let $\xi\colon E \to X$ be a Hurewicz fibration over a compact metric space with a fixed section $s\colon X \to E$. Suppose ξ is the pullback of a CW fibration $\xi_0\colon E_0 \to B$ for some map $h\colon X \to B$. Then $\Gamma(\xi;s)$ is of CW homotopy type [**19**, Proposition 3.2] Given an inverse system of finite complexes X_j for X with compatible maps $g_j\colon X \to X_j$ as above, choose an index m so that $h\colon X \to B$ factors as $h = h_m \circ g_m$ for some map $h_m\colon X_m \to B$. Given $j \geq m$, write $\xi_j\colon E_j \to X_j$ for the pullback of the fibration ξ_0 via the map $h_j = h_m \circ g_{mj}\colon X_j \to B$. Restricting again to indices $j \geq m$ gives a direct system of spaces of sections $\Gamma(\xi_j)$ with structure maps $\gamma_{ij}\colon \Gamma(\xi_i) \to \Gamma(\xi_j)$ and compatible maps $\gamma_j\colon \Gamma(\xi_j) \to \Gamma(\xi)$ induced by g_{ij} and g_j for $j \geq i \geq m$.

After further reindexing, the direct system $\Gamma(\xi_j)$ gives rise to a corresponding system of connected components $\Gamma(\xi_j;s_j)$. To see this, consider the diagram:

$$\begin{array}{ccccccc}
E & & & \xrightarrow{\widetilde{h}} & & & E_0 \\
{\scriptstyle s}\uparrow\downarrow & & & & & \nearrow^{\sigma_m} & \downarrow{\scriptstyle p} \\
X & \longrightarrow & X_{m'} & \xrightarrow[g_{mm'}]{} & X_m & \xrightarrow[h_m]{} & B.
\end{array}$$

Since E_0 is a CW complex, $\widetilde{h} \circ s$ factors through X_m for some m. Since $\varinjlim_j [X_j, B] = [X, B]$, there exists an index $m' \geq m$ such that $p \circ \sigma_m \circ g_{mm'} \simeq h_{m'}$ and so we get a section $s_{m'} \in \Gamma(\xi_{m'})$ with $\gamma_{m'}(s_{m'}) \simeq s \in \Gamma(\xi)$. Restricting now to indices $j \geq m'$, we have a direct system of connected spaces $\Gamma(\xi_j; s_j)$ with compatible maps $\gamma_j \colon \Gamma(\xi_j; s_j) \to \Gamma(\xi; s)$ for $j \geq m'$. We note that, in fact,

$$\varinjlim \pi_0(\Gamma(\xi_j)) \cong \pi_0(\Gamma(\xi)).$$

This result is a particular case of [**19**, Theorem 6.5] which gives this isomorphism in all degrees. Quoting this theorem in higher degrees, we have:

THEOREM 2.3. ([**19**, Theorem 6.5]) *Let $\xi \colon E \to X$ be a Hurewicz fibration over X a compact metric space with a fixed section $s \colon X \to E$. Suppose ξ is the pullback of a CW fibration. Then, for all $n \geq 1$, the maps γ_j above induce an isomorphism*

$$\varinjlim_j \pi_n(\Gamma(\xi_j; s_j)) \cong \pi_n(\Gamma(\xi; s)).$$

□

By the work of May [**26**], a fibration $\xi \colon E \to X$ of CW complexes with nilpotent fibre F admits a *fibrewise* $\mathbb{P}$*-localization* which is a fibration $\xi_{(\mathbb{P})} \colon E_0 \to X$ and a map $g \colon E \to E_0$ over X such that g induces $\mathbb{P}$-localization $F \to F_{\mathbb{P}}$ on fibres. This fibrewise $\mathbb{P}$-localization of ξ is unique up to fibre-homotopy equivalence by Llerena [**22**, Theorem 6.1].

We may directly extend this construction to non-CW fibrations which are nevertheless the pullback of an appropriate CW fibration. Specifically, let $\xi \colon E \to X$ be the pullback of a fibration $\xi_0 \colon E_0 \to B$ of CW complexes with nilpotent fibre via a map $h \colon X \to B$. We take $\xi_{(\mathbb{P})} = h^{-1}((\xi_0)_{(\mathbb{P})})$, the pullback of the fibrewise $\mathbb{P}$-localization of ξ_0. We note that uniqueness is no longer assured.

THEOREM 2.4. *Let $\xi \colon E \to X$ be a fibration of spaces with nilpotent fibre and compact metric base. Suppose ξ is a pullback of a CW fibration and that, for some section s of ξ, the component $\Gamma(\xi; s)$ is a nilpotent space. Let $\xi \to \xi_{(\mathbb{P})}$ be a fibrewise $\mathbb{P}$-localization and s_0 the induced section. Then*

$$\Gamma(\xi; s)_{\mathbb{P}} \simeq \Gamma(\xi_{(\mathbb{P})}; s_0).$$

PROOF. Choose an inverse system X_j of finite complexes for X and let ξ_j be the corresponding fibrations over X_j with fibre F and compatible sections s_j. Consider the commutative diagram, as in the proof of Theorem 2.2:

$$\begin{array}{ccc} \varinjlim_j \pi_n(\Gamma(\xi_j; s_j)) & \xrightarrow{\cong} & \pi_n(\Gamma(\xi; s)) \\ \downarrow & & \downarrow \\ \varinjlim_j \pi_n(\Gamma((\xi_j)_{(\mathbb{P})}; (s_j)_0)) & \xrightarrow{\cong} & \pi_n(\Gamma(\xi_{(\mathbb{P})}; s_0)) \end{array}$$

The horizontal maps are isomorphisms by Theorem 2.3. The left vertical map is $\mathbb{P}$-localization by [**24**, Theorem 5.3]. Thus the right vertical map is a $\mathbb{P}$-localization, as well. □

Theorem 2.4 implies:

PROOF OF THEOREM 3. Recall we are assuming $\zeta\colon E \to X$ is fibrewise grouplike over a compact metric space X with fibre G a connected CW grouplike space. Further, we assume ζ is the pullback of CW fibration. By [**19**, Proposition 3.2], $\Gamma(\zeta)_\circ$ is of CW type and so a nilpotent space since grouplike. The result is thus a direct consequence of Theorem 2.4. $\square$

We now turn to the proof of Theorems 1 and 2. Fix a compact metric space X and a finite CW complex F. We write $\xi_\infty\colon E_\infty \to B\mathrm{aut}(F)$ for the universal F-fibration. (Here $E_\infty \simeq B\mathrm{aut}_*(F)$.) Given a map $h\colon X \to B\mathrm{aut}(F)$, let $\xi\colon E \to X$ be the pullback. In [**2**], the authors define a fibration

$$\xi \,\square_1 \xi_\infty\colon E \,\square E_\infty \to X$$

in which the total space is the set of all homotopy equivalences $f_{a,b}\colon E_a \to (E_\infty)_b$ between fibres of ξ and fibres of ξ_∞ suitably topologized. The projection is given by $f_{a,b} \mapsto a \in X$ and so the fibre over a basepoint of X may be identified with $\mathrm{aut}(F)$. Next, the authors define a fibration

$$\Phi\colon \Gamma(\xi \,\square_1 \xi_\infty) \to \mathrm{map}(X, B\mathrm{aut}(F))$$

with fibre $\Phi^{-1}(h) \approx \mathrm{aut}(\xi)$. We observe that the construction of Φ and the identification of the fibre both hold in our case. For the former, the key is [**2**, Proposition 2.1] whose proof depends on the use of exponential laws for functional fibrations which hold for X compact by [**3**]. The identification of the fibre over h with $\mathrm{aut}(\xi)$ requires that ξ be an $\mathcal{F}$-fibration in the sense of [**25**, Definition 2.1] where $\mathcal{F}$ is the category of "fibres" homotopic to F. This fact is a consequence of our assumption that ξ is the pullback of a CW fibration (use [**25**, Lemma 3.3 and Proposition 3.4]).

LEMMA 2.5. *The space* $\Gamma(\xi \,\square_1 \xi_\infty)$ *is contractible.*

PROOF. Observe that $\xi \,\square_1 \xi_\infty$ is the pullback of the fibration

$$\xi_\infty \,\square_1 \xi_\infty\colon E_\infty \,\square E_\infty \to B\mathrm{aut}(F)$$

by $h\colon X \to B\mathrm{aut}(F)$. The fibration $\xi_\infty \,\square_1 \xi_\infty$ is a CW fibration by Schön [**30**] since both the base $B\mathrm{aut}(F)$ and the fibre $\mathrm{aut}(F)$ are CW since F is finite CW. Thus $\Gamma(\xi \,\square_1 \xi_\infty)$ is of CW type [**19**, Proposition 3.2]. It thus suffices to show $\Gamma(\xi \,\square_1 \xi_\infty)$ is weakly contractible.

For this, we apply Theorem 2.3. Following the procedure described before this result, write $X = \varprojlim_j X_j$ with each X_j a finite complex. For $j \geq m$ as above, let $\xi_j\colon E_j \to X_j$ be the corresponding fibration over X_j. Let $\xi_j \,\square_1 \xi_\infty\colon E_j \,\square E_\infty \to X_j$ the associated construction. Functoriality of this construction gives fibre maps $\xi_j \,\square_1 \xi_\infty \to \xi_k \,\square_1 \xi_\infty$ for $j \geq k \geq m$ and compatible fibre maps $\xi \,\square_1 \xi_\infty \to \xi_j \,\square_1 \xi_\infty$. We thus obtain a direct system $\Gamma(\xi_j \,\square_1 \xi_\infty)$ of sections with compatible maps $\gamma_j\colon \Gamma(\xi_j \,\square_1 \xi_\infty) \to \Gamma(\xi \,\square_1 \xi_\infty)$ satisfying the conditions of Theorem 2.3. By [**2**, Proposition 3.1], each space $\Gamma(\xi_j \,\square_1 \xi_\infty)$ is contractible. Thus

$$\pi_n(\Gamma(\xi \,\square_1 \xi_\infty)) \cong \varinjlim_j \pi_n(\Gamma(\xi_j \,\square_1 \xi_\infty)) = 0,$$

for $n \geq 0$, as needed. $\square$

We obtain, as a consequence:

PROOF OF THEOREM 1. Let X be a compact metric space, F a finite CW complex and $h\colon B \to B\mathrm{aut}(F)$ a map. Let $\xi\colon E \to X$ be the corresponding F-fibration. By Lemma 2.5, the connecting map

$$\delta\colon \Omega\mathrm{map}(X, B\mathrm{aut}(F); h)) \to \Phi^{-1}(h) \approx \mathrm{aut}(\xi)$$

in the Barratt-Puppe sequence for $\Phi\colon \Gamma(\xi \square_1 \xi_\infty) \to \mathrm{map}(X, B\mathrm{aut}(X))$ is a homotopy equivalence. By [**2**, Theorem 3.3], δ is a multiplicative map and so gives the desired H-equivalence. □

Using Theorem 1 we obtain:

PROOF OF THEOREM 2. Recall we are assuming X is a simply connected, compact metric space and $\xi\colon E \to X$ is the pullback of the universal F-fibration via $h\colon X \to B\mathrm{aut}(X)$. Since X is simply connected, h lifts to a map $\tilde{h}\colon X \to B\mathrm{aut}(X)_\circ$ to the universal cover. Further, we have an H-equivalence

$$\Omega_\circ\mathrm{map}(X, B\mathrm{aut}(F); h) \simeq \Omega_\circ\mathrm{map}(X, B\mathrm{aut}(F)_\circ; \tilde{h})$$

as these spaces are evidently weakly H-equivalent and both are CW complexes [**27**]. The result now follows from Theorem 2.2 and Theorem 1 □

3. Consequences in Rational Homotopy Theory

There are good algebraic models for the rational homotopy theory of the monoid $\mathrm{aut}(\xi)$ and the space of sections $\Gamma(\zeta)$. (See [**11**] for the former and [**16, 4**] for the latter.) However, these models require finiteness and/or nilpotence conditions on the fibration and so are not directly applicable to the case of compact domain. In our main results above, we require that ξ and ζ occur as pullbacks of CW fibrations. In this section, we observe that, in certain special circumstances, the structure of the rationalization of these CW fibrations allow for a description of the rational homotopy theory of these grouplike spaces of maps.

For the monoid $\mathrm{aut}(\xi)$, the CW fibration in question is the universal F-fibration. The following result is the $\mathbb{P}$-local version of [**2**, Proposition 6.1] in our setting.

THEOREM 3.1. *Let X be a simply connected, compact metric space and F a finite complex. Suppose the $\mathbb{P}$-localization of $B\mathrm{aut}(F)_\circ$ is a grouplike space. Then, for all fibrations ξ corresponding to maps in $[X, B\mathrm{aut}(F)]$, we have an H-equivalence*

$$(\mathrm{aut}(\xi)_\circ)_\mathbb{P} \simeq \mathrm{map}(X, (\mathrm{aut}(F)_\circ)_\mathbb{P}; 0)$$

where the latter is the space of null maps into the grouplike space $(\mathrm{aut}(F)_\circ)_\mathbb{P}$ which is a grouplike space with pointwise multiplication.

PROOF. The proof is the same as that given in [**11**, Example 4.7]. We give the details for completeness. Using the homotopy inverse for $(B\mathrm{aut}(F)_\circ)_\mathbb{P}$, we obtain a homotopy equivalence between $\mathrm{map}(X, (B\mathrm{aut}(F)_\circ)_\mathbb{P}; \tilde{h}_\mathbb{P})$ and $\mathrm{map}(X, (B\mathrm{aut}(F)_\circ)_\mathbb{P}; 0)$. Looping this equivalence and applying Theorem 2 yields H-equivalences:

$$(\mathrm{aut}(\xi)_\circ)_\mathbb{P} \simeq \Omega_\circ\mathrm{map}(X, (B\mathrm{aut}(F)_\circ)_\mathbb{P}; \tilde{h}_\mathbb{P}) \simeq \Omega_\circ\mathrm{map}(X, (B\mathrm{aut}(F)_\circ)_\mathbb{P}; 0)$$

At the null component, we may pull $\Omega_\circ$ inside to get

$$\Omega_\circ\mathrm{map}(X, (B\mathrm{aut}(F)_\circ)_\mathbb{P}; 0) \simeq \mathrm{map}(X, \Omega_\circ(B\mathrm{aut}(F)_\circ)_\mathbb{P}; 0) \simeq \mathrm{map}(X, (\mathrm{aut}(F)_\circ)_\mathbb{P}; 0),$$

as needed. □

The condition on $(B\mathrm{aut}(F)_\circ)_{\mathbb{P}}$ is, of course, very strong and will be rarely satisfied. However, as we explain now, there is a famous class of spaces in rational homotopy theory, identified by Halperin in [**15**], that conjecturally (and in many known cases) do have this property after rationalization. In the rational case, observe that it is sufficient to check $(B\mathrm{aut}(F)_\circ)_{\mathbb{Q}}$ is an H-space to apply Theorem 3.1 since, rationally, all H-spaces are grouplike (see, e.g., [**29**]).

We say a space F is an F_0-*space* if F is a simply connected elliptic space (i.e., a finite complex with $\pi_*(F)\otimes\mathbb{Q}$ finite-dimensional) and satisfying $H^{\mathrm{odd}}(F;\mathbb{Q})=0$. Halperin's conjecture (see [**15**]) translated to our setting is:

CONJECTURE 3.2. (Halperin) *Let F be an F_0-space. Then the monoid* $\mathrm{aut}(F)_\circ$ *has vanishing even degree rational homotopy groups.*

Examples of F_0-spaces include products of even-dimensional spheres and complex projective spaces for which Conjecture 3.2 is easily confirmed. Homogeneous spaces G/H of equal rank, compact pairs $H\subseteq G$ are also F_0-spaces and satisfy Conjecture 3.2 by [**31**].

We note that, if F is an F_0-space satisfying Conjecture 3.2, then $(B\mathrm{aut}(F)_\circ)_{\mathbb{Q}}$ is a rational H-space since the Sullivan model of a nilpotent space with no odd rational homotopy has trivial differential. Further, the (odd) rational homotopy groups of $\mathrm{aut}(F)_\circ$ can be, in practice, directly computed from the minimal model $(\wedge V;d)$ of F using Sullivan's identity

$$\pi_k(\mathrm{aut}(F)_\circ)\otimes\mathbb{Q}\cong H_k(\mathrm{Der}(\wedge V;d)).$$

(See, e.g., [**14**].) Here the latter space is the homology of the DG vector space of negative degree derivations of $(\wedge V;d)$.

The following result extends [**10**, Theorem 4] and [**11**, Example 2.7]. We write $\check{H}^*(X;\mathbb{Q})$ for rational Čech cohomology.

THEOREM 3.3. *Let X be a simply connected, compact metric space and F an F_0-space satisfying Conjecture 3.2. Let $h\colon X\to B\mathrm{aut}(F)$ be a map and ξ the corresponding fibration over X. Then:*

(1) *The rational H-homotopy type of the monoid* $\mathrm{aut}(\xi)_\circ$ *is independent of the classifying map h.*

(2) *The monoid* $\mathrm{aut}(\xi)_\circ$ *is rationally H-commutative and equivalent to a product of Eilenberg-Mac Lane spaces with homotopy groups given by*

$$\pi_n(\mathrm{aut}(\xi)_\circ)\otimes\mathbb{Q}\cong\bigoplus_{k\geq n}\check{H}^{k-n}(X;\mathbb{Q})\otimes(\pi_k(\mathrm{aut}(F)_\circ)\otimes\mathbb{Q})$$

for $n\geq 1$.

PROOF. The first statement follows directly from the preceding paragraph and Theorem 3.1. As for (2), we note $\pi_*(\mathrm{aut}(F)_\circ)\otimes\mathbb{Q}$ is oddly graded and so, for degree reasons, does not admit any Samelson products. It follows that $\mathrm{aut}(F)_\circ$ is rationally H-commutative [**29**]. Thus $\mathrm{aut}(\xi)_\circ\simeq_{\mathbb{Q}}\mathrm{map}(X,\mathrm{aut}(F)_\circ;0)$ is rationally H-commutative, as well by [**23**, Theorem 4.10].

The computation of the rational homotopy groups of $\mathrm{map}(X,\mathrm{aut}(F);0)$ reduces to the problem of computing homotopy groups of the space of maps into an Eilenberg-Mac Lane space. Here we have the identity of Thom [**33**]:

$$\pi_q(\mathrm{map}(X,K(\pi,n)))\cong H^{n-q}(X;\pi)$$

which holds, with ordinary cohomology when X is CW. For compact spaces, the same identity holds with rational Čech cohomology by writing $X = \varprojlim X_j$ and using the continuity of Čech cohomology (see the proof of [**23**, Theorem 5.6].) □

We prove a related result for grouplike spaces of sections. We say a fibration $\xi\colon E \to X$ of connected CW complexes is *nilpotent* if the spaces E and X are nilpotent. In this case, F is also nilpotent by [**17**, Theorem II.2.2] and term-by-term $\mathbb{P}$-localization gives a fibration $\xi_{\mathbb{P}}\colon E_{\mathbb{P}} \to X_{\mathbb{P}}$ with fibre $F_{\mathbb{P}}$ [**17**, Theorem II.3.12]. Note that the term-by-term $\mathbb{P}$-localization $\xi_{\mathbb{P}}\colon E_{\mathbb{P}} \to X_{\mathbb{P}}$ is not the same as the fibrewise $\mathbb{P}$-localization $\xi_{(\mathbb{P})}\colon E_0 \to X$. In fact, we have:

LEMMA 3.4. *Let $\xi\colon E \to X$ be a nilpotent fibration with term-by-term $\mathbb{P}$-localization $\xi_{\mathbb{P}}$ and fibrewise $\mathbb{P}$-localization $\xi_{(\mathbb{P})}$. Then there is a fibre-homotopy equivalence*

$$\xi_{(\mathbb{P})} \simeq \ell_X^{-1}(\xi_{\mathbb{P}})$$

where $\ell_X\colon X \to X_{\mathbb{P}}$ is a $\mathbb{P}$-localization map.

PROOF. Using the uniqueness theorem for fibrewise localization [**22**, Theorem 6.1] we obtain a fibrewise map $\xi_{(\mathbb{P})} \to (\ell_X)^{-1}(\xi_{\mathbb{P}})$ over X. By the 5-lemma and [**7**, Theorem 3.3], this is a fibre homotopy equivalence. □

As a direct consequence, we have:

THEOREM 3.5. *Let $\zeta\colon E \to X$ be a fibrewise grouplike space over a compact metric space X with fibre G a connected CW grouplike space. Suppose ζ is the pullback of nilpotent fibration ξ with $\mathbb{P}$-localization $(\xi)_{\mathbb{P}}$ fibre-homotopically trivial. Then*

$$(\Gamma(\zeta)_\circ)_{\mathbb{P}} \simeq \mathrm{map}(X, G_{\mathbb{P}}; 0).$$

PROOF. By the remarks preceding the proof of Theorem 3, the fibrewise localization of ζ may be taken as the pullback of the fibrewise localization $\xi_{(\mathbb{P})}$ of ξ. Since the latter is the pullback of $\xi_{\mathbb{P}}$ which is, by hypothesis, fibre-homotopically trivial, we see $\zeta_{(\mathbb{P})}$ is fibre-homotopically trivial, as well. Now use Theorem 3. □

We apply this last result to a natural generalization of the gauge group. Given a group G let $PG = G/Z(G)$ denote the projectification. Given a map $h\colon X \to BPG$ write $P_h\colon E \to X$ for the corresponding principal PG-bundle and

$$\mathrm{Pad}(P_h)\colon E \times_{PG} G^{\mathrm{ad}} \to X$$

the associated G-bundle where PG acts on $G^{\mathrm{ad}} = G$ by the adjoint action. This is a fibrewise group. We call the group of sections $\Gamma(\mathrm{Pad}(P_h))$ the *projective gauge group*. When $G = U(n)$, the projective gauge group corresponds to the group of unitaries of a complex matrix bundle (see [**19**, Example 3.7]).

The classification of rational H-types of projective gauge groups for G a compact, connected Lie group and X a compact metric space is given by [**19**, Theorem F]. The following result extends this to all connected Lie groups but at the expense of the H-structure.

THEOREM 3.6. *Let X be a compact metric space and G a connected Lie group. Then:*

(1) *The rational homotopy type of $\Gamma(\mathrm{Pad}(P_h))_\circ$ is independent of $h \in [X, BPG]$.*

(2) *The rational homotopy groups of* $\Gamma(\mathrm{Pad}(P_h))_\circ$ *are given by*

$$\pi_n(\Gamma(\mathrm{Pad}(P_h))_\circ)\otimes\mathbb{Q}\cong\bigoplus_{k\geq n}\check{H}^{k-n}(X;\mathbb{Q})\,\otimes\,(\pi_k(G)\otimes\mathbb{Q}),$$

for $n\geq 1$.

PROOF. We note that $\mathrm{Pad}(P_h)$ is the pullback by h of the universal projective adjoint bundle $\mathrm{Pad}(\eta_{PG})\colon EPG\times_{PG}G^{\mathrm{ad}}\to BPG$. Here $\eta_{PG}\colon EPG\to BPG$ is the universal principal PG-bundle. By [**19**, Lemma 5.11], the total space $EPG\times_{PG}G^{\mathrm{ad}}$ is a nilpotent space. (This result assumes G is compact Lie. However, all that is used there is that $Z(G)$ have vanishing higher rational homotopy groups, which is true for G connected Lie.) Thus $\mathrm{Pad}(\eta_{PG})$ is nilpotent since BPG is simply connected. Below we show the rationalization of $\mathrm{Pad}(\eta_{PG)})$ is fibre-homotopically trivial. The result (1) then follows from Theorem 3.5. The result (2) is proved using the arguments given in the proof of Theorem 3.3 (2).

To show $\mathrm{Pad}(\eta_{PG})_\mathbb{Q}\colon(EPG\times_{PG}G^{\mathrm{ad}})_\mathbb{Q}\to(BPG)_\mathbb{Q}$ is fibre homotopically trivial, we first show the total space is an H-space. Let $EG\times_G G^{\mathrm{ad}}$ denote the total space of the universal G-adjoint bundle. Then $EG\times_G G^{\mathrm{ad}}$ is also a nilpotent space. In fact, we have

$$EG\times_G G^{\mathrm{ad}}\simeq\mathrm{map}(S^1,BG;0)$$

the free loop space of the classifying space (see, e.g., [**19**, Lemma 9.1]). Since BG is simply connected, $\mathrm{map}(S^1,BG;0)$ is nilpotent by [**17**, Theorem II.3.11]. Further, by this last result again,

$$(EG\times_G G^{\mathrm{ad}})_\mathbb{Q}\simeq\mathrm{map}(S^1,(BG)_\mathbb{Q};0).$$

Now since G is a connected Lie group, it has oddly graded rational homotopy groups. It follows, as above, that $(BG)_\mathbb{Q}$ is an H-space. Thus $\mathrm{map}(X,(BG)_\mathbb{Q};0)$ is an H-space, as well. Next, by [**19**, Lemma 5.9], the natural map

$$\pi\colon EG\times_G G^{\mathrm{ad}}\to EPG\times_{PG}G^{\mathrm{ad}}$$

induces a surjection on rational homotopy groups. (Again, the lemma is stated for G compact Lie but the proof uses only that $Z(G)$ is rationally aspherical.) Thus, by [**23**, Lemma 5.8], since $EG\times_G G^{\mathrm{ad}}$ is a rational H-space so is $EPG\times_{PG}G^{\mathrm{ad}}$. We have shown that $\mathrm{Pad}(\eta_{PG})$ is a sectioned, nilpotent fibration of rational H-spaces with simply connected base. By Lemma 3.7 below, $(\mathrm{Pad}(\eta_{PG}))_\mathbb{Q}$ is fibre-homotopically trivial, as needed. □

LEMMA 3.7. *Let* $\xi\colon E\to B$ *be a fibration of nilpotent spaces with* E,B *and the fibre* F *each a rational H-space of finite type. Suppose* B *is simply connected and the linking homomorphism* $\partial\colon\pi_k(B)\longrightarrow\pi_{k-1}(E)$ *is trivial after rationalization for all* $k\geq 2$. *Then* $\xi_\mathbb{Q}$ *is fibre homotopically trivial.*

PROOF. By [**15**, Theorem 4.6], ξ is a rational fibration as in [**15**, Definition 4.5]. In general, this means that $\xi_\mathbb{Q}$ admits a Koszul-Sullivan model which is a sequence

$$(\wedge V_3;d_3)\xrightarrow{P}\wedge(V_2;d_2)\xrightarrow{J}(\wedge V_1;d_1).$$

Here $(\wedge V_j;d_j)$ is a free DG algebra over $\mathbb{Q}$ for $j=1,2,3$ and a Sullivan model for F,E,B, respectively. The maps P and J are Sullivan models for the projection and fibre inclusion. The models $(\wedge V_1;d_1)$ and $(\wedge V_3;d_3)$ are minimal, meaning $d_j(V_j)$ is contained in the decomposables of $\wedge V_j$ for $j=1,3$. Since B and F are assumed to be rational H-spaces, this forces $d_3 = d_1 = 0$. The model $(\wedge V_2; d_2)$ is not, in general, minimal. However, by [**15**, Theorem 4.12], the vanishing of the rational linking homomorphism implies $(\wedge V_2; d_2)$ is a minimal model for E, as well. Since $H^*(E;\mathbb{Q}) \cong H^*(\wedge V_2; d_2)$ is free we conclude $d_2 = 0$. Now we may directly obtain a DG algebra map $\widetilde{J}\colon \wedge(V_2;0) \to \wedge(V_3;0)$ with $\widetilde{J} \circ P = 1_{\wedge V_3}$. The maps $\widetilde{J}, J$ induce an isomorphism $A\colon \wedge(V_2;0) \to (\wedge V_1;0) \otimes (\wedge V_3;0)$ of DG algebras satisfying $\pi_2 \circ A = P$ where $\pi_2\colon (\wedge V_1;0) \otimes (\wedge V_3;0) \to \wedge(V_3;0)$ is the projection. Using the correspondence between (homotopy classes of) maps between minimal DGAs and maps between rational spaces, we obtain $\alpha\colon F_{\mathbb{Q}} \times B_{\mathbb{Q}} \longrightarrow E_{\mathbb{Q}}$, the needed fibre homotopy equivalence. □

4. The Based Case

We deduce versions of our main results in the basepoint preserving cases. First, let $\xi\colon E \to X$ be a fibration over a based space X and let F be the fibre over a fixed basepoint. We denote by $\mathrm{aut}^F(\xi)$ the submonoid of $\mathrm{aut}(\xi)$ consisting of equivalences inducing the identity on F. This is the natural based version of $\mathrm{aut}(\xi)$. Note that $\mathrm{aut}^F(\xi)$ is the fibre over the identity map 1_F of the restriction map $\mathrm{res}\colon \mathrm{aut}(\xi) \to \mathrm{aut}(F)$. We have the following based version of the identity (1) in our setting. Write $\mathrm{map}_*(X,Y)$ for the space of basepoint perserving maps from X to Y.

THEOREM 4.1. *Let X be a compact metric space, F a finite CW complex and $h\colon B \to B\mathrm{aut}(F)$ a map. Let $\xi\colon E \to X$ be the corresponding F-fibration. Then there is an H-equivalence*

$$\mathrm{aut}^F(\xi) \simeq \Omega\mathrm{map}_*(X, B\mathrm{aut}(F); h).$$

PROOF. First, note that, by [**30**], $\mathrm{aut}^F(\xi)$ is of CW type since $\mathrm{aut}(\xi)$ and $\mathrm{aut}(F)$ are. Now compare the long exact homotopy sequence of the restriction fibration above to that of loops on the evaluation fibration $\omega\colon \mathrm{map}(X, B\mathrm{aut}(F); h) \to B\mathrm{aut}(F)$ with fibre $\mathrm{map}_*(X, B\mathrm{aut}(F); h)$. Note that we obtain a map between these fibrations by the functoriality of the construction $\Phi\colon \Gamma(\xi \,\square_1 \xi) \to \mathrm{map}(X, B\mathrm{aut}(F))$ applied to the inclusion of the basepoint $* \to X$. The result follows from Theorem 1 and the 5-lemma. □

As before, this gives a corresponding localization result for $\mathrm{aut}^F(\xi)_\circ$:

THEOREM 4.2. *Let X be a compact metric space, F a finite CW complex and $h\colon B \to B\mathrm{aut}(F)$ a map. Let $\xi\colon E \to X$ be the corresponding F-fibration. Then $\mathrm{aut}^F(\xi)_\circ$ is a nilpotent space and we have an H-equivalence*

$$(\mathrm{aut}^F(\xi)_\circ)_{\mathbb{P}} \simeq \Omega_\circ\mathrm{map}_*(X, (B\mathrm{aut}(F)_\circ)_{\mathbb{P}}; (\tilde{h})_{\mathbb{P}}).$$

□

Next, given a fibration $\zeta\colon E \to X$ write $\Gamma_*(\zeta)$ for the space of basepoint preserving sections of ζ. We have:

THEOREM 4.3. *Let $\zeta\colon E \to X$ be a fibrewise grouplike space with connected, CW grouplike fibre G and base X a compact metric space. Suppose ζ is the pullback of a CW fibration. Then $\Gamma_*(\zeta)_\circ$ is a nilpotent space and the map*

$$\Gamma_*(\zeta)_\circ \to \Gamma_*(\zeta_{(\mathbb{P})})_\circ$$

induced by a fibrewise $\mathbb{P}$-localization $\zeta \to \zeta_{(\mathbb{P})}$ is a $\mathbb{P}$-localization map.

PROOF. Here we may repeat the entire argument in the based case to achieve the needed result. We note that Möller's theorem [**24**, Theorem 5.3], for X a finite complex, is proved in the based setting. The rest of the argument thus proceeds as before. □

Finally, we remark that our applications Theorems 3.3 and 3.6 also hold in the respective based settings but with ordinary Čech cohomology replaced by reduced Čech cohomology in the rational homotopy calculations.

References

[1] M. F. Atiyah and R. Bott, *The Yang-Mills equations over Riemann surfaces*, Philos. Trans. Roy. Soc. London Ser. A **308** (1983), 523–615. MR **85k:**14006

[2] P. Booth, P. Heath, C. Morgan and R. Piccinini, *H-spaces of self-equivalences of fibrations and bundles,* Proc. London Math. Soc. (3) **49** (1984), no. 1, 111–127. MR **0743373 (85k:**55013)

[3] P. Booth, P. Heath and R. Piccinini, *Fibre preserving maps and functional spaces,* Lecture Notes in Math., 673, Springer, Berlin, pp. 158–167, 1978. MR **0517090 (80g:**55011)

[4] U. Buijs, Y. Félix and A. Murillo, *Lie models for the components of the space of sections of a nilpotent fibration,* Trans. Amer. Math. Soc. 361 (2009), 5601–5614. MR 2515825

[5] M. C. Crabb and I. M. James, *Fibrewise homotopy theory*, Springer-Verlag, 1998. MR 1646248

[6] M. C. Crabb and W. A. Sutherland, *Counting homotopy types of gauge groups*, Proc. London. Math. Soc. (3) **81** (2000), 747-768. MR **1781154 (2001m:**55024)

[7] A. Dold, *Partitions of unity in the theory of fibrations,* Ann. of Math. (2) **78** 1963, 223–255. MR 0155330 (27 #5264)

[8] A. Dold and R. Lashof *Principal quasi-fibrations and fibre homotopy equivalence of bundles*, Ill. J. Math. (1959), 285-30. MR 0101521 (21 #331)

[9] S. Eilenberg and N. Steenrod, *Foundations of algebraic topology*, Princeton University Press, Princeton, 1952. MR 0050886 (14,398b)

[10] Y. Félix and J.-C. Thomas, *Nilpotent subgroups of the group of fibre homotopy equivalences,* Publ. Mat. **39** (1995), no. 1, 95–106 MR 1336359

[11] Y. Félix, G. Lupton, S. B. Smith, *Rational homotopy type of the groups of self-equivalences of a fibration,* preprint, arXiv:0903.1470

[12] D. H. Gottlieb, *On fibre spaces and the evaluation map*, Ann. Math. **87** (1968), 42–55. MR 0221508 (36 #4560)

[13] ——— *The total space of universal fibrations*, Pacific J. Math. **46** (1973), 415–417 MR 0331384 (48 #9717)

[14] P.-P. Grivel, *Algèbres de Lie de dérivations de certaines algèbres pures*, J. Pure Appl. Algebra **91** (1994), no. 1-3, 121–135. MR **1255925 (95a:**55024)

[15] S. Halperin, *Rational fibrations, minimal models, and fibrings of homogeneous spaces*, Trans. Amer. Math. Soc. **244** (1978), 199–224. MR 58 #24264

[16] A. Haefliger, *Rational homotopy of the space of sections of a nilpotent bundle*, Trans. Amer. Math. Soc. **273** (1982), no. 2, 609–620. MR **0667163 (84a:**55010)

[17] P. Hilton, G. Mislin, and J. Roitberg, *Localization of nilpotent groups and spaces*, North-Holland Publishing Co., Amsterdam, 1975, North-Holland Mathematics Studies, No. 15. MR 57 #17635

[18] P. Hilton, G. Mislin, J. Roitberg, R. Steiner, *On free maps and free homotopies into nilpotent spaces,* Lecture Notes in Math., vol. 673, Springer-Verlag, New York, 1978. MR **0517093 (80c:**55007)

[19] J. R. Klein, C. L. Schochet and S. B. Smith, *Continuous trace C^*-algebras, gauge groups and rationalization,* J. Topol. Analysis **1** (2009), 261-288.

[20] A. Kono, *A note on the homotopy type of certain gauge groups*, Proc. Roy. Soc. Edinburgh Sect. A **117** (1991), no. 3-4, 295–297 MR 1103296

[21] A. Kono and S. Tsukuda, *4-manifolds X over $B\mathrm{SU}(2)$ and the corresponding homotopy types* $\mathrm{Map}(X, B\mathrm{SU}(2))$. J. Pure Appl. Algebra **151** (2000), no. 3, 227–237 MR 1776430

[22] I. Llerena, *Localization of fibrations with nilpotent fibre*, Math. Z. **188** (1985), 397-410. MR **0771993 (86e:**55014)

[23] G. Lupton, N. C. Phillips, C. L. Schochet and S. B. Smith, *Banach algebras and rational homotopy theory*, Trans. Amer. Math. Soc. **361** (2009), no. 1, 267–295. MR 2439407

[24] J. M. Möller, *Nilpotent spaces of sections,* Trans. Amer. Math. Soc. **303** (1987), 733–741. MR 0902794

[25] J. P. May, *Classifying spaces and fibrations*, Memoir Amer. Math. Soc. **155** (1975). MR 0370579 (51 #6806)

[26] ———, *Fibrewise localization and completion*, Trans. Amer. Math. Soc. **258** (1980), 127-146.. MR **0554323 (81f:**55005)

[27] J. Milnor, *On spaces having the homotopy type of CW-complex*, Trans. Amer. Math. Soc. **90** (1959), 272-280. MR 20 #6700

[28] H. Scheerer, *Localisation des espaces de sections et de rétractions,* C. R. Acad. Sci. Paris Sr. A-B **287** (1978), no. 13, A851–A853. MR **0551762 (81i:**55010)

[29] ———, *On rationalized H- and co-H-spaces. With an appendix on decomposable H- and co-H-spaces*, Manuscripta Math. **51** (1985), 63–87. MR **88k:**55007

[30] R. Schön, *Fibrations over a CWh-base,* Proc. Amer. Math. Soc. **62** (1976), no. 1, 165–166. MR 0431163 (55 #4165)

[31] H. Shiga and M. Tezuka, *Rational fibrations, homogeneous spaces with positive Euler characteristics and Jacobians,* Ann. Inst. Fourier (Grenoble) **37** (1987), no. 1, 81–106. MR **0894562 (89g:**55019)

[32] J. Stasheff, *A classification theorem for fibre spaces*, Topology **2** (1963), 239–246 MR 0154286 (27 #4235)

[33] R. Thom, *L'homologie des espaces fonctionnels*, Colloque de topologie algébrique, Louvain, 1956, Georges Thone, Liège, 1957, pp. 29–39. MR 19,669h

Department of Mathematics, Wayne State University, Detroit MI 48202
E-mail address: `claude@math.wayne.edu`

Department of Mathematics, Saint Joseph's University, Philadelphia PA 19131
E-mail address: `smith@sju.edu`

Contemporary Mathematics
Volume **519**, 2010

Non-integral central extensions of loop groups

Christoph Wockel

Abstract. It is well-known that the central extensions of the loop group of a compact, simple and 1-connected Lie group are parametrised by their level $k \in \mathbb{Z}$. This article concerns the question how much can be said for arbitrary $k \in \mathbb{R}$ and we show that for each k there exists a Lie groupoid which has the level k central extension as its quotient if $k \in \mathbb{Z}$. By considering categorified principal bundles we show, moreover, that the corresponding Lie groupoid has the expected bundle structure.

Introduction

In this paper we generalise a construction of the universal central extension $\widehat{\Omega K}$ of the loop group ΩK of a compact simple and 1-connected Lie group K, going back to Mickelsson and Murray [**Mic87**], [**Mur88**]. They construct $\widehat{\Omega K}$ as a quotient of a central extension $U(1) \times_{\kappa} P_e\Omega K$ of the based path group $P_e\Omega K$. For this construction one has the freedom to choose a real number $k \in \mathbb{R}$ (after having fixed all normalisations appropriately), which is usually referred to as the *level*. The construction from [**Mic87**] and [**Mur88**] then yields a normal subgroup $N_k \trianglelefteq U(1) \times_{\kappa} P_e\Omega K$ if and only if $k \in \mathbb{Z}$ and constructs $\widehat{\Omega K}$ as $(U(1) \times_{\kappa} P_e\Omega K)/N_1$.

The point of this article is that, although the construction of N_k works if and *only* if $k \in \mathbb{Z}$, for general k there still exists an infinite-dimensional Lie group $\mathcal{K}$ acting on $U(1) \times_{\kappa} P_e\Omega K$ and the quotient of this action coincides with $(U(1) \times_{\kappa} P_e\Omega K)/N_k$ if $k \in \mathbb{Z}$. This then gives rise to an action Lie groupoid. By passing to a Morita equivalent Lie groupoid we show that this Lie groupoid has the structure of a generalised principal bundle.

The results that we get here are closely related to the general extension theory of infinite-dimensional Lie groups by categorical Lie groups from [**Woc08**]. However, this article concerns more the global and differential point of view to those extensions for the particular case of loop groups, while [**Woc08**] provides a more detailed perspective from the side of cocycles.

Notation: Throughout this article, G denotes a (possibly infinite-dimensional) connected Lie group with Lie algebra $\mathfrak{g}$, modelled on a locally convex space, $\mathfrak{z}$ is a sequentially complete locally convex space and $\Gamma \subseteq \mathfrak{z}$ is a discrete subgroup of (the additive group of) $\mathfrak{z}$. Moreover, we set $Z := \mathfrak{z}/\Gamma$.

1991 *Mathematics Subject Classification.* Primary 22E67; Secondary 58H05, 22A22.
Key words and phrases. Loop group, central extension, Lie groupoid, principal 2-bundle.

©2010 American Mathematical Society

Most of the time, G will be the pointed loop group

$$\Omega K := \{\gamma : \mathrm{R} \to K : \gamma(0) = e, \gamma(x+n) = \gamma(x)\, \forall n \in \mathrm{N}\}$$

of smooth and pointed loops in a compact, simple and 1-connected Lie group K, endowed with the usual Fréchet topology and point-wise multiplication. The Lie algebra of ΩK is then

$$\Omega\mathfrak{k} := \{\gamma : \mathrm{R} \to \mathfrak{k} : \gamma(0) = 0, \gamma(x+n) = \gamma(x)\, \forall n \in \mathrm{N}\}$$

with $\mathfrak{k} := L(K)$. In this case, $\mathfrak{z}$ will be R, Γ will be Z and thus $Z = \mathrm{R}/\mathrm{Z} =: U(1)$. We circumvent all normalisation issues by choosing this quite unnatural realisation of the circle group. Moreover, we denote by exp the canonical quotient map $\exp : \mathrm{R} \to \mathrm{R}/\mathrm{Z}$.

We will be a bit sloppy in our conventions concerning the precise model for S^1 and B^2. Instead, we collect the things that we want to assume:

- B^2 and S^1 are manifolds with corners such that S^1 may be identified with a submanifold of B^2 (which we denote by ∂B^2) and the base-point of B^2 is contained in ∂B^2.
- $C^\infty_*(S^1, G)$ may be identified with the kernel of the evaluation map $\mathrm{ev} : P_eG \to G$, where P_eG denotes the space of smooth maps $f : [0,1] \to G$ with $f(0) = e$.
- The map $C^\infty_*(B^2, G) \to C^\infty_*(S^1, G)_e$, $f \mapsto f|_{\partial B^2}$ is surjective.

Here, the subscript $_*$ denotes pointed maps and the subscript $_e$ denoted the connected component of the identity.

1. Generalities on central extensions of infinite-dimensional Lie groups

We briefly review essentials on central extensions of infinite-dimensional Lie group, established by Neeb in [**Nee02**]. There, the second locally smooth group cohomology $H^2_{\mathrm{loc}}(G, Z)$ is defined to be the set of functions $f : G \times G \to Z$ such that

- f is smooth on $U \times U$ for $U \subseteq G$ some open identity neighbourhood
- $f(g,h) + f(gh,k) = f(g,hk) + f(h,k)$ for all $g,h,k \in G$
- $f(g,e) = f(e,g) = 0$ for all $g \in G$,

(called *locally smooth group cocycles* in this paper) modulo the equivalence relation

$$(f \sim f') :\Leftrightarrow f(g,h) - f'(g,h) = b(g) - b(gh) + b(h) \tag{1.1}$$

for some $g : G \to Z$ which is smooth on some identity neighbourhood and satisfies $b(e) = 0$. Similarly, we define $H^2_{\mathrm{glob}}(G, Z)$ to be defined in the same way except that we require f (respectively b) to be smooth on $G \times G$ (respectively G). We shall call such a f a *globally smooth group cocycle*. Then in [**Nee02**] it is shown that $H^2_{\mathrm{loc}}(G, Z)$ corresponds bijectively to the equivalence classes of central extensions

$$Z \to \widehat{G} \to G \tag{1.2}$$

of Lie groups such that (1.2) is a *locally* trivial principal bundle and that $H^2_{\mathrm{glob}}(G, Z)$ corresponds bijectively to equivalence classes of central extensions of Lie groups such that (1.2) is a *globally* trivial principal bundle.

The bulk of the work in [**Nee02**] concerns the integration issue for central extensions, i.e., how to derive a continuous Lie algebra cocycle $D(f) : \mathfrak{g} \times \mathfrak{g} \to \mathfrak{z}$ from a locally smooth group cocycle and to determine whether for a given continuous Lie algebra cocycle ω there exists a locally smooth group cocycle f such that $[Df] = [\omega] \in H^2_c(\mathfrak{g}, \mathfrak{z})$ (where the subscript $_c$ means continuous Lie algebra cohomology). In the latter case we say that ω *integrates*. The main result in [**Nee02**] is an exact sequence

$$\operatorname{Hom}(\pi_1(G), Z) \to H^2(G, Z) \xrightarrow{D} H^2_c(\mathfrak{g}, \mathfrak{z}) \xrightarrow{P} \operatorname{Hom}(\pi_2(G), Z) \oplus \operatorname{Hom}(\pi_1(G), \operatorname{Lin}_c(\mathfrak{g}, \mathfrak{z})),$$

where Lin_c denotes continuous linear maps and

$$\operatorname{per}_\omega := \operatorname{pr}_1(P([\omega])) : \pi_2(G) \to Z, \quad [\sigma] \mapsto \left[\int_\sigma \omega^l\right]$$

for σ a smooth representative of $[\sigma] \in \pi_2(G)$ and ω^l the left-invariant 2-form on G with $\omega^l(e) = \omega$. In particular, when G is simply connected, then the sequence reduces to a shorter exact sequence

$$0 \to H^2(G, Z) \xrightarrow{D} H^2_c(\mathfrak{g}, \mathfrak{z}) \xrightarrow{\operatorname{per}} \operatorname{Hom}(\pi_2(G), Z).$$

Thus a given cocycle ω integrates in this case if and only if the corresponding *period homomorphism* $\operatorname{per}_\omega$ vanishes.

2. The topological type of central extensions

In [**Mic85**], Mickelsson derives a Čech 1-cocycle for $\widehat{\Omega\,\mathrm{SU}_2}$. In this section we shall describe how to derive the topological type of the principal bundle

$$Z \to \widehat{G} \to G$$

for a central extension coming from a locally smooth cocycle $f : G \times G \to Z$. This description is much more general than the one from [**Mic85**] and it will become apparent from this construction that for a globally smooth cocycle the corresponding bundle is automatically trivial. For this we will make use of the following fact.

THEOREM 2.1. *Let H be a group $W \subseteq H$ be a subset containing e and let W be endowed with a manifold structure. Moreover, assume that there exists an open neighbourhood $Q \subseteq W$ of e with $Q^{-1} = Q$ and $Q \cdot Q \subseteq W$ such that*

- *$Q \times Q \ni (g, h) \mapsto gh \in W$ is smooth,*
- *$Q \ni g \mapsto g^{-1} \in Q$ is smooth and*
- *Q generates H as a group.*

Then there exists a manifold structure on H such that Q is open in H and such that group multiplication and inversion is smooth. Moreover, for each other choice of Q, satisfying the above conditions, the resulting smooth structures on H coincide.

PROOF. The proof is well-known and straight-forward, cf. [**Woc08**, Thm. II.1], [**Bou98**, Prop. III.1.9.18]. □

We now derive a central extension from a locally smooth group cocycle $f : G \times G \to Z$. First, we define a twisted group structure on the set-theoretical direct product $Z \times G$ by $(a, g) \cdot (b, h) := (a + b + f(g, h), gh)$. Then the requirement on f to define a group cocycle implies that this defines a group multiplication with neutral element $(0, e)$ and $(a, g)^{-1} = (-a - f(g, g^{-1}), g^{-1})$. We denote this group by $Z \times_f G$. If f is smooth on $U \times U$ and $V \subseteq U$ is an open identity neighbourhood with $V \cdot V \subseteq W$ and $V^{-1} = V$, then $Z \times U$ carries the product manifold structure and $Z \times V$ is open in $Z \times U$. Since G is assumed to be connected, $Z \times_f G$ is generated by $Z \times V$ and the preceding theorem yields a Lie group structure on $Z \times_f G$. Clearly, the sequence

$$Z \to Z \times_f G \to G$$

is a locally trivial principal bundle for we have the smooth section

$$U \ni x \mapsto (0, x) \in Z \times U \subseteq Z \times_f G.$$

LEMMA 2.2. *The assignment*

$$\tau(f)_{g,h} : gV \cap hV \to Z, \quad x \mapsto f(g, g^{-1}x) - f(h, h^{-1}x)$$

defines a Čech 1-cocycle $\tau(f)$ on the open cover $(gV)_{g \in G}$ and thus an element of $\check{Z}^1(G, Z)$. If $[f] = [f']$ in $H^2_{\text{loc}}(G, Z)$, then $[\tau(f)] = [\tau(f')]$ in $\check{H}^1(G, Z)$.

PROOF. We first note that $gh^{-1} \in V \cdot V \subseteq W$ if $gV \cap hV \neq \emptyset$. From this it follows that

$$x \mapsto f(g, g^{-1}x) - f(h, h^{-1}x) = f(g^{-1}h, h^{-1}x) - f(g, g^{-1}h)$$

is smooth on $gV \cap hV$, for f is smooth on $U \times U$. From the definition it is also clear that $\tau(f)_{g,h} - \tau(f)_{g,k} + \tau(f)_{h,k}$ vanishes.

If $[f] = [f']$, then $f(g, h) - f'(g, h) = b(g) - b(gh) + b(h)$ for $b : G \to Z$. We assume without loss of generality that f and f' are smooth on $U \times U$ and b is smooth on U (presumably, the identity neighbourhoods may be distinct for f and f' and b). Then

$$\tau(b)_g : gV \to Z, \quad x \mapsto b(g) + b(g^{-1}x)$$

defines a Cech cochain with $\tau(f) - \tau(f') = \check{\delta}(\tau(b))$. □

We thus have a map $\tau : H^2_{\text{loc}}(G, Z) \to \check{H}^1(G, Z)$, which clearly is a group homomorphism.

PROPOSITION 2.3. *The principal bundle*

$$Z \to Z \times_f G \to G \tag{2.1}$$

is classified by $[\tau(f)] \in \check{H}^1(G, Z)$.

PROOF. From the construction of the topology on $Z \times_f G$ it follows immediately that $\sigma_e(x) := (0, x)$ defines a smooth section on V. Thus the assignment

$$\sigma_g : gV \to Z \times_f G, \quad x \mapsto (f(g, g^{-1}x), x) = \left(\lambda_{(0,g)} \circ \sigma_e \circ \lambda_{g^{-1}}\right)(x)$$

is smooth, where $\lambda_{g^{-1}}$ denotes left multiplication in G with g^{-1} and $\lambda_{(0,g)}$ denotes left multiplication in $Z \times_f G$ with $(0, g)$. Consequently, $(\sigma_g : gV \to Z \times_f G)_{g \in G}$ defines a system of sections for the principal bundle (2.1) and since $\tau(f)$ satisfies $\sigma_g(x) = \sigma_h(x) \cdot \tau(f)_{g,h}(x)$, this already shows the claim. □

COROLLARY 2.4. *A locally smooth cocycle $f : G \times G \to Z$ is equivalent to a globally smooth cocycle if and only if the principal bundle, underlying $Z \to Z \times_f G \to G$, is topologically trivial.*

PROOF. If the bundle is topologically trivial, then there exists a smooth section $\sigma : G \to Z \times_f G$ and

$$f'(g,h) := \sigma(g)\sigma(h)\sigma(gh)^{-1}$$

defines a Z-valued cocycle. Since Z acts freely on $Z \times_f G$, we have $(0,g) = \sigma(g)\cdot b(g)$ for $b : G \to Z$, smooth on a identity neighbourhood and satisfying (1.1). The "only if" part is clear from the construction of $\tau(f)$. □

We thus obtain a sequence

$$0 \to H^2_{\text{glob}}(G,Z) \to H^2_{\text{loc}}(G,Z) \xrightarrow{\tau} \check{H}^1(G,Z)$$

which is obviously exact. It would be interesting to determine for which groups and which coefficients the map τ is *not* surjective. Note that the case of Z being connected is the interesting one, since for a discrete group A, each principal A-bundle over G is a covering and thus admits a compatible Lie group structure.

3. The universal central extension of Loop groups

The results described in the preceding section applies to loop groups ΩK in the following way. If $\langle\cdot,\cdot\rangle : \mathfrak{k}\times\mathfrak{k} \to \mathrm{R}$ denotes the Killing form (which is non-degenerate and negative definite in our case), then

$$\omega : \Omega\mathfrak{k} \times \Omega\mathfrak{k} \to \mathrm{R}, \quad (f,g) \mapsto \int_{S^1} \langle f(t), g'(t)\rangle dt$$

defines a continuous Lie algebra cocycle. If we normalise $\langle\cdot,\cdot\rangle$ in such way that the left-invariant extension ω^l of ω satisfies $\int_\sigma \omega^l = 1$ for σ a generator[1] of $\pi_2(\Omega K) \cong \pi_3(K) \cong \mathrm{Z}$, then the calculations in [**MN03**] show that for $Z = \mathrm{R}$ we have $\mathrm{per}_\omega(\pi_2(\Omega K)) = \mathrm{Z}$. Thus the cocycle $k\cdot\omega$ integrates to a locally smooth group cocycle $f_k : \Omega K \times \Omega K \to U(1)$, defining a central extension

$$U(1) \to \widehat{\Omega K}_k \to \Omega K$$

if and only if $k \in \mathrm{Z}$. Moreover, the central extension for $k = \pm 1$ is universal, as it is shown in [**MN03**]. This means that for each other central extension $Z \to \widetilde{\Omega K} \to \Omega K$ there exist unique morphisms $U(1) \to Z$ and $\widehat{\Omega K}_{\pm 1} \to \widetilde{\Omega K}$ making the diagram

$$\begin{array}{ccccc} U(1) & \longrightarrow & \widehat{\Omega K}_{\pm 1} & \longrightarrow & \Omega K \\ \downarrow & & \downarrow & & \downarrow \\ Z & \longrightarrow & \widetilde{\Omega K} & \longrightarrow & \Omega K \end{array}$$

commute.

There also exist more ad-hoc constructions of $\widehat{\Omega K}_k$, cf. [**PS86**], [**Mic87**], [**Mur88**], [**MS01**] or [**MS03**] which, more or less, all construct $\widehat{\Omega K}_k$ first constructing a central extension $\widehat{P_e\Omega K} \to P_e\Omega K$, corresponding to the pull-back of $\widehat{\Omega K}_k$ along $\mathrm{ev} : P_e\Omega K \to \Omega K$, and then considering an appropriate quotient of $\widehat{P_e\Omega K}$. Since $P_e\Omega K$ is contractible, the results of the preceding sections imply that

[1] The ambiguity in the sign that one still has for the normalisation of $\langle\cdot,\cdot\rangle$ will play no role in the sequel.

the pull-back of the central extension of $L(\Omega K) \cong \Omega\mathfrak{k}$ along the evaluation homomorphism $L(\text{ev}) : L(P_e\Omega K) \to L\mathfrak{k}$ to $L(P_e\Omega K) \cong P_e\Omega\mathfrak{g}$ integrates to a central extension

$$U(1) \to \widehat{P_e\Omega K} \to P_e\Omega K,$$

given by a globally smooth cocycle $\kappa : PG \times PG \to U(1)$. The constructions of $\widehat{\Omega K}$ cited above all deal with an explicit description of a normal subgroup $N_k \trianglelefteq \widehat{P_e\Omega K}$ in order to obtain an induced central extension

$$U(1) \to \underbrace{\widehat{P_e\Omega K}/N_k}_{\cong \widehat{\Omega K}_k} \to \underbrace{P_e\Omega K/\Omega(\Omega K)}_{\cong \Omega K}$$

(cf. also [**GN03**, Section III]).

4. Central extensions of loop groups from Lie groupoids

We shall put more structure on the ad-hoc construction of $\widehat{\Omega K}$ from the previous section. In particular, we show that $\widehat{P_e\Omega K}/N_k$ may be obtained as the quotient of an action Lie groupoid, which also exists for non-integral values of k.

REMARK 4.1. We briefly recall the essential notions for Lie groupoids. A Lie groupoid is a category object in the category of locally convex manifolds, such that source and target maps admit local inverses[2]. More precisely, it consists of two locally convex manifolds M_1 and M_0, together with smooth maps $\text{id} : M_0 \to M_1$ and $s, t : M_1 \to M_0$, admitting local inverses, and a smooth map[3] $\circ : M_1 \,{}_s\!\times_t M_1 \to M_0$ satisfying the usual relations of identity, source, target and composition map of a small category. Moreover, we require that each morphism of this category is invertible an that the map $M_1 \to M_1$, assigning to each morphism its inverse, is smooth. The quotient of such a Lie groupoid is defined to be the set of equivalence classes of isomorphic objects. The smooth structure on M_0 may or may not induce a smooth structure on the quotient, depending on how badly the quotient actually is behaved.

A typical example of a Lie groupoid, called *action groupoid*, is obtained from a smooth right action of a Lie group G on a manifold M. With this data given, we set $M_0 := M$, $M_1 := M \times G$, $\text{id}(m) := (m, e)$, $s(m, g) := m$, $t(m, g) := m.g$ and $(m.g, h) \circ (m, g) := (m, g \cdot h)$. Clearly, the inverse of (m, g) is $(m.g, g^{-1})$. The quotient of the Lie groupoid clearly is given by M/G. If it admits a smooth structure such that the quotient map $M \to M/G$ is smooth an admits local inverses, then the action groupoid is Morita equivalent[4] to the Lie groupoid with $M_0 = M_1 = M/G$ and all structure maps the identity. If M/G does not carry a smooth structure, then the action groupoid is an appropriate replacement for M/G.

[2]We shall use concepts from the usual theory of Lie groupoids by replacing the term "surjective submersion" at each occurrence by the term "admits local inverses". This is equivalent in the finite-dimensional case but may not be in the infinite-dimensional one.

[3]Note that the existence of local inverses for s and t ensures the existence of a manifold structure on $M_1 \,{}_s\!\times_t M_1$.

[4]Morita equivalent Lie groupoids are the correct replacement for the concept of equivalent categories. In fact, Morita equivalent Lie groupoids are equivalent as categories and possess the same amount of "differential" information, cf. [**MM03**].

In order to motivate our procedure we recall that N_k is defined to be the subset

$$\{(z,\gamma)\in U(1)\times P_e\Omega K : \gamma\in C^\infty_*(S^1,\Omega K), z=\exp(-k\cdot\int_{D_\gamma}\omega^l)\},$$

where $D_\gamma : B^2\to\Omega K$ is a smooth map with $D_\gamma|_{\partial B^2}=\gamma$ and ω^l is the left-invariant 2-from on ΩK with $\omega^l(e)=\omega$. Since ω^l is an integral 2-from on ΩK, the value of $\exp(k\cdot\int_{D_\gamma}\omega^l)$ does not depend on the choice of D_γ if $k\in\mathbb{Z}$. The groupoid that we will construct carries some more information, namely not only the boundary value of D_γ but also the homotopy type of it relative to ∂B^2. This information is contained in the group

$$C^\infty_*(B^2,\Omega K)/C^\infty_*(S^2,\Omega K)_e,$$

(where we identify $C^\infty_*(S^2,\Omega K)$ with the normal subgroup in $C^\infty_*(B^2,\Omega K)$ of functions that vanish on ∂B^2) which we shall now endow with a Lie group structure. The following proof shall make use of the fact that smooth and continuous homotopies of functions with values in locally convex manifold agree, we refer to [**Woc09a**] for details on this.

LEMMA 4.2. *If G is a connected locally convex Lie group with Lie algebra $\mathfrak{g}$, then the quotient group*

$$C^\infty_*(B^2,G)/C^\infty_*(S^2,G)_e$$

carries a Lie group structure, modelled on $C^\infty_*(S^1,\mathfrak{g})$.

PROOF. We shall make use of the Lie group structure on $C^\infty_*(M,G)$ (with respect to point-wise group operations), which exists for each compact manifold M, possibly with corners [**Woc06**]. If U is open in G, then

$$C^\infty(M,U):=\{f\in C^\infty_*(M,G):f(M)\subseteq U\}$$

is open in $C^\infty(M,G)$ and, likewise, if U' is open in $\mathfrak{g}$, then $C^\infty(M,U')$ is open in $C^\infty(M,\mathfrak{g})$. If $U\subseteq G$ is an open identity neighbourhood and $\varphi:U\to\varphi(U)\subseteq\mathfrak{g}$ is a chart with $\varphi(U)$ open and convex and satisfying $\varphi(e)=0$, then a chart for the manifold structure, underlying $C^\infty_*(M,G)$, is given by

$$C^\infty(M,U)\ni f\mapsto\varphi\circ f\in C^\infty(M,\varphi(U)).$$

Clearly, this induces a map

$$\widetilde{\varphi}:q(C^\infty_*(B^2,U))\to C^\infty_*(S^1,\varphi(U)),\quad [f]\mapsto\varphi\circ(f|_{\partial B^2}),$$

where $q: C^\infty_*(B^2,G)\to C^\infty_*(B^2,G)/C^\infty_*(S^2,G)_e$ denotes the canonical quotient map. This map is bijective since each map $f\in C^\infty_*(S^1,\varphi(U))$ has a homotopy to the map which is constantly 0, defining an extension of f to a map $F:B^2\to\varphi(U)$ with $F|_{\partial B^2}=f$ and $[\varphi^{-1}\circ F]$ is mapped to f under $\widetilde{\varphi}$. Similarly, we deduce that $\widetilde{\varphi}$ is injective, since each two maps in $C^\infty_*(B^2,U)$, which restrict to the same value on ∂B^2, are homotopic.

We are now ready to verify that the conditions of Theorem 2.1 are satisfied, which we want to apply to the subset $W:=q(C^\infty_*(B^2,U))$. On this we have a smooth structure, induced by the bijection $\widetilde{\varphi}$. Moreover, if $V\subseteq U$ is an open identity neighbourhood of G with $V^2\subseteq U$ and $V^{-1}=V$, then $q(C^\infty_*(B^2,V))$ is open in $q(C^\infty_*(B^2,U))$. The structure maps on the mapping group under consideration are all given by the point-wise group structure in G, and so it follows that the coordinate representation of the structure maps on $q(C^\infty_*(B^2,V))$ coincides with the coordinate representation of the structure maps of $C^\infty_*(S^1,G)$. Since the latter are smooth it follows that the structure maps on $q(C^\infty_*(B^2,V))$ are smooth. Finally, $q(C^\infty_*(B^2,V))$ generates $C^\infty_*(B^2,G)/C^\infty_*(S^2,G)_e$, because G is connected. □

Note that there is a natural homomorphism

$$\mathcal{K}:=C^\infty_*(B^2,\Omega K)/C^\infty_*(S^2,\Omega K)_e\to C^\infty_*(S^1,\Omega K),\quad [f]\mapsto f|_{\partial B^2},$$

which obviously is smooth and surjective, because $\pi_1(\Omega K)$ vanishes. The kernel of this map is $C^\infty_*(S^2,\Omega K)/C^\infty_*(S^2,\Omega K)_e\cong\pi_2(\Omega K)$ and we thus obtain a central extension

$$\pi_2(\Omega K)\to\mathcal{K}\to C^\infty_*(S^1,\Omega K).$$

That $\pi_2(\Omega K)$ is in fact central follows from the fact that it is a discrete normal subgroup of the connected group $\mathcal{K}$. For general, not necessarily simply connected G, we only obtain a crossed module

$$C^\infty_*(B^2,G)/C^\infty_*(S^2,G)_e\to C^\infty_*(S^1,G).$$

Since the image of this morphism is precisely $C^\infty_*(S^1,G)_e$, this in turn gives rise to the four term exact sequence

$$\pi_2(G)\to C^\infty_*(B^2,G)/C^\infty_*(S^2,G)_e\to C^\infty_*(S^1,G)\to\pi_1(G).$$

This sequence has a characteristic class in in $H^3(\pi_1(G),\pi_2(G))$, which has first been constructed in [**EM46**].

The second smooth map, naturally associated to $\mathcal{K}$ is given by $\mathcal{K}\to\mathrm{R}$, $[f]\mapsto\int_f\omega^l$, where the integral only depends on the homotopy class of f because ω^l is closed.

For the following lemma we define a generalisation of the the cocycle κ by

$$\kappa_k:(P_e\Omega K)\times(P_e\Omega K)\to U(1),$$
$$(\gamma,\eta)\mapsto\exp\left(k\cdot\int_0^1\int_0^1\langle\gamma(s)^{-1}\gamma'(s),\eta'(t)\eta(t)^{-1}\rangle\,ds\,dt\right),$$

which is for $k=1$ the cocycle κ from [**Mur88**] (cf. also [**BCSS07**]).

PROPOSITION 4.3. *For each $k\in\mathrm{R}$, the Lie group $\mathcal{K}$ acts smoothly from the right on $U(1)\times P_e\Omega K$ by*

$$(z,\gamma).[f]:=(z\cdot\exp(-k\cdot\int_f\omega^l)\cdot\kappa_k(\gamma,f|_{\partial B^2}),\gamma\cdot f|_{\partial B^2}).\tag{4.1}$$

PROOF. It is clear that the action map is smooth on the product

$$(U(1)\times P_e\Omega K)\times\mathcal{K},$$

because the restriction map $\mathcal{K}\to C^\infty_*(S^1,\Omega K)$ and the integration map $\mathcal{K}\to U(1)$ are smooth.

In order to show that (4.1) actually defines a group action we have to verify that $(z,\gamma).[f\cdot g]=((z,\gamma).[f]).[g]$, which is equivalent to

$$\exp(-k\cdot\int_{f\cdot g}\omega^l)\cdot\kappa_k(\gamma,(f\cdot g)|_{\partial B^2})=$$
$$\exp(-k\cdot\int_f\omega^l)\cdot\kappa_k(\gamma,f|_{\partial B^2})\cdot\exp(-k\cdot\int_g\omega^l)\cdot\kappa_k(\gamma\cdot f|_{\partial B^2},g|_{\partial B^2}).$$

But this in turn follows immediately from the cocycle condition for κ_1, because

$$\kappa_1(f|_{\partial B^2}, g|_{\partial B^2}) = \exp(\int_f \omega^l)\cdot\exp(\int_g \omega^l)\cdot\exp(-\int_{f\cdot g}\omega^l)$$

follows from $f|_{\partial B^2}, g|_{\partial B^2} \in C^\infty_*(S^1, \Omega K)$ (cf. [**Mur88**, Sect. 6]). □

For each $k \in \mathrm{R}$, the action (4.1) now defines an action Lie groupoid

$$(U(1)\times P_e\Omega K \times \mathcal{K} \rightrightarrows_k U(1)\times P_e\Omega K), \tag{4.2}$$

i.e., $s(z,\gamma,[f]) = (z,\gamma)$, $t(z,\gamma,[f]) = (z,\gamma).[f]$ and

$$((z,\gamma).[f],[f'])\circ((z,\gamma),[f]) = (z,\gamma,[f\cdot f']).$$

From formula (4.1) we see in particular, that the action of $\mathcal{K}$, and thus the Lie groupoid (4.2) is not proper for arbitrary k. In fact, the subgroup $\mathrm{Z}\cong\pi_2(\Omega K)\subseteq\mathcal{K}$ acts on $U(1)\times P_e\Omega K$ by

$$a.(z,\gamma) = (z\cdot\exp(-k\cdot a),\gamma),$$

since we assumed that ω is normalised so that $\int_\sigma \omega^l = 1$ for a generator $[\sigma]$ of $\pi_2(\Omega K)\cong\pi_3(K)$. Of course, the interesting range for k in the previous proposition is $k\in[0,1]$, for then the quotient of the action "interpolates" between the trivial and the universal extension:

$$\text{for } k=0: \ U(1)\times P_e\Omega K/\mathcal{K}\cong U(1)\times\Omega K$$

$$\text{for } k=1: \ U(1)\times P_e\Omega K/\mathcal{K}\cong\widehat{\Omega K}_1$$

Moreover, we see that for each $k\in\mathrm{Q}$ the quotient $U(1)\times P_e\Omega K/\mathcal{K}$ exists as a manifold, we shall give a precise argument for this at the end of the next section. However, the group structure on $U(1)\times P_e\Omega K$, given by the cocycle κ, only induces a group structure on the quotient in the case $k\in\mathrm{Z}$.

5. Lie groupoids as principal 2-bundles

In this section we show that the Lie groupoids, derived in the previous section, possess the structure of a principal 2-bundle. For this we give at first a very short and condensed introduction to principal 2-bundles. The details can be found in [**Woc09b**].

A strict Lie 2-group is a category object in the category of locally convex Lie groups i.e., it consists of two locally convex Lie groups G_0 and G_1, together with morphisms $s,t: G_1\rightrightarrows G_0$, a morphism $i: G_0\to G_1$ and a morphism $c: G_1\,{}_s\!\times_t G_1\to G_1$ (assuming that the pull-back $G_1\,{}_s\!\times_t G_1$ exists), such that (G_0,G_1,s,t,i,c) constitutes a small category. In short, we write $(G_1\rightrightarrows G_0)$ for this (cf. [**BL04**] and [**Por08**]). A smooth 2-space is simply a Lie groupoid and similar to the case of Lie groups and manifolds, one defines a (right) $(G_1\rightrightarrows G_0)$-2-space to be a 2-space $(M_1\rightrightarrows M_0)$, together with a smooth functor

$$(\rho_1,\rho_0):(M_1\rightrightarrows M_0)\times(G_1\rightrightarrows G_0)\to(M_1\rightrightarrows M_0),$$

such that ρ_1 defines a (right) G_1-action on M_1 and ρ_0 defines a (right) G_0-action on on M_0. Similarly, one defines a morphism of $(G_1\rightrightarrows G_0)$-2-spaces $(M_1\rightrightarrows M_0)$ and $(N_1\rightrightarrows N_0)$ to be a smooth functor $(\varphi_1\times\varphi_0):(M_1\times M_0)\to(N_1\times N_0)$ such that φ_1 (respectively φ_0) defines a morphism of G_1 (respectively G_0)-spaces. A 2-morphism $\alpha:\varphi\Rightarrow\psi$ between two morphisms $\varphi,\psi:(M_1\times M_0)\to(N_1\times N_0)$

of $(G_1 \rightrightarrows G_0)$-spaces consists of a smooth map $\alpha : M_0 \to N_1$ such that α defines a natural transformation between the functors φ and ψ and, moreover, satisfies $\alpha(m.g) = \alpha(m).\operatorname{id}_g$ for each $m \in M_0$ and $g \in G_0$.

With this said one defines a principal $(G_1 \rightrightarrows G_0)$-2-bundle over the smooth manifold M (viewed as a smooth 2-space with only identity morphisms, we write $\underline{M}$ for this 2-space) as follows. It is a smooth $(G_1 \rightrightarrows G_0)$-2-space $(P_1 \rightrightarrows P_0)$, together with a smooth functor $\pi : (P_1 \rightrightarrows P_0) \to (M \rightrightarrows M)$, commuting with the action functor ρ, such that there exist

- an open cover $(U_i)_{i\in I}$ of M
- morphisms
$$\Phi_i : \pi^{-1}(\underline{U_i}) \to \underline{U_i} \times (G_1 \rightrightarrows G_0) \quad \text{and}$$
$$\overline{\Phi}_i : \underline{U_i} \times (G_1 \rightrightarrows G_0) \to \pi^{-1}(\underline{U_i})$$
of $(G_1 \rightrightarrows G_0)$-2-spaces
- 2-morphisms
$$\tau_i : \Phi_i \circ \overline{\Phi}_i \Rightarrow \operatorname{id}_{\underline{U_i}\times(G_1 \rightrightarrows G_0)}$$
$$\overline{\tau}_i : \overline{\Phi}_i \circ \Phi_i \Rightarrow \operatorname{id}_{\pi^{-1}(\underline{U_i})}$$
between morphisms of $(G_1 \rightrightarrows G_0)$-2-spaces,

such that π, Φ_i and $\overline{\Phi}_i$ commute in the usual way with the projection functor $\operatorname{pr} : \underline{U}_i \times (G_1 \rightrightarrows G_0) \to \underline{U}_i$.

We are now aiming at showing that the action Lie groupoid
$$(U(1) \times P_e\Omega K \times \mathcal{K} \rightrightarrows_k U(1) \times P_e\Omega K)$$
possesses the structure of a principal 2-bundle (we used the subscript $_k$ to denote the value of k in the action map (4.1)). The structure 2-group of this bundle shall be given by $(U(1) \times \pi_2(\Omega K) \rightrightarrows_k U(1))$ with $s(z, [\sigma]) = z$, $t(z, [\sigma]) = z \cdot \exp(-k \cdot \int_\sigma \omega^l)$ and $(z \cdot \exp(-k \cdot \int_\sigma \omega^l), [\sigma']) \circ (z, [\sigma]) = (z, [\sigma' \cdot \sigma])$.

Before showing the claim of this section, we have to pass from the action Lie groupoid (4.2) to a Morita equivalent one, which we will denote by $(P_1 \rightrightarrows_k P_0)$. For this we choose a system $(\sigma_i : U_i \to P_e\Omega K)_{i\in I}$ of smooth local sections of the principal bundle $\operatorname{ev} : P_e\Omega K \to \Omega K$. For technical reasons, that will become apparent later, we choose this system so that there exists smooth maps $\sigma_{ij} : U_i \cap U_j \to \mathcal{K}$ such that $\sigma_i(x) = \sigma_j(x) \cdot \sigma_{ij}(x)|_{\partial B^2}$. Then we set
$$P_0 := \coprod_{i\in I} (U(1) \times \{\sigma_i(x) : x \in U_i\}),$$
which we endow with the smooth structure induced from $U(1) \times P_e\Omega K$. The set of morphisms we set to be
$$P_1 := \{(z, \gamma, \eta, [f]) \in U(1) \times P_0 \times P_0 \times \mathcal{K} : \operatorname{ev}(\gamma) = \operatorname{ev}(\eta), \gamma = \eta \cdot f|_{\partial B^2}\}.$$
For a fixed choice of γ and η, the possible different choices of $[f]$ are parametrised by $\pi_2(\Omega K)$, and so P_1 has a natural manifold structure, modelled on $C^\infty_*(S^1, \Omega K)$. Source and target maps are induced by the two projections from P_1 to P_0 and composition is induced by multiplication in $\mathcal{K}$.

We define a smooth functor from $(P_1 \rightrightarrows_k P_0)$ to $(U(1) \times P_e\Omega K \times \mathcal{K} \rightrightarrows_k U(1) \times P_e\Omega K)$ by inclusion on objects and on morphisms by $(z, \gamma, \eta, [f]) \mapsto (z, \gamma, [f])$.

One easily checks that this functor actually defines a Morita equivalence.

There exists an obvious $(U(1) \times \pi_2(\Omega K) \rightrightarrows_k U(1))$-2-space structure on $(P_1 \rightrightarrows_k P_0)$, given by

$$(z,\gamma).w = (z\cdot w,\gamma) \quad \text{on objects and by} \tag{5.1}$$

$$(z,\gamma,\eta,[f]).(w,[\sigma]) = (z\cdot w\cdot \exp(-k\cdot \int_\sigma),\gamma,\eta,[f\cdot\sigma]) \quad \text{on morphisms.} \tag{5.2}$$

Moreover, there exists a natural smooth functor $\pi : (P_1 \rightrightarrows_k P_0) \to \underline{\Omega K}$, given on objects by $(z,\gamma)\mapsto \mathrm{ev}(\gamma)$ and on morphisms by $(z,\gamma,\eta,[f])\mapsto \mathrm{ev}(\gamma)$. We are now ready to prove the main result on this section.

PROPOSITION 5.1. *The $(U(1) \times \pi_2(\Omega K) \rightrightarrows_k U(1))$-2-space structure on $(P_1 \rightrightarrows_k P_0)$, given by* (5.1) *and* (5.2), *along with the smooth functor π, defines a principal 2-bundle.*

PROOF. We observe that (z,γ) is an object of $\pi^{-1}(\underline{U_i})$ if and only if $\mathrm{ev}(\gamma)\in U_i$ and $\gamma=\sigma_i(\mathrm{ev}(\gamma))$. From $\mathrm{ev}(\gamma)=\mathrm{ev}(\gamma\cdot f|_{\partial B^2})$ for each $[f]\in\mathcal{K}$ it follows that a morphisms has source in $\pi^{-1}(\underline{U_i})$ if and only if it has target in $\pi^{-1}(\underline{U_i})$, so that $\pi^{-1}(\underline{U_i})$ is in fact a full subcategory.

We now define local trivialisations Φ_i by

$$(z,\gamma)\mapsto(\mathrm{ev}(\gamma),z) \qquad \text{on objects and by}$$

$$(z,\gamma,\eta,[f])\mapsto(z,\sigma_{ij}(\gamma,\eta)\cdot[f]^{-1}) \quad \text{on morphisms.}$$

This is smooth due to the requirements that we put on the choice of $(\sigma_i : U_i \to P_e\Omega K)_{i\in I}$ and that it actually defines a functor follows from the fact that $\pi_2(\Omega K)$ is central in $\mathcal{K}$. The "inverse" trivialisations $\overline{\Phi}_i$ we define by

$$(l,z)\mapsto(z,\sigma_i(l)) \qquad \text{on objects and by}$$

$$(l,(z,[\sigma]))\mapsto(z,\sigma_i(l),\sigma_i(l),e) \quad \text{on morphisms.}$$

These obviously define smooth functors commuting with the $(U(1) \times \pi_2(\Omega K) \rightrightarrows_k U(1))$-action, and we have $\Phi_i\circ\overline{\Phi}_i=\mathrm{id}$. We then define $\overline{\tau}_i : \overline{\Phi}_i\circ\Phi_i\Rightarrow\mathrm{id}$ by

$$(z,\gamma)\mapsto(z,\sigma_i(x),\sigma_j(x),\sigma_{ij}(x)) \text{ if } \gamma=\sigma_j(x) \text{ for } x\in U_{ij}.$$

It is easily checked that $\overline{\tau}_i$ actually defines a natural transformation and satisfies $\overline{\tau}_i((z,\gamma).z')=\overline{\tau}_i((z,\gamma)).(z',e)$. □

Note that the functors Φ_i and the natural transformations $\overline{\tau}_i$ in the previous proof were smooth for they only need to be defined if (γ,η) can be written as $(\sigma_i(x),\sigma_j(x))$ for $x=\mathrm{ev}(\gamma)=\mathrm{ev}(\eta)$. If one tried to define a 2-bundle structure on the whole action groupoid (4.2) in a similar way, then one would need a smooth global section of $\mathcal{K}\to C^\infty_*(S^1,\Omega K)$, which does not exist. Thus the passage to the Morita equivalent groupoid $(P_1 \rightrightarrows_k P_0)$ was necessary to ensure the smoothness properties of the local trivialisations.

COROLLARY 5.2. *If $k\in\mathbb{Q}$, then the quotient P_k of the groupoid $(P_1 \rightrightarrows_k P_0)$ can be endowed with the structure of a smooth manifold. Moreover, the action* (5.1) *induces on P_k the structure of a smooth $(U(1)/k)$-principal bundle.*

PROOF. This is exactly the construction of the band of a principal 2-bundle from [**Woc09b**]. □

The previous result can also be obtained as in Section 2 by considering the Lie group $U(1)/k = \mathrm{R}/(\mathrm{Z}+k\mathrm{Z})$. This shows actually that P_k can also be endowed with a Lie group structure, turning

$$(U(1)/k) \to P_k \to \Omega K$$

into a central extension of Lie groups. However, the group structure on P_k is not induced by the one on $U(1) \times_{\kappa_1} P_e\Omega K$ any more.

References

[BCSS07] John C. Baez, Alissa S. Crans, Danny Stevenson, and Urs Schreiber, *From loop groups to 2-groups*, Homology, Homotopy Appl. **9** (2007), no. 2, 101–135. MR MR2366945

[BL04] John C. Baez and Aaron D. Lauda, *Higher-dimensional algebra. V. 2-groups*, Theory Appl. Categ. **12** (2004), 423–491 (electronic). MR **MR2068521 (2005m:**18005)

[Bou98] Nicolas Bourbaki, *Lie groups and Lie algebras. Chapters 1–3*, Elements of Mathematics (Berlin), Springer-Verlag, Berlin, 1998, Translated from the French, Reprint of the 1989 English translation. MR **MR1728312 (2001g:**17006)

[EM46] Samuel Eilenberg and Saunders MacLane, *Determination of the second homology and cohomology groups of a space by means of homotopy invariants*, Proc. Nat. Acad. Sci. U. S. A. **32** (1946), 277–280. MR MR0019307 (8,398b)

[GN03] Helge Glöckner and Karl-Hermann Neeb, *Banach-Lie quotients, enlargibility, and universal complexifications*, J. Reine Angew. Math. **560** (2003), 1–28. MR **MR1992799 (2004i:**22007)

[Mic85] Jouko Mickelsson, *Two-cocycle of a Kac-Moody group*, Phys. Rev. Lett. **55** (1985), no. 20, 2099–2101. MR **MR811843 (87b:**22036)

[Mic87] ______, *Kac-Moody groups, topology of the Dirac determinant bundle, and fermionization*, Comm. Math. Phys. **110** (1987), no. 2, 173–183. MR **MR887993 (89a:**22036)

[MM03] Ieke Moerdijk and Janez Mrčun, *Introduction to foliations and Lie groupoids*, Cambridge Studies in Advanced Mathematics, vol. 91, Cambridge University Press, Cambridge, 2003. MR **MR2012261 (2005c:**58039)

[MN03] Peter Maier and Karl-Hermann Neeb, *Central extensions of current groups*, Math. Ann. **326** (2003), no. 2, 367–415. MR **MR1990915 (2005e:**22019)

[MS01] Michael K. Murray and Danny Stevenson, *Yet another construction of the central extension of the loop group*, Geometric analysis and applications (Canberra, 2000), Proc. Centre Math. Appl. Austral. Nat. Univ., vol. 39, Austral. Nat. Univ., Canberra, 2001, pp. 194–200. MR **MR1852705 (2002h:**22028)

[MS03] ______, *Higgs fields, bundle gerbes and string structures*, Comm. Math. Phys. **243** (2003), no. 3, 541–555. MR **MR2029365 (2004i:**53027)

[Mur88] Michael K. Murray, *Another construction of the central extension of the loop group*, Comm. Math. Phys. **116** (1988), no. 1, 73–80. MR **MR937361 (89e:**22035)

[Nee02] Karl-Hermann Neeb, *Central extensions of infinite-dimensional Lie groups*, Ann. Inst. Fourier (Grenoble) **52** (2002), no. 5, 1365–1442. MR **MR1935553 (2003j:**22025)

[Por08] Sven S. Porst, *Strict 2-Groups are Crossed Modules*, 2008.

[PS86] Andrew Pressley and Graeme Segal, *Loop groups*, Oxford Mathematical Monographs, The Clarendon Press Oxford University Press, New York, 1986, Oxford Science Publications. MR **MR900587 (88i:**22049)

[Woc06] Christoph Wockel, *Smooth extensions and spaces of smooth and holomorphic mappings*, J. Geom. Symmetry Phys. **5** (2006), 118–126. MR **MR2269885 (2007g:**58011)

[Woc08] Christoph Wockel, *Categorified central extensions, étale Lie 2-groups and Lie's Third Theorem for locally exponential Lie algebras*, 2008.

[Woc09a] ______, *A Generalisation of Steenrod's Approximation Theorem*, Arch. Math. (Brno) **45** (2009), no. 2, 95–104.

[Woc09b] ______, *Principal 2-bundles and their gauge 2-groups*, to appear in Forum Math. (2009), 40 pp.

Mathematisches Institut, Georg-August-Universität Göttingen, Bunsenstr. 3-5, 37073 Göttingen, Germany

E-mail address: `christoph@wockel.eu`

Problem List

Contemporary Mathematics
Volume **519**, 2010

Problems on mapping spaces and related subjects

Yves Félix

ABSTRACT. The present list is based on problem sessions organized during the workshop in Oberwolfach. This paper has been realized in collaboration with Andrey Lazarev, Greg Lupton, John Oprea, Sam Smith, Jeff Strom, Daniel Tanré, Antonio Viruel, Christoph Wockel and Toshihiro Yamaguchi.

1. The Gottlieb groups. (G. Lupton)

An element $\alpha \in \pi_n(X)$ is a *Gottlieb element* if there exists an extension A (called *an associated map*) in the diagram

$$\begin{array}{ccc} S^n \vee X & \xrightarrow{(\alpha, \mathrm{id}_X)} & X \\ \downarrow & \nearrow_{A} & \\ S^n \times X & & \end{array}$$

The set of all Gottlieb elements in $\pi_n(X)$ is a subgroup of $\pi_n(X)$ denoted $G_n(X)$. The exponential correspondence (see [**Whi**], I.4.21) applied to the diagram defining a Gottlieb element shows that $G_n(X)$ is equal to

$$\mathrm{Image}\,(\mathrm{ev}_*\colon \pi_n(\mathrm{Map}(X, X; 1_X)) \to \pi_n(X))\,,$$

where $\mathrm{ev}\colon X^X \to X$ is evaluation $\mathrm{ev}(f) = f(x_0)$ at a specified basepoint $x_0 \in X$. Moreover, as a consequence of the existence of a classifying fibration for fibrations with fibre X, it is known that yet another equivalent definition is: $G_n(X)$ is equal to

$$\bigcup \mathrm{Im}(\partial_*\colon \pi_{n+1}(B) \to \pi_n(X)),$$

where the union is over all fibrations $X \to E \to B$ having X as fibre and ∂_* is the connecting homomorphism.

These groups were discovered and studied by Gottlieb in the early 1960's (see [**Go1, Go2**]) and have led to many interesting results in homotopy theory and fixed point theory (see, for example, [**Op**]). Some interesting questions about Gottlieb groups still remain however.

At the level of homotopy theory, if X is a simply connected finite CW complex, then ([**FH**]) $G_{2n}(X) \otimes \mathbb{Q} = 0$, all n and $\sum_n \mathrm{rank}\, G_{2n+1}(X) \leq \mathrm{cat}(X)$. In particular, $G_n(X)$ is a finite group for n large enough. The Gottlieb groups have a lot of implications. We mention here two applications. First, the vanishing of the

©2010 American Mathematical Society

Gottlieb groups in high degrees implies that for a n-dimensional simply connected CW complex X and for r large enough, we have ([**La**]),

$$\sum_{i=r+1}^{r+n-1} \dim \pi_i(X)\otimes\mathbb{Q} \geq \frac{\dim \pi_r(X)\otimes\mathbb{Q}}{\dim H^*(X;\mathbb{Q})}\,.$$

On the other hand, the Gottlieb groups may be used to distinguish between homotopy types of components of mapping spaces ([**LS3**]). For instance ([**Yo**]),

$$\mathrm{Map}(S^n, X; f) \cong \mathrm{Map}(S^n, X; 0) \text{ if and only if } [f]\in G_n(X)\,.$$

Problem 1. ([**FHT**]) Let X be a simply connected complex of dimension n. Is it true that $G_{\geq 2n}(X)\otimes\mathbb{Q}=0$?

Problem 2. Compute $G_*(X)$ when X is a symplectic manifold, a configuration space or a mapping space.
For instance for an odd dimensional sphere S^n, we have $2\pi_k(S^n)\subset G_k(S^n)$ ([**GM**]).

Problem 3. If $\alpha : S^n\to X$ is a continuous map, what are the relations between $G_*(X\cup_\alpha e^{n+1})$ and $G_*(X)$?

Problem 4. (B. Jessup, G. Lupton). For an elliptic space, is $\dim H^*(X;\mathbb{Q}) \geq 2^{\dim G_*(X)\otimes\mathbb{Q}}$?
This question is very close to the so-called torus rank conjecture (TRC),

(TRC) : If an r-torus acts continuously with finite isotropy groups on a closed simply connected manifold, then $\dim H^*(X;\mathbb{Q})\geq 2^r$.

TRC has been proved for a large family of spaces including homogeneous spaces (Allday and Haperin, [**AH**]) and homology Kähler manifolds (Allday and Puppe, [**AP**]). Examples in [**JL**] show that the torus rank may be smaller or larger than the rank of $G(X)$.

Let X be a finite complex with $\chi(X)\neq 0$. By a result of Gottlieb ([**Go2**]), $G_1(X)$ is contained in the kernel of the Hurewicz map, and for n odd, $G_n(X)$ is contained in the kernel of the rational Hurewicz map.

Problem 5. (D. Gottlieb) When $\chi(X)\neq 0$, are all the Gottlieb groups $G_n(X)$ contained in the kernel of the Hurewicz map ?
Remark that the condition $\chi(X)\neq 0$ is very strong. For instance if $\chi(X)\neq 0$ and $\pi_{2n}(X)\otimes\mathbb{Q}=0$ for all $n\leq m$, then $G_r(X)\otimes\mathbb{Q}=0$ for $r\leq 2m$ ([**FH2**]).

Let $f : M\to N$ be a map between m-dimensional manifolds and let ev : $\mathrm{Map}(M,N;f)\to N$ be the evaluation map. Denote by Λ_f the Lefschetz number of f. In [**Go3**] D. Gottlieb proves that if $\Lambda_f\neq 0$, then $\pi_1(ev)=0$.

Problem 6. If $\Lambda_f\neq 0$, can you say something about $H^*(ev;\mathbb{Q})$?
This question is motivated by the fact that in the absolute case, $M=N$ and $f=id$, the hypothesis $\chi(M)\neq 0$ implies $H^*(ev;\mathbb{Q})=0$ ([**FL**]).

Question. Find conditions that imply the existence of nonzero Gottlieb groups.

Finite cocategory is not a good condition for the existence of non trivial Gottlieb group. Here is an example of an infinite CW complex with finite cocategory with $G_*(X)\otimes\mathbb{Q}=0$. Consider the Sullivan relative minimal model

$$(\wedge(a_1,a_2,\dots,b_1,b_2,\dots),0)\to(\wedge(a_1,a_2,\dots,b_1,b_2,\dots,x_1,x_2,\dots,y_1,y_2,\dots),d)$$

with $|a_i| = 2$, $|b_i| = 4 \cdot 2^{i-1} - 1$, $|x_i| = 4 \cdot 2^{i-1}$, $|y_i| = 4 \cdot 2^i - 2$, $d(a_i) = d(b_i) = 0$, $d(x_i) = a_i b_i$ and $d(y_i) = x_i b_i - b_{i+1}$. This is trivially a space of cocategory ≤ 3. A minimal model of the space is $(\wedge(b_1, a_i, x_i), d)$, with $d(x_1) = a_1 b_1$, $d(x_2) = a_2 x_1 b_1$, $d(x_3) = a_3 x_2 x_1 b_1$ and $d(x_n) = a_n x_{n-1} x_{n-2} \dots x_1 b_1$. This space has no Gottlieb element.

2. Group actions on manifolds. (J. Oprea)

Suppose we have a group action $A\colon G \times X \to X$, where G is a compact Lie group. The Borel fibration associated to the action is

$$X \to EG \times_G X \to BG,$$

with connecting map (in the Puppe sequence) $\partial\colon \Omega BG \simeq G \to X$. It is known that ∂ is homotopic to the orbit map given by $g \mapsto g \cdot x_0$ for any $x_0 \in X$. Therefore, the image of $\pi_*(G)$ under an orbit map is contained in the Gottlieb group of X. Note that this also means that $G_1(X) \subseteq \mathcal{Z}(\pi_1(X)$, the center of the fundamental group.

Question. Is there anything special about the set of Gottlieb elements coming from group actions?

In particular, we can consider circle actions on manifolds as a starting point. Of course, this will only possibly give elements in G_1, but it is a place to start. It is known by work of Conner and Raymond that an effective S^1 action on a $K(\pi, 1)$-manifold M induces an injection on the fundamental group level, so this provides an element of $G_1(M)$. By a factorization result of Browder and Hsiang [**BH**], we can generalize this to certain symplectic manifolds.

Let (M, ω) be a compact symplectic manifold. Then we write $\omega_{|\pi_2 M} = 0$ if ω is zero on the image of the Hurewicz map $h_2\colon \pi_2(M) \to H_2(M)$. Here we are taking $\omega \in H^2(M; \mathbb{R})$ and thinking of it as a homomorphism $\omega\colon H_2(M) \to \mathbb{R}$. Such a symplectic manifold is said to be *symplectically aspherical.* Symplectically aspherical manifolds are exactly those for which the original Arnold conjecture has been proven.

Theorem [**LO, Mc, On**]. An effective action of S^1 on a symplectically aspherical manifold (M, ω) has an orbit map $\alpha\colon S^1 \to M$ with $[\alpha] \in \pi_1(M)$ of infinite order. Hence, the Gottlieb group $G_1(M)$ is infinite.

Problem 7. What are the restrictions on groups that can be Gottlieb groups $G_1 M$ for symplectically aspherical manifolds (M, ω)?

The interest in such a problem comes from several different directions. First, R. Gompf showed that any finitely presented group can be realized as the fundamental group of a (even 4-dimensional) compact symplectic manifold. D. Kotschick has also shown the analogous result for almost complex manifolds. Yet symplectically aspherical manifolds cannot realize all groups. For instance, $\mathbb{Z}$ cannot be the fundamental group of a symplectically aspherical manifold. A study of which groups can occur is made in [**IKRT**]. On the other hand, in [**OS**], it is shown that any finitely presented abelian group occurs as $G_1(M)$ for some manifold M. So exactly how special *are* symplectically aspherical manifolds? How does the condition of being symplectically aspherical restrict the topology of the manifold? In particular, if we can understand $G_1(M)$ in some uniform way for these manifolds, then we shall also understand what types of symmetries (i.e. group actions) they support.

3. Realization of Gottlieb groups.

Problem 8. (J. Strom). Determine the pairs (G, n) such that there exists a finite complex X with $G_n(X) = G$.
Remark that, if n is even, then G must be finite. On the other hand, since $G_n(X \times Y) = G_n(X) \times G_n(Y)$, we can focus on $G = \mathbb{Z}/m\mathbb{Z}$.
A first answer is given by J. Oprea and J. Strom ([**OS**]). They prove that for any finitely generated abelian group G there is a compact manifold M with $G_1(M) \cong G$. The paper [**OS**] also gives some preliminary results about realizing $G_n(X)$, where X is a manifold or a finite complex, but there is still work to be done here. Therefore, we can ask

Question. Can every finitely presented abelian group be realized as $G_n(X)$ for a finite complex (or manifold) X?

4. Homotopy type of mapping spaces. (S. Smith)

The determination of the homotopy type of mapping spaces is a very hard problem. The first results on the subject are due to Milnor and Thom. By a result of Milnor ([**Mi**]), if X is a compact metric space and Y is a CW complex, the mapping space $\mathrm{Map}(X, Y)$ and the pointed mapping space $\mathrm{Map}_*(X, Y)$ are CW complexes. Then, Thom proves that for an abelian group π, we have an homotopy equivalence ([**Th**])

$$\mathrm{Map}(X, K(\pi, n)) \cong \prod_q K(G_q, q)\,,$$

with $G_q = H^{n-q}(X; \pi)$.

The first results on the localization of mapping spaces are due to Hilton, Mislin and Roitberg ([**HMR**]). Let X be a finite CW complex and Y be a nilpotent CW complex. Denote by $\ell_Y : Y \to Y_P$ the P-localization. Then,

(1) $\mathrm{Map}(X, Y; f)$ and $\mathrm{Map}_*(X, Y; f)$ are nilpotent spaces,
(2) The maps induced by ℓ_Y, $\mathrm{Map}(X, Y; f) \to \mathrm{Map}(X, Y_P; f_P)$ and $\mathrm{Map}_*(X, Y; f) \to \mathrm{Map}_*(X, Y_P; f_P)$ are P-localizations.

The rational homotopy of mapping spaces has been first described by A. Haefliger in [**Hae**], then by E. Brown and R. Szczarba in [**BS**]. A survey on the models is given in [**FOT**].
The following general problem is classical.

Problem 9. Classify the path components of $\mathrm{Map}(X, Y)$ (resp. $\mathrm{Map}_*(X, Y)$) up to homotopy type.

See the survey ([**Sm4**]) for a discussion of this problem. Among partial results in that direction, let us mention

- the classification of the components of $\mathrm{Map}(M, S^n)$ by the degree when M is a compact manifold of dimension m (Hansen [**Ha1**], Sutherland [**Su**]).
- The classification of the components of $\mathrm{Map}(\mathbb{C}P^n, \mathbb{C}P^m)$ by the degree (Møller [**Mo**]).

Specializing to the case $Y = BG$, Problem 9 is related to the classification of gauge groups. Let $\mathcal{G}(p)$ be the group of automorphisms of a principal G-bundle $p\colon E \to X$ over X. Then $B\mathcal{G}(p) \simeq \mathrm{Map}(X, BG; h)$, where $h\colon X \to BG$ is the classifying map.

Problem 10. Classify the path components of $\mathrm{Map}(M, BG)$ (resp. $\mathrm{Map}_*(M, BG)$) up to homotopy type.
Denote by $\mathrm{Ad}(p) : E \times_G G^{ad} \to X$ the associated adjoint bundle where G acts on G by conjugation. Then $\mathcal{G}(p)$ is the space of sections of $\mathrm{Ad}(p)$ ([**AB**]). In [**CS**] Crabb and Sutherland prove that $\mathrm{Ad}(p)$ is trivial after fibrewise rationalization and that the set of homotopy types of $\mathcal{G}(p)$, when p ranges over all principal G-bundles over X, is finite. In [**KT**] Kono and Tsukuda generalize the result of Crabb and Sutherland and prove that $\mathrm{Ad}(p)$ is trivial after localization at a prime q if q is large enough, and in this case $\mathcal{G}(p)_q \cong \mathrm{Map}(B, G)_q$.

After rationalization, Problem 9 has been studied using the Haefliger model for function spaces. Here we have:

- the classification by Møller and Raussen ([**MR**]) of the components of the rationalization of the space $\mathrm{Map}(X, \mathbb{C}P^n)$ where X is simply connected and rationally $2n + 1$ coconnected.
- the classification by Smith ([**Sm1**]) of the "simple" components of the rationalization of the space $\mathrm{Map}(X, Y)$ for X, Y various flag-manifolds.

Problem 11. (S. Smith) Let G_1 and G_2 compact Lie groups and T_i maximal torus. Are all the components of $\mathrm{Map}(G_1/T_1, G_2/T_2)$ formal ?

We next have another general problem.

Problem. Compute homotopy invariants of $\mathrm{Map}(X, Y, f)$.
For instance, consider L.S. category. In most of the cases, $\mathrm{cat}_0(\mathrm{Map}(X, Y; 0))$ is infinite ([**Fe**]).

Problem 12. What can you say about the Gottlieb groups of the mapping space $\mathrm{Map}_*(X, Y; f)$?

Problem 13. (G. Lupton) Can $\mathrm{Map}_*(X, Y; f)$ be homotopically trivial for X and Y simply connected finite dimensional and Y hyperbolic?
This is not possible when $f = 0$, because in that case

$$\pi_q(\mathrm{Map}_*(X, Y; 0)) \otimes \mathbb{Q} \cong \mathrm{Hom}(\widetilde{H}_*(X; \mathbb{Q}), \pi_{*+q}(Y) \otimes \mathbb{Q}) .$$

The hypothesis simply connected is necessary. In [**Go3**] D. Gottlieb proves that if $\deg f \neq 0$ and $\chi(N) \neq 0$, then $\mathrm{Map}(M, N, f)$ is contractible when N is aspherical.

Now recall that the Whitehead length of a space Z, $\mathrm{WL}(Z)$, is the maximum n of the length of Whitehead brackets in $\pi_*(Z)$. By a result of Ganea ([**Ga**]) $\mathrm{WL}\,\mathrm{Map}_*(X, Y; 0) \leq \mathrm{cl}X$, and $\mathrm{WL}\,\mathrm{Map}_*(X, Y; 0) \leq \mathrm{cocat}Y$. Here $\mathrm{cl}Y$ is the cone-length of X and $\mathrm{cocat}Y$ is the cocategory of Y. The first inequality extends to general components after rationalization, where rational Whitehead length is interpreted as the longest bracket in the higher rational homotopy groups [**LS4**].

Problem 14 (G. Lupton and S. Smith) Determine whether the second inequality above holds rationally for arbitrary path-components.

Remark that the first inequality can be replaced by $\mathrm{WL}\,\mathrm{Map}_*(X,Y;0) < \mathrm{cl}X$ when $\mathrm{WL}\,(Y) = 1$ as proved by U. Buijs ([**Bu**]). We can also ask

Problem 15 (Y. Félix) Denote by e_0 the rational Toomer invariant of X, i.e., the minimal integer n such that the Ganea fibration $q_n : G_n(X) \to X$ is surjective in rational homology. Is the following inequality true for any component : $\mathrm{WL}\,\mathrm{Map}_*(X,Y;f) \leq e_0(X)$?

5. Miller Spaces. (J. Strom)

A space X is a Miller space if for every finite nilpotent complex K the space $\mathrm{map}_*(X,K)$ of pointed maps from X to K is weakly contractible. They are named for Haynes Miller, who proved the fundamental result that $B\mathbb{Z}/p$ is a Miller space [**Mil**] (this is not how he phrased it). To date, all known examples of Miller spaces flow easily from Miller's theorem.

Problem 16. Without using Miller's theorem or the T-functor, exhibit a Miller space. Note that by [**St**] if $\mathrm{map}_*(X,S^n) \sim *$ for all sufficiently large n, then X is a Miller space.

A strongly closed class is a class of pointed spaces that is closed under extensions by pointed homotopy colimits and extensions by fibrations. If X is a space, then $\overline{\mathcal{C}(X)}$ is the smallest strongly closed class containing X. It is not hard to see that if X is a Miller space, then every space in $\overline{\mathcal{C}(X)}$ is also a Miller space.

Problem 17. Does $\overline{\mathcal{C}(B\mathbb{Z}/p)}$ contain all the p-local Miller spaces?

6. The group $\mathcal{E}(X) = \pi_0(aut\,X)$. (A. Viruel)

Before to begin this section, recall that a list of problems on self-equivalences can be found in [**Ark**]. This section is a complement to that list.

Problem 18. Let G be a finitely generated nilpotent group. What are the conditions on G to be the group, $\mathcal{E}(X)$, of homotopy classes of homotopy self-equivalences of a finite CW complex X ?

Partial answers exist. Given a finite group G, an integer k, and a rational (faithful) representation $G < GL_n(\mathbb{Q})$, there exists a k-connected rational space X such that $\mathcal{E}(X) = N_{GL_n(\mathbb{Q})}(G)$ (the normalizer of G in $GL_n(\mathbb{Q})$) ([**Vi**]). This implies in particular that every finite group shows up as a normal subgroup in some $\mathcal{E}(X)$.

Problem 19. Given G and a distinguished subgroup or quotient $\hat{\mathcal{E}}(X)$ of $\mathcal{E}(X)$, find X such that $\hat{\mathcal{E}}(X) = G$.
Here $\hat{\mathcal{E}}(X)$ may be the subgroup $\mathcal{E}_\#(X)$ of self-equivalences that induce the identity on the homotopy groups, the subgroup $\mathcal{E}_*(X)$ consisting of self-equivalences that induce the identity in cohomology, or the derived subgroup, or $\hat{\mathcal{E}}(X)$ may be a quotient $\mathcal{E}(X)/\mathcal{E}_\#(X)$ or $\mathcal{E}(X)_{ab}$.

Given a group G and a distinguished subgroup or quotient of equivalences $\hat{\mathcal{E}}(X)$, define the $\hat{\mathcal{E}}$-genus of G as the set of homotopy types of spaces X such that $G < \hat{\mathcal{E}}(X)$.

Problem 20. Does the $\hat{\mathcal{E}}$-genus tell finite groups apart ? That is, if G and H are finite groups such that for every space $G < \hat{\mathcal{E}}(X)$ if and only if $H < \hat{\mathcal{E}}(Y)$, is then $G = H$?

The problem has a negative answer for infinite groups: if G and H are both free groups (of different rank), both groups have the same $\hat{\mathcal{E}}$-genus. Notice that since G and H are both free, then there are monomorphisms $G < H$ and $H < G$. This problem is somehow an homotopy version of the classical isomorphism problem in representation theory, as one may think of $G < \hat{\mathcal{E}}(X)$ as X being a $G-\hat{\mathcal{E}}$-module. Then the problem above can be rewritten as: if two groups G and H have the same set of $\hat{\mathcal{E}}$-modules, are G and H isomorphic ? Of course, the realization problem for finite groups will provide a positive answer for problem 20, although this problem may be proven without mentioning the realization problem.

The next problem is the rational version of the realization problem.

Problem 21. Let G be the rationalization of a nilpotent finitely generated group. What are the conditions for G to be the group of homotopy self-equivalences of the rationalization of a simply connected finite CW complex ?
Partial answers are given in [**Pi**] and [**FF**]. In particular 2-solvable nilpotent groups are realizable.

7. The space of homotopy self-equivalences

Let X be a simply connected finite CW complex and aut_1X the identity component of self-homotopy equivalences of X. The Dold-Lashof classifying space, $Baut_1X$, is the classifying space for orientable fibrations with fiber the homotopy type of X. We are interested in the rational homotopical properties. Milnor showed that, when $X = S^n$, $Baut_1X_0 \cong K(\mathbb{Q}, 2n)$ if n is even and $Baut_1X \cong K(\mathbb{Q}, n+1)$ if n is odd. Recall now the classical Halperin conjecture

Halperin Conjecture. Suppose that X is rationally elliptic and that $\chi(X) > 0$, then the rational homology of aut_1X is finite dimensional.

This is equivalent to say that aut_1X has finite rational category or that $Baut_1X_0$ is a finite product $\prod_i K(\mathbb{Q}, 2n_i)$. The Halperin conjecture covers many other forms. For instance if $F \to E \xrightarrow{p} B$ is a fibration where F is an elliptic space with $\chi(F) > 0$, the Halperin conjecture states that the Serre spectral sequence of the fibration collapses at the E_2-term. We refer to [**Lu**] for other forms of the Halperin conjecture. The conjecture has been solved in many geometric situations, for instance by Shiga and Tezuka ([**ST**]) for quotients G/H where G is a compact connected Lie group and H a subgroup of maximal rank.
Further examples of function space components of finite rational L.S. category related to the Halperin conjecture are given in [**Sm2**]. As we mention above Halperin conjecture states that under some hypothesis, $\mathrm{Map}(X, X; 1_X)$ has finite rational category. More generally we can ask :

Problem 22. (S. Smith) Determine which components of $\mathrm{Map}(X, Y)$ are of finite and which of infinite rational L.S. category, for X and Y given elliptic spaces with $\chi(X), \chi(Y) > 0$.

We discuss now on the existence of an H-space structure on the classifying space. If Halperin conjecture is true, then when X is elliptic with $\chi(X) > 0$, $Baut_1X$ is a rational H-space. But $Baut_1X$ is not a (rational) H-space in general, as discussed in [**Sm3**]. It is complicated even when X is a product of spheres [**Sm3**]. For example, it is a rational H-space if X is $S^3 \times S^3$, but it is not if X is $S^3 \times S^4$.

Recall that a space X is an $H(n)$-space if there exists a map $\mu_n : G_n(X \times X) \to X$ such that $\mu_n \circ i_n^\ell = \mu_n \circ i_n^r = p_n : G_n(X) \to X$ ([**FTa**]). Here $p_n : G_n(X) \to X$ is the n-th Ganea fibration and i_n^ℓ, $i_n^r : G_n(X) \to G_n(X \times X)$ are the canonical applications induced by the standard injections of X in $X \times X$. A space Z is a rational H(2)-space if and only if the Lie algebra $\pi_*(\Omega Z) \otimes \mathbb{Q}$ is abelian.

Problem 23. (T. Yamaguchi) Is $Baut_1X$ a rational H-space if it is a rational H(2)-space ?
Note that when $Baut_1X$ is coformal, it satisfies the above condition.

Problem 24. Determine the rational category of aut_1X when X is a simply connected rationally hyperbolic space.

Consider now the homotopy nilpotency of aut_1X. For an homotopy associative H-space X we denote by Hnil$X$ the maximal n (or ∞) such that the n-fold commutator map $c_n : X^n \to X$ is null homotopic. In [**Ra**] V. Rao proves that a connected Lie group is homotopy nilpotent if and only if it has no torsion in homology. Concerning self-maps, in [**Sa**] P. Salvatore proves that Hnil$(aut_1(S^2)) \geq 3$ and that Hnil$(aut_1(X)) = \infty$ when X is a wedge of at least two spheres S^{2n+1}.

Problem 25. Let X be a simply connected CW complex rationally hyperbolic. Is Hnil(aut_1X) always infinite ?

8. Mapping spaces and spaces of sections.

Let $f : X \to Y$ be a pointed map between simply connected spaces and suppose X finite. Denote by $\varphi : (\wedge V, d) \to (B, d)$ a Sullivan model of f with $(\wedge V, d)$ minimal, and by $\psi : (\mathbb{L}(V), d) \to (\mathbb{L}(W), d)$ a minimal Quillen model for f. Denoting by $\mathrm{Der}(\wedge V, B, \varphi)$ the complex of φ-derivations of $\wedge V$ into B and by $\mathrm{Der}(\mathbb{L}(V), \mathbb{L}(W), \psi)$ the complex of ψ-derivations of $\mathbb{L}(V)$ in $\mathbb{L}(W)$, we have ([**LS1**], [**LS2**], [**BM**], [**BL**])

$$\begin{cases} \pi_n(\mathrm{Map}(X, Y; f)) \otimes \mathbb{Q} \cong H_n(\mathrm{Der}(\wedge V, B, \varphi)), \\ \pi_n(\mathrm{Map}_*(X, Y; f)) \otimes \mathbb{Q} \cong H_n(\mathrm{Der}(\wedge V, B_+, \varphi)), \\ \pi_n(\mathrm{Map}_*(X, Y; f)) \otimes \mathbb{Q} \cong H_n(\mathrm{Der}(\mathbb{L}(V), \mathbb{L}(W), \psi)). \end{cases}$$

Since homology of complexes of derivations is André-Quillen homology, we get an homological formulation of the rational homotopy groups of mapping spaces (Block and Lazarev, [**BL**]),

$$\pi_n(\mathrm{Map}(X, Y; f)) \otimes \mathbb{Q} \cong HAQ^{-n}(C^*(Y), C^*(X)).$$

On the other hand, in [**GK**] Gatsinzi and Kwashira construct an explicit isomorphism

$$\pi_n(\Omega\mathrm{Map}(X, Y; f)) \otimes \mathbb{Q} \cong \mathrm{Ext}^n_{(T(V),d)}(\mathbb{Q}, \mathbb{L}(W)),$$

where the right hand term is a part of the Hochschild homology

$$\mathrm{Ext}^n_{(T(V),d)}(\mathbb{Q},(T(W),d)) \cong HH_*(T(V),d),(T(W),d)) .$$

Problem 26. Is it possible to describe the rational homotopy groups of a section space in terms of derivations, of André-Quillen homology, or in terms of differential Ext.

9. Models for mapping spaces (A. Lazarev)

In what follows we will adopt the shorthand cdga for 'commutative differential graded algebra' and dgla for 'differential graded Lie algebra'.

Let X and Y be based spaces and $f : X \to Y$ be a based map. Denote by $\mathrm{Map}(X,Y)$ the space of all continuous maps $X \to Y$, not necessarily preserving the base point. The space $\mathrm{Map}(X,Y))$ has a base point given by f; the basic problem that we are interested in is constructing an algebraic model (Sullivan or Lie) for this mapping space. Note that there is also a mapping space $\mathrm{Map}_*(X,Y)$ consisting of *based* maps $X \to Y$ and discussion below could be modified to include this case as well; however we will stick to the space of unbased maps for definiteness. There have been a number of results, cf. for example [**BS, BM, BL, BFM, BFM1**] in this direction which we will now briefly survey. For these results to make sense we need to impose some restrictions on the spaces X and Y; for example it is natural to assume that both X and Y are rational finite type nilpotent spaces.

Let B and A be the Sullivan (cdga) model for X and Y respectively. The map f induces a cdga map $A \to B$; thus B becomes an A-module and we can form $\mathrm{HAQ}^*(A,B)$ the André-Quillen cohomology of A with coefficients in B. Then for $n > 0$ we have:

$$\pi_n(\mathrm{Map}(X,Y),f) \cong \mathrm{HAQ}^{-n}(A,B).$$

A similar result holds for the Lie models. The map f induces a dgla map $L \to M$; thus M becomes an L-module and we can form $\mathrm{HCE}^*(A,B)$, the Chevalley-Eilenberg cohomology of L with coefficients in M. Then for $n > 0$ we have

$$\pi_n(\mathrm{Map}(X,Y),f) \cong \mathrm{HCE}^{1-n}(L,M).$$

In light of these results we pose the following problem.

Problem 27. Generalize the above isomorphisms by constructing dgla models for $\mathrm{Map}(X,Y)$ in terms of André-Quillen and Chevalley-Eilenberg cohomology.

The obvious conjecture is that the complexes $\mathrm{AQ}^*(A,B)$ and $\Sigma\,\mathrm{CE}(L,M)$ are themselves dgla models for the space $\mathrm{Map}(X,Y))$ as a whole (or, rather, its connected component containing f). In fact, the constructions in [**BFM, BFM1**] come close, but do not quite reach, the desired statement. In order to make sense of this we need to supply $\mathrm{AQ}^*(A,B)$ and $\Sigma\,\mathrm{CE}(L,M)$ with a Lie bracket which is compatible with the differential. Let us describe how this is done in the André-Quillen case; the other case is similar. We refer to [**HLa**] for a general detailed discussion of André-Quillen (also known as Harrison) and Chevalley-Eilenberg cohomology theories.

So let A and B be two cdga's over a $\mathbb{Q}$ and $f : A \to B$ be a map of cdga's. Associated to f is a certain element $\tilde{f} \in A^*\hat{\otimes}B \cong \mathrm{Hom}(A,B)$. Here the hat over $\otimes$ means the completed tensor product which reduces to the usual notion if either A or B are finite-dimensional. Via the suspension isomorphism $\Sigma A^* \cong A^*$ this element can be viewed as belonging to $\Sigma A^*\hat{\otimes}B$ and further, as an element in $\hat{L}(\Sigma A^*)\hat{\otimes}B$ via the inclusion $\Sigma A^*\hat{\otimes}B \subset \hat{L}(\Sigma A^*)\hat{\otimes}B$. Here $\hat{L}(-)$ stands for the completed free Lie algebra functor on $-$.

Note that $\hat{L}(\Sigma A^*)$ can be identified with the subspace of primitive elements in the (completed) tensor algebra $\hat{T}(\Sigma A^*)$. The latter is in fact a complex; the differential being the sum of the internal differential induced by that in A^* and the cobar-differential. This differential restricts to $\hat{L}(\Sigma A^*)$ making it into a dgla. Since B is a cdga the completed tensor product $\hat{L}(\Sigma A^*)\hat{\otimes}B$ is a dgla; this is the Harrison complex of A with coefficients in B. One can check that $\tilde{f}$ is a Maurer-Cartan element in this dgla. That means that it satisfies the Maurer-Cartan equation $d\tilde{f} + \frac{1}{2}[\tilde{f},\tilde{f}] = 0$. Using $\tilde{f}$ we twist the differential in $\hat{L}(\Sigma A^*)\hat{\otimes}B$ in a standard fashion; the twisted differential $d_{\tilde{f}}$ has the form $d_{\tilde{f}}(-) = d(-) + [-,\tilde{f}]$; the Maurer-Cartan condition implies that $d_{\tilde{f}}$ squares to zero.

One can check that $\mathfrak{g} := \hat{L}(\Sigma A^*)\hat{\otimes}B$ supplied with the differential $d_{\tilde{f}}$ is the standard Harrison (or André-Quillen) complex computing $\mathrm{AQ}^*(A,B)$. The advantage of this presentation is that it is now manifestly a dgla.

One problem with the dgla $\mathfrak{g}$ is that it is non-zero in negative degrees, and may well have non-trivial homology there. Therefore we should replace it by an appropriate truncation $\bar{\mathfrak{g}}$ concentrated in non-negative degrees and having there the same homology as $\mathfrak{g}$ in non-negatiove degrees. We conjecture that this dgla is a Lie model for the space $\mathrm{Map}(X,Y)$.

Let us emphasize that this model (or a similar Chevalley-Eilenberg model) is quite different from the other known models of mapping spaces in that it uses essentially only the classical homological algebra ingredients rather than simplicial methods. The Chevalley-Eilenberg complex is defined in terms of classical *abelian* derived functors and the André-Quillen cohomology is just a direct summand of the Hochschild cohomology which is also defined in terms of abelian derived functors. It also makes possible to compute purely algebraically the *cohomology* of mapping spaces; indeed if the conjecture is true then the $\mathrm{H}^*\,\mathrm{Map}(X,Y) \cong \mathrm{HCE}^*(\bar{\mathfrak{g}})$.

10. Lusternik-Schnirelmann category (D. Tanré)

Denote by $F_nX \to G_nX \overset{q_n}{\to} X$ the nth Ganea fibration of X. The Lusternik-Schnirelmann category of X is the least integer n such that q_n admits a section. In [**SST**], H. Scheerer, D. Stanley and D. Tanré introduce a variant of the category called the Q-category and designed for the study of the Ganea conjecture. Let Q be a base-point free version of the functor $\Omega^\infty\Sigma^\infty$. By applying the functor Q in a fibrewise way to the Ganea fibration, we get a new fibration $QF_nX \to \widetilde{G_nX} \overset{q'_n}{\to} X$, and Q-cat X is the least integer n such that q'_n admits a section.

The classifying spaces for fibrations with fibers F_nX and QF_nX are respectively denoted by $M_{aut\,F_nX} \to B_{aut\,F_nX}$ and $M_{Aut\,QF_nX} \to B_{Aut\,QF_nX}$. The natural transformation $\mathrm{id} \to Q$ induces a commutative diagram

$$(*)\qquad \begin{array}{ccc} M_{aut\,F_nX} & \to & M_{aut\,QF_nX} \\ p_n \downarrow & & \downarrow p'_n \\ B_{aut\,F_nX} & \overset{f}{\to} & B_{aut\,QF_nX}. \end{array}$$

Now denote by $\varphi : X \to B_{aut\,F_nX}$ the classifying map of the Ganea fibration. Then $\operatorname{cat} X \leq n$ if and only if there is a map $\psi : X \to M_{aut\,F_nX}$ such that $p_n \circ \psi \simeq \varphi$ and $Q\operatorname{cat} X \leq n$ if and only if there is a map $\psi' : X \to M_{aut\,QF_nX}$ such that $p'_n \circ \psi' \simeq f \circ \varphi$.

Problem 28. Describe a model of the square (*) in terms of algebraic models of X.

With the previous notation, observe that the existence of ψ' implies the existence of φ, which means $Qcat(X) \leq cat(X)$.

Problem 29. By using the algebraic models for (*), prove that the reverse way is true in rational homotopy theory, i.e. the existence of the lifting $\varphi_{\mathbb{Q}}$ implies the existence of the lifting $\psi'_{\mathbb{Q}}$.

In [**SS**], H. Scheerer and M. Stelzer proved that, for a rational space Y, the invariant $Qcat(Y)$ is equal to the invariant $Mcat(Y)$ defined by S. Halperin and J.-M. Lemaire in [**HL**]. A solution to the previous problem would give an alternative proof of the K. Hess theorem, [**He**],$Mcat(Y) = cat(Y)$, which is the cornerstone of many rational homotopy results.

11. *G*-bundles and Gauge groups. (C. Wockel)

Let G be a compact and connected Lie group and X a compact manifold. Then the space of $\mathcal{C}^k$-maps from X to G, $C^k(X,G)$, is an infinite dimensional Lie group for $0 \leq k \leq \infty$. The associated Lie algebra is $C^k(X, L(G))$, where $L(G)$ is the Lie algebra associated to G [**Gl**, Sect. 3]] It is well-known that the real cohomology of G is isomorphic to the cohomology of the invariant forms [**FOT**, Thm. 1.28], which can in turn be identified with the real cohomology of its Lie algebra $L(G)$ [**Ne**, Lem. 3.10]. In particular, the rational homotopy type of G determines the Lie algebra cohomology of $L(G)$.

Problem 30. Do there exist similar statements for $C^k(X,G)$? In particular, does the rational homotopy type of $C^k(X,G)$ determine the cohomology of the algebra of invariants forms $\Omega_L(C^k(X,G))$ (there really is a difference to the Lie algebra cohomology of $C^k(X,L(G))$, cf. [**Ne**, Sect. 8])?

The most important case would be the case of gauge groups, $\operatorname{Gau}(P)$ for a principal G-bundle $P \to X$, where [**FO**] and [**Wo1**, Prop. 1.20] show that the rational homotopy type of $\operatorname{Gau}(P)$ is that of $C^k(X,G)$. The geometric calculations in [**MN**] and [**NW**] show that the central extensions of $\operatorname{Gau}(P)$ and of $C^k(X,G)$ are "essentially of the same type", so they might give rise to an explicit isomorphism

$$H^2(L(\operatorname{Gau}(P)),\mathbb{R}) \cong H^2(C^k(X,G),\mathbb{R}).$$

References

[AH] C. Allday and S. Halperin, *Lie group actions on spaces of finite rank*, Quart. J. Math. Oxford Ser. **29** (1978), 63-76.

[AP] C. Allday and V. Puppe, *Bounds on the torus rank*, in : Transformation groups, Poznam 1985, Lecture Notes in Math. 1217 (1986), 1-10.

[Ark] M. Arkowitz, *Problems on self-equivalences*, in Groups of homotopy self-equivalences and related topics, Contemporary Math. 274, AMS, (2001), 309-315.

[AB] M. Atiayh and R. Bott, *The Yang-Mills equation over Riemann surfaces*, Phil. Trans. Roy. Soc. London **308** (1983), 523-615.

[BL] J. Block and A. Lazarev, *André-Quillen cohomology and rational homotopy of function spaces*, Adv. Math. **193** (2005), 18-39.

[BS] E. H. Brown and R. H. Szczarba, *On the rational homotopy type of function spaces*, Trans. Amer. Math. Soc. **349** (1997), 4931–4951.

[BH] W. Browder and W-C. Hsiang, *G-Actions and the Fundamental Group*, Invent. Math. **65** (1982), 411-424.

[Bu] U. Buijs, *Upper bounds for the Whitehead-length of mapping spaces*, this proceedings.

[BFM] U. Buijs, Y. Félix, A. Murillo, *Lie models for the components of sections of a nilpotent fibration*, Trans. Amer. Math. Soc. **361** (2009), no. 10, 5601–5614.

[BFM1] U. Buijs, Y. Félix, A. Murillo, *L_∞-models for mapping spaces*, preprint.

[BM] U. Buijs and A. Murillo, *The rational homotopy Lie algebra of function spaces*, Comment. Math. Helvetici **83** (2008), 723-739.

[CS] M. Crabb and W. Sutherland, *Counting homotopy types of gauge groups*, Proc. London Math. Soc. **81** (2000), 747-768.

[FF] J. Federinov and Y. Félix, *Realization of 2-solvable nilpotent groups as groups of classes of homotopy self-equivalences*, Top. and its Appl. **154** (2007), 2425-2433.

[Fe] Y. Félix, *Rational category of the space of sections of a nilpotent bundle*, Comment. Math. Helvetici **65** (1990), 615-622.

[FH] Y. Félix and S. Halperin, *Rational L.S. category and its applications*, Trans. Amer. Math. Soc. **273** (1982), 1-38.

[FH2] Y. Félix and S. Halperin, *The rational homotopy of spaces with nonzero Euler-Poincaré characteristic*, Bull. Soc. Math. Belgique **38** (1986), 171-173.

[FHT] Y. Félix, S. Halperin and J.C. Thomas, *Rational Homotopy Theory*, Graduate Texts in Mathematics 205, 2001.

[FL] Y. Félix and G. Lupton, *Evaluation maps in rational homotopy*, Topology **46** (2007), 493-506.

[FO] Y. Félix and J. Oprea, *Rational homotopy of gauge groups*, Proc. Amer. Math. Soc. **137** (2009), 1519-1527.

[FOT] Y. Félix, J. Oprea, and D. Tanré, *Algebraic models in geometry*, volume 17 of *Oxford Graduate Texts in Mathematics*. Oxford University Press, Oxford, 2008.

[FTa] Y. Félix and D. Tanré, *H-space structure on pointed mapping spaces*, Algebraic Geometric Topology **5** (2005), 713-724.

[Ga] T. Ganea, *Lusternik-Schnirelmann category and cocategory*, Proc. London Math. Soc. **10** (1960), 623-639.

[GK] J.-B. Gatsinzi and R. Kwashira, *Rational homotopy groups of function spaces*, this proceedings.

[Gl] H. Glöckner, *Lie group structures on quotient groups and universal complexifications for infinite-dimensional Lie groups*, J. Funct. Anal. **194** (2002), 347-409.

[GM] M. Golasinski and J. Mukai, *Gottlieb groups of spheres*, Topology **47** (2008), 399-430.

[Go1] D. Gottlieb, *A certain subgroup of the fundamental group*, Amer. J. Math. **87** (1965), 840-856.

[Go2] D. Gottlieb, *Evaluation subgroups of homotopy groups*, Amer. J. of Math. **91** (1969), 729-756.

[Go3] D. Gottlieb, *Self coincidence numbers and the fundamental group*, Fixed Point Theory Appl. **2** (2007), 73-83.

[GT] J. Grbic and S. Terzic, *The integral Pontrjagin homology of the based loop space on a flag manifold*, preprint 2009.

[Hae] A. Haefliger, *Rational homotopy of the space of sections of a nilpotent bundle*, Trans. Amer. Math. Soc. **273** (1982), 609-620.

[HL] S. Halperin and J.M. Lemaire, *Notions of category in differential algebra*, In : Algebraic topology - rational homotopy, Louvain-la-Neuve 1986, Lect. Notes in Math. 1318 (1988), 138-154

[Ha] V. Hansen, *The homotopy problem for the components in the space of maps on the n-sphere*, Quart. J. Math. Oxford **25** (1974), 313-321.

[Ha1] V. Hansen, *On spaces of maps of n-manifolds into the n-sphere*, Trans. Amer. Math. Soc. **265** (1981), 273-281.

[HLa] A. Hamilton, A. Lazarev, *Cohomology theories for homotopy algebras and noncommutative geometry*, Algebraic and Geometric Topology **9**, (2009) 15031583.

[He] K. Hess, *A proof of Ganea's conjecture for rational spaces*, Topology **30** (1991), 205-214

[HMR] P. Hilton, G. Mislin and J. Roitberg, *Localization of nilpotent groups and spheres*, North-Holland Mathematical Studies 15, North-Holland 1975.

[IKRT] R. Ibez, J. Kędra, Y. Rudyak, and A Tralle, *On fundamental groups of symplectically aspherical manifolds*, Math. Z. **248** (2004), 805-826.

[JL] B. Jessup and G. Lupton, *Free Torus Actions and Two-Stage Spaces*, Math Proc. Camb Phil.Soc. **137** (2004), 191-207.

[Ko] A. Kono, *A note on the homotopy type of certain gauge groups*, Proc. Roy. Soc. Edinburgh **117** (1991), 295-297.

[KT] A. Kono and S. Tsukuda, *Notes on the triviality of adjoint bundles*, this proceedings.

[La] P. Lambrechts, *Analytic properties of Poincaré series of spaces*, Topology **37** (1998), 1363-1370.

[Lu] G. Lupton, *Variations on a conjecture of Halperin*, Homotopy and Geometry (Warsaw 1997), Banach Center Publ. 45 (1998), 115-135.

[LO] G. Lupton and J. Oprea, *Cohomologically symplectic spaces: toral actions and the Gottlieb group*, Trans. Amer. Math. Soc. **347** (1995), 261-288.

[LS1] G. Lupton and S. Smith, *Rationalized evaluation subgroups of a map I : Sullivan models, derivations and G-sequences*, Journal of Pure and Applied Algebra **209** (2007), 159-171.

[LS2] G. Lupton and S. Smith, *Rationalized evaluation subgroups of a map II : Quillen models and adjoint maps*, Journal of Pure and Applied Algebra **209** (2007), 173-188.

[LS3] G. Lupton and S. Smith, *Criteria for components of a function space to be homotopy equivalent*, Math. Proc. Cambridge Philos. Soc. **145** (2008), 95-106.

[LS4] G. Lupton and S. Smith, *Whitehead products in function spaces: Quillen model formulae*, to appear in J. Japan. Math. Soc.

[Mc] D. McDuff, *The Moment Map for Circle Actions on Symplectic Manifolds*, J. Geom. Phys. **5** (1988), 149-160.

[MN] P. Maier and K.-H. Neeb, *Central extensions of current groups*, Math. Ann., 326 (2003), 367-415.

[Mil] H. Miller, *The Sullivan conjecture on maps from classifying spaces*, Ann. of Math. **120** (1984), 39–87.

[Mi] J. Milnor, *On spaces having the homotopy type of a CW complex*, Trans. Amer. Math. Soc. **90** (1959), 272-280.

[Mo] J. Møller, *On spaces of maps between complex projective spaces*, Proc. Amer. Math. Soc. **91** (1984), 471-476.

[MR] J. Møller and M. Raussen, *Rational homotopy of spaces of maps into spheres and complex projective spaces*, Trans. Amer. Math. Soc. **292** (1985), 721-732.

[Ne] K.-H. Neeb, *Central extensions of infinite-dimensional Lie groups*, Ann. Inst. Fourier (Grenoble), **52** (2002), 1365-1442.

[NW] K.-H. Neeb and C. Wockel, *Central extensions of groups of sections*, to appear in Ann. Glob. Anal. Geom., 2009, arXiv:0711.3437.

[On] K. Ono, *Obstruction to Circle Group Actions Preserving Symplectic Structure*, Hokkaido Math. J. **21** (1992), 99-102.

[Op] J. Oprea, *Gottlieb groups, group actions, fixed points and rational homotopy*, Lecture Notes Series, 29. Seoul National University, Research Institute of Mathematics, Global Analysis Research Center, Seoul, 1995.

[OS] J. Oprea and J. Strom, *On the realizability of Gottlieb groups*, this proceedings.

[Pi] S. Piccarreta, *Rational nilpotent group as subgroups of self-homotopy equivalences*, Cahiers de topologie et géométrie différentielle catégorique **42** (2001), 137-151.

[Ra] V. Rao, *Homotopy nilpotent Lie groups have no torsion in homology*, Manuscripta Math. **92** (1997), 455-462.

[Sa] P. Salvatore, *Rational homotopy nilpotency of self-equivalences*, Topology and its applications **77** (1997), 37-50.

[SST] H. Scheerer, D. Stanley and D. Tanré, *Fibrewise construction applied to Lusternik-Schnirelmann category*, Israel J. Math. **131** (2002), 333-359

[SS] H. Scheerer and M. Stelzer, *Fibrewise infinite symmetric products and M-category*, Bull. Korean Math. Soc. **36** (1999), 671-682

[ST] H. Shiga and M. Tezuka, *Rational fibrations, Homogeneous spaces with positive Euler characteristics and Jacobians*, Ann. Inst. Fourier **37** (1987), 81-106.

[Sm1] S. Smith, *Rational classification of simple function space components for flag-manifolds,* Can. J. Math. **49** (1997), 855-866.

[Sm2] S. Smith, *Rational L.S. category of function spaces for F_0-spaces,* Bull. Belg. Math. Soc. **6**, (1999), 295-304.

[Sm3] S. Smith, *Rational type of classifying space for fibrations*, Contemporary Math. **274** (2001), 299-307.

[Sm4] S. Smith *The homotopy theory of function spaces: a survey,* these proceeedings

[St] J. Strom, *Miller Spaces and Spherical Resolvability of Finite Complexes, Fund. Math.* ***178*** *(2003), no. 2, 97–108.*

[Su] W.A. Sutherland, Path-components of function spaces, Quart. J. Math. Oxford **34** *(1983), 223-233.*

[Th] R. Thom, L'homologie des espaces fonctionnels, Colloque de topologie algébrique, Louvain 1956, (1957), 29-39.

[Ts] S. Tsukuda, Comparing the homotopy types of the components of map($S^4, BSU(2)$), J. Pure Appl. Algebra **161** *(2001), 235-243.*

[Vig] M. Vigué-Poirrier, Sur l'homotopie rationnelle des espaces fonctionnels, Manuscripta Math. **56** *(1986), 177-191.*

[Vi] A. Viruel, Invariant theory and the realisation problem for self homotopy equivalences, article in preparation, private communication.

[Whi] G. W. Whitehead, Elements of Homotopy Theory, Graduate Texts in Mathematics **61**, *Springer-Verlag, 1978.*

[Wo1] C. Wockel. Lie group structures on symmetry groups of principal bundles, J. Funct. Anal., **251** *(2007), 254-288.*

[Wo2] C. Wockel, The Samelson product and rational homotopy for gauge groups, preprint.

[Yo] Y.S. Yoon, On n-cyclic maps, J. Korean Math. Soc. **26** *(1989), 17-25.*

DÉPARTEMENT DE MATHÉMATIQUE, 2, CHEMIN DU CYCLOTRON, 1348 LOUVAIN-LA-NEUVE, BELGIUM

Titles in This Series

527 **Ricardo Castaño-Bernard, Yan Soibelman, and Ilia Zharkov, Editors,** Mirror symmetry and tropical geometry, 2010

526 **Helge Holden and Kenneth H. Karlsen, Editors,** Nonlinear partial differential equations and hyperbolic wave phenomena, 2010

525 **Manuel D. Contreras and Santiago Díaz-Madrigal, Editors,** Five lectures in complex analysis, 2010

524 **Mark L. Lewis, Gabriel Navarro, Donald S. Passman, and Thomas R. Wolf, Editors,** Character theory of finite groups, 2010

523 **Aiden A. Bruen and David L. Wehlau, Editors,** Error-correcting codes, finite geometries and cryptography, 2010

522 **Oscar García-Prada, Peter E. Newstead, Luis Álverez-Cónsul, Indranil Biswas, Steven B. Bradlow, and Tomás L. Gómez, Editors,** Vector bundles and complex geometry, 2010

521 **David Kohel and Robert Rolland, Editors,** Arithmetic, geometry, cryptography and coding theory 2009, 2010

520 **Manuel E. Lladser, Robert S. Maier, Marni Mishna, and Andrew Rechnitzer, Editors,** Algorithmic probability and combinatorics, 2010

519 **Yves Félix, Gregory Lupton, and Samuel B. Smith, Editors,** Homotopy theory of function spaces and related topics, 2010

518 **Gary McGuire, Gary L. Mullen, Daniel Panario, and Igor E. Shparlinski, Editors,** Finite fields: Theory and applications, 2010

517 **Tewodros Amdeberhan, Luis A. Medina, and Victor H. Moll, Editors,** Gems in experimental mathematics, 2010

516 **Marlos A.G. Viana and Henry P. Wynn, Editors,** Algebraic methods in statistics and probability II, 2010

515 **Santiago Carrillo Menéndez and José Luis Fernández Pérez, Editors,** Mathematics in finance, 2010

514 **Arie Leizarowitz, Boris S. Mordukhovich, Itai Shafrir, and Alexander J. Zaslavski, Editors,** Nonlinear analysis and optimization II, 2010

513 **Arie Leizarowitz, Boris S. Mordukhovich, Itai Shafrir, and Alexander J. Zaslavski, Editors,** Nonlinear analysis and optimization I, 2010

512 **Albert Fathi, Yong-Geun Oh, and Claude Viterbo, Editors,** Symplectic topology and measure preserving dynamical systems, 2010

511 **Luise-Charlotte Kappe, Arturo Magidin, and Robert Fitzgerald Morse, Editors,** Computational group theory and the theory of groups, II, 2010

510 **Mario Bonk, Jane Gilman, Howard Masur, Yair Minsky, and Michael Wolf, Editors,** In the Tradition of Ahlfors-Bers, V, 2010

509 **Primitivo B. Acosta-Humánez and Francisco Marcellán, Editors,** Differential algebra, complex analysis and orthogonal polynomials, 2010

508 **Martin Berz and Khodr Shamseddine, Editors,** Advances in p-Adic and non-archimedean analysis, 2010

507 **Jorge Arvesú, Francisco Marcellán, and Andrei Martínez-Finkelshtein, Editors,** Recent trends in orthogonal polynomials and approximation theory, 2010

506 **Yun Gao, Naihuan Jing, Michael Lau, and Kailash C. Misra, Editors,** Quantum affine algebras, extended affine Lie algebras, and their applications, 2010

505 **Patricio Cifuentes, José García-Cuerva, Gustavo Garrigós, Eugenio Hernández, José María Martell, Javier Parcet, Alberto Ruiz, Fernándo Soria, José Luis Torrea, and Ana Vargas, Editors,** Harmonic analysis and partial differential equations, 2010

504 **Christian Ausoni, Kathryn Hess, and Jérôme Scherer, Editors,** Alpine perspectives on algebraic topology, 2009

503 **Marcel de Jeu, Sergei Silvestrov, Christian Skau, and Jun Tomiyama, Editors,** Operator structures and dynamical systems, 2009

502 **Viviana Ene and Ezra Miller, Editors,** Combinatorial Aspects of Commutative Algebra, 2009

501 **Karel Dekimpe, Paul Igodt, and Alain Valette, Editors,** Discrete groups and geometric structures, 2009

500 **Philippe Briet, François Germinet, and Georgi Raikov, Editors,** Spectral and scattering theory for quantum magnetic systems, 2009

499 **Antonio Giambruno, César Polcino Milies, and Sudarshan K. Sehgal, Editors,** Groups, rings and group rings, 2009

498 **Nicolau C. Saldanha, Lawrence Conlon, Rémi Langevin, Takashi Tsuboi, and Pawel Walczak, Editors,** Foliations, geometry and topology, 2009

497 **Maarten Bergvelt, Gaywalee Yamskulna, and Wenhua Zhao, Editors,** Vertex operator algebras and related areas, 2009

496 **Daniel J. Bates, GianMario Besana, Sandra Di Rocco, and Charles W. Wampler, Editors,** Interactions of classical and numerical algebraic geometry, 2009

495 **G. L. Litvinov and S. N. Sergeev, Editors,** Tropical and idempotent mathematics, 2009

494 **Habib Ammari and Hyeonbae Kang, Editors,** Imaging microstructures: Mathematical and computational challenges, 2009

493 **Ricardo Baeza, Wai Kiu Chan, Detlev W. Hoffmann, and Rainer Schulze-Pillot, Editors,** Quadratic Forms—Algebra, Arithmetic, and Geometry, 2009

492 **Fernando Giráldez and Miguel A. Herrero, Editors,** Mathematics, Developmental Biology and Tumour Growth, 2009

491 **Carolyn S. Gordon, Juan Tirao, Jorge A. Vargas, and Joseph A. Wolf, Editors,** New developments in Lie theory and geometry, 2009

490 **Donald Babbitt, Vyjayanthi Chari, and Rita Fioresi, Editors,** Symmetry in mathematics and physics, 2009

489 **David Ginzburg, Erez Lapid, and David Soudry, Editors,** Automorphic Forms and L-functions II. Local aspects, 2009

488 **David Ginzburg, Erez Lapid, and David Soudry, Editors,** Automorphic forms and L-functions I. Global aspects, 2009

487 **Gilles Lachaud, Christophe Ritzenthaler, and Michael A. Tsfasman, Editors,** Arithmetic, geometry, cryptography and coding theory, 2009

486 **Frédéric Mynard and Elliott Pearl, Editors,** Beyond topology, 2009

485 **Idris Assani, Editor,** Ergodic theory, 2009

484 **Motoko Kotani, Hisashi Naito, and Tatsuya Tate, Editors,** Spectral analysis in geometry and number theory, 2009

483 **Vyacheslav Futorny, Victor Kac, Iryna Kashuba, and Efim Zelmanov, Editors,** Algebras, representations and applications, 2009

482 **Kazem Mahdavi and Deborah Koslover, Editors,** Advances in quantum computation, 2009

481 **Aydın Aytuna, Reinhold Meise, Tosun Terzioğlu, and Dietmar Vogt, Editors,** Functional analysis and complex analysis, 2009

480 **Nguyen Viet Dung, Franco Guerriero, Lakhdar Hammoudi, and Pramod Kanwar, Editors,** Rings, modules and representations, 2008

479 **Timothy Y. Chow and Daniel C. Isaksen, Editors,** Communicating mathematics, 2008

478 **Zongzhu Lin and Jianpan Wang, Editors,** Representation theory, 2008

477 **Ignacio Luengo, Editor,** Recent Trends in Cryptography, 2008

For a complete list of titles in this series, visit the
AMS Bookstore at **www.ams.org/bookstore/**.